International Scientific Practical Conference on Materials Science, Shape-Generating Technologies and Equipment (ICMSSTE 2020)

Held online due to COVID-19

IOP Conference Series: Materials Science and Engineering
Volume 889

Yalta, Crimea
25 - 29 May 2020

ISBN: 978-1-7138-3176-1
ISSN: 1757-8981

Printed from e-media with permission by:

Curran Associates, Inc.
57 Morehouse Lane
Red Hook, NY 12571

Some format issues inherent in the e-media version may also appear in this print version.

Institute of Physics
Dirac House, Temple Back
Bristol BS1 6BE UK

Phone: 44 1 17 929 7481
Fax: 44 1 17 920 0979

techtracking@iop.org

Additional copies of this publication are available from:

Curran Associates, Inc.
57 Morehouse Lane
Red Hook, NY 12571 USA
Phone: 845-758-0400
Fax: 845-758-2633
Email: curran@proceedings.com
Web: www.proceedings.com

TABLE OF CONTENTS

MATERIALS SCIENCE AND PHYSICAL CHEMISTRY OF MATERIALS

STUDY OF DIFFUSION OXIDE HARDENING ON INTERMETALLIC MATERIALS 1
Elena Chekalova, Andrey Zhuravlev

THE X-RAY DIFFRACTION STUDY OF THE COMPOSITION OF THE POWDER MATERIALS FROM THE WASTE STEEL X13 8
E.V. Ageeva, S.V. Khardikov, A. N. Novikov

X-RAY DIFFRACTION ANALYSIS OF SINTERED PRODUCTS FROM ELECTRO-EROSION MATERIALS, OBTAINED FROM CR17 ALLOY WASTE 12
A A Sysoev, SA Grashkov, NA Pivovar

TEXTURE AND ANISOTROPY OF THE MECHANICAL PROPERTIES OF 2 MM SHEETS OF MG-9LI-3AL-0.4CE ALLOY 16
S Ya Betsofen, I A Grushin, A A Petrov, K A Speransky

METALLIC MATERIALS: STRUCTURE, PROPERTIES, EQUIPMENT

OXIDE FILMS RESISTANCE TO HYDROGEN PENETRATION IN VT6 TITANIUM ALLOY 20
O N Gvozdeva, A V Shalin, A S Stepushin

IMPROVING THE RELIABILITY OF METAL STRUCTURES OF BUILDINGS AND STRUCTURES DURING RECONSTRUCTION 27
A Mezentseva, Z Gel'manova, A Konakbayeva

INCREASING THE OPERATIONAL RELIABILITY OF CAR VARIATORS DUE TO CREATING REGULAR SURFACE MICRORELIEF BY LASER ABLATION 32
Elmar Yagyaev, Leonid Shron, Dmitriy Meniuk

CONSTRUCTIVE-TECHNOLOGICAL PECULIARITIES OF THE MACHINERY FOR IMPLEMENTING THE TECHNOLOGY OF THE SURFACE MODIFICATION BY HIGH-ENTROPIC MATERIALS WITH SME UNDER THE CONDITIONS OF HIGH-VOLTAGE PULSE ELECTROMECHANICAL INFLUENCES 39
R L Plomodialo, Zh M Blednova, D V Dmitrenko

IMPROVING THE ADHESIVE STRENGTH OF COATINGS FROM MULTICOMPONENT MATERIALS WITH THERMOELASTIC PHASE TRANSFORMATIONS BY EXTERNAL HIGH-ENERGY INFLUENCES AT VARIOUS PROCESSING STAGES 46
Zh M Blednova, G V Kurapov, E Yu O Balaev

DEVELOPMENT AND EXPERIMENTAL STUDIES OF NEW TYPES OF SHEET METAL STAMPING EQUIPMENT 56
A Yu Botashev, R A Bayramukov, N U Bisilov, R S Malsugenov

NON-METAL MATERIALS: STRUCTURE, PROPERTIES, EQUIPMENT

THEORETICAL CALCULATIONS OF MECHANOTERMIC DESTRUCTION PROCESS WITH REGARD TO GENERAL RUBBER GOODS 62
I E Ametov, C V Abkhairova

BIOPOLYMER BINDINGS FOR CASTING MANUFACTURING PROCESSES ON THE TECHNOLOGICAL LIGNIN BASIS 68
Yu A Svinoroev, V V Dyadichev, O A Ternovskoy, I V Dyadicheva, A Yu Melnik, S Ye Chornobay

COMPOSITE AND POWDER MATERIALS

INTERACTION OF COMPONENTS FEATURES OF THE REACTION MIXTURE $ZRO_{2(M)}$ - AL-C SYSTEMS WHEN HEATED 75
S I Dzhelialov

OPTIMIZATION METHOD OF THE EXTRUSION MIXTURE COMPONENTS FORMULATION AND CONSUMPTION IN THE PROCESSING OF THE SECONDARY HETEROGENEOUS POLYMERIC MATERIALS 82
V V Dyadichev, A V Kolesnikov, A V Dyadichev, S G Menyuk, O F Babushkina, S Ye Chornobay

THE STRUCTURE AND PROPERTIES OF LAMINATED ALUMINUM-GLASS REINFORCED PLASTICS 88
M I Gordeeva, Yu A Knyazeva

COATING

OPTIMISATION OF ELECTRO-SPARK COATING METHOD 92
N Yu Loginov, V A Gulyaev, O Yu Khudyakova

THE STUDY OF CONTACT ELECTRICAL RESISTANCE OF COATINGS ON THE BASIS OF REFRACTORY METAL ALLOYS CO-W, CO-MO, NI-W, NI-MO 98
G P Gololobov, D V Suvorov, M A Serpova, A S Arefiev

QUANTITATIVE PHASE ANALYSIS OF MG-AL-ZN AND MG-ZN-ZR ALLOYS 104
Yu V Chernysheva, I A Grushin, I I Voskresenskaya, D S Savostin

EFFECT OF BIAS VOLTAGE ON THE TEXTURE OF THE TIN AND ZRN COATINGS DEPOSITED BY VACUUM ION-PLASMA METHOD 108
A A Ashmarin, S Ya Betsofen, L M Petrov, M A Lebedev

INVESTIGATION OF HYDROABRASIVE RESISTANCE OF INTERNAL ANTI-CORROSION COATINGS USED IN THE OIL AND GAS INDUSTRY 112
M Kovalev, E Alekseeva, N Shaposhnikov

OPERATIONAL PROPERTIES OF DRILL PIPES MADE OF COATED ALUMINUM ALLOY 2024 117
O V Shvetsov, S Y Kondrat'ev, V A Yakhimovich, M K Kurakin

NANOMATERIALS

POLYMERIC NANOMATERIALSBASED ON IRON (III) TRIS-ACETYLACETONATE: SYNTHESIS, STRUCTURE AND PROPERTIES 128
N P Shapkin, I G Khalchenko, E K Papynov

EFFECTS OF «SIMPLEST» POST-TREATMENT TECHNIQUES ON THE COMPOSITION OF POROUS SILICON SURFACE 135
A S Lenshin, O V Chernousova, P V Seredin, K A Barkov, D A Minakov

BUILDING MATERIALS, PROPERTIES, TECHNOLOGIES, EQUIPMENT

TECHNOLOGIES OF REUSING MATERIALS OBTAINED DURING STEP-BY-STEP DEMOLITION OF BUILDINGS 142
S N Zolotukhin, V P Volokitin, N M Kovalev

THERMAL TRANSFORMATIONS IN POLYMINERAL TECHNOGENIC CEMENT RAW 149
O A Miryuk

CALCULATION UPDATE OF STRENGTH OF EXPANDED CLAY FIBER-REINFORCED CONCRETE ELEMENTS WITH MIXED REINFORCEMENT 157
D R Mailyan, A M Blyagoz, K M Kretinin

INVESTIGATION ON BOND PERFORMANCE BETWEEN DIFFERENT TYPES OF FRP - REINFORCEDREBARS AND CONCRETE 165
A I Komarov, N V Makarova, V G Tsuprik

BIOLOGICAL MATERIALS: STRUCTURE, PROPERTIES, TECHNOLOGIES, EQUIPMENT

ANTIOXIDANT IN A MODEL BIOMEMBRANE – ASTAXANTHIN AND ITS ESTERS MIXED WITH DPPC IN LANGMUIR FILMS 172
E A Kulikov, A A Stupnikov, Y N Malakhova

ETIDRONIC ACID ADDUCT AND BIS (2-PYRIDYL-1,2,4-TRIAZOLYL-3) BUTANE: PROPERTIES AND EFFECT ON PAIN SENSITIVITY 177
I V Cheretaev, M Yu Ravaeva, E N Chuyan

COORDINATING METAL CO^{2+} AND ZN^{2+} COMPOUNDS WITH ACETYLSALICYLIC ACID: THE INFLUENCE ON THE CARDIORESPIRATORY SYSTEM 182
M Yu Ravaeva, E N Chuyan, I V Cheretaev

MODELING THE STRUCTURE, PROPERTIES OF MATERIALS

IMPROVEMENT OF HIGH-SPEED STEEL CUTTING TOOL PRODUCTION TECHNOLOGY USING THE METHOD OF NANOSTRUCTURE GENERATION BY LASER ABLATION 188
Elmar Yagyaev, Seran Akimov

METHODS OF CONTROL AND ANALYSIS OF MATERIALS

TEMPERATURE DEPENDENCE OF $GE_2SB_2TE_5$ CONDUCTIVITY AND ITS CHANGE DURING AGING 194
Y V Vorobyov, A O Yakubov, A V Ermachikhin

OTHER TOPICS OF MATERIALS SCIENCE, SHAPE-GENERATING TECHNOLOGIES AND EQUIPMENT

HOPPING CONDUCTION IN SINGLE CRYSTALS OF THE DILUTED MAGNETIC SEMICONDUCTOR $(ZN_{1-X}FE_X)_3AS_2$ (X=0.005) 200
V S Zakhvalinskii, E A Pilyuk, T B Nikulicheva

EXPERIMENTAL EVALUATION OF THE EFFICIENCY OF LUBRICATING PROCESS MEDIA DURING THE RUNNING-IN PERIOD OF A METAL-CUTTING TOOL 205
Vladimir Skakun, Ruslan Dzhemalyadinov, Ilimdar Abdulkerimov

BERRY PHASE IN SINGLE CRYSTALS OF THE DILUTE MAGNETIC SEMICONDUCTOR $(CD_{0.7} ZN_{0.28} MN_{0.02})_3AS_2$ 211
V S Zakhvalinskii, T B Nikulicheva, EA Pilyuk

Author Index

IOP Conf. Series: Materials Science and Engineering **889** (2020) 011001 doi:10.1088/1757-899X/889/1/011001

Preface

Valery Dyadichev*

[1]V.I. Vernadsky Crimean Federal University, Simferopol, Crimea, Russia

*mr.dyadichev@mail.ru

International Scientific Practical Conference «Materials science, shape-generating technologies and equipment 2020» (ICMSSTE 2020) was held in Yalta, Russia from 25 to 29 May 2020.

ICMSSTE 2020 is organized by the: Federal State Autonomous Educational Institution of Higher Education «V. I. Vernadsky Crimean Federal University»; Federal State Autonomous Educational Institution of Higher Education «Sevastopol State University»; Federal State Autonomous Educational Institution of Higher Education «Polzunov Altai State Technical University»; Limited Liability Company "Engineering Centre The Best Available Technologies".

The Conference provides scientists, industry experts and business representatives from various countries with the opportunity to summarize and share innovative and latest scientific research developments in materials science, material properties, new materials and products manufacturing technologies, industrial equipment and tools, to discuss the development of nanomaterials and nanotechnologies, new equipment and technologies.

In addition, the Conference aim is to establish new scientific relations between conference participants with the purpose to integrate science and business of the various countries, in order to conduct further joint researches, to implement some new joint research projects and introduce the scientific results into the real production industry and business.

ICMSSTE 2020 proceedings cover these topics: Materials science and physical chemistry of materials; Metallic materials: structure, properties, equipment; Non-metal materials: structure, properties, equipment; Composite and powder materials; Coating; Nanomaterials; Building materials, properties, technologies, equipment; Biological materials: structure, properties, technologies, equipment; Modeling the structure, properties of materials; Methods of control and analysis of materials; Other topics of Materials science, shape-generating technologies and equipment.

This year the organizing committee was forced to lead the conference online, in the virtual form, in connection with the current situation with the epidemiological situation in the world and the impossibility of the personal presence of participants from Russia and other countries in the city of Yalta, because of travel restrictions. The organizing committee decided to hold a conference with online reports of the participants, and not to cancel or a postpone the holding of conference in order to enable scientists to present the obtained topical scientific results and share them with the colleagues, since the science development is important and especially urgent in times of general isolation and problems with personal international communication.

So, the Conference was held on the online platform – "Zoom", with the online reports of authors, with the video and visual presentation of reports, with the opportunity to ask the questions to each other and discuss the scientific results.

The organizers was located in Yalta and some of them in Simferopol, where is located the main unit of V.I. Vernadsky Crimean Federal University. Some of organizers and participant was located at home, because of remote work at home.

The Conference model was as for full-time participation, there were three speeches in the plenary session, all reports were held in accordance with the sections of the conference, each author had from 10 to 15 minutes for his report and the answer to questions from other participants. Also there was time at the ending dates of the conference for general discussion and consideration of poster presentations of participants who were unable to personally submit a video report.

IOP Conf. Series: Materials Science and Engineering **889** (2020) 011001 doi:10.1088/1757-899X/889/1/011001

Overall, the 206 authors submitted papers for ICMSSTE 2020, from 6 countries: Russia, Ukraine, Chech Republic, Iraq, Vietnam, Kazakhstan. The Conference Program, video from online reports and presentations of the research articles are available on the Conference website.

Virtual form of reports did not prevent authors from actively participating and establishing new relations and experience exchange. Authors and researches from different countries and organizations, including companies of real sector of the economy, showed the great interest and took an active part in the conference. We look forward to further personal meetings and face-to-face reports by our fellow researchers during the next Conference ICMSSTE 2020.

Editors want to thank the international scientific committee and the reviewers for their great contribution to the work and organization of the Conference. After the peer review process, the submitted papers were selected on the basis of originality, significance, and clarity for the purpose of the conference.

The organizing committee is very grateful to all the participants and authors of the conference, who presented the results of their research in various fields of materials science, technologies and equipment on our Conference.

Prof., Conference Chair, Valery Dyadichev, V.I. Vernadsky Crimean Federal University, Russia

Conference foto – online reports of conference participants:

IOP Conf. Series: Materials Science and Engineering **889** (2020) 011001 doi:10.1088/1757-899X/889/1/011001

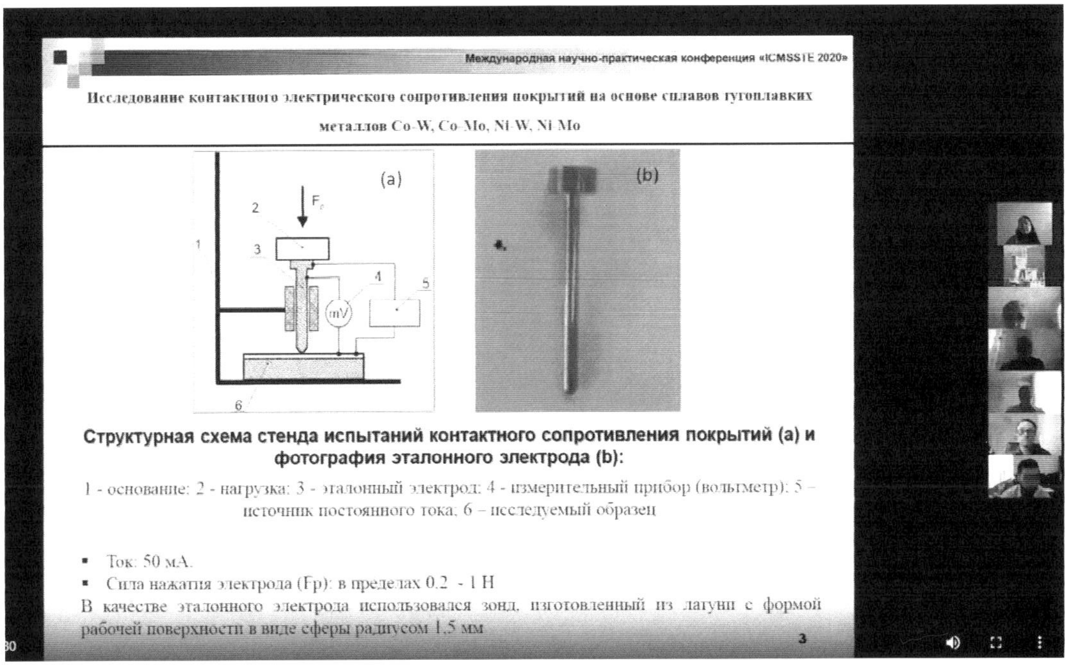

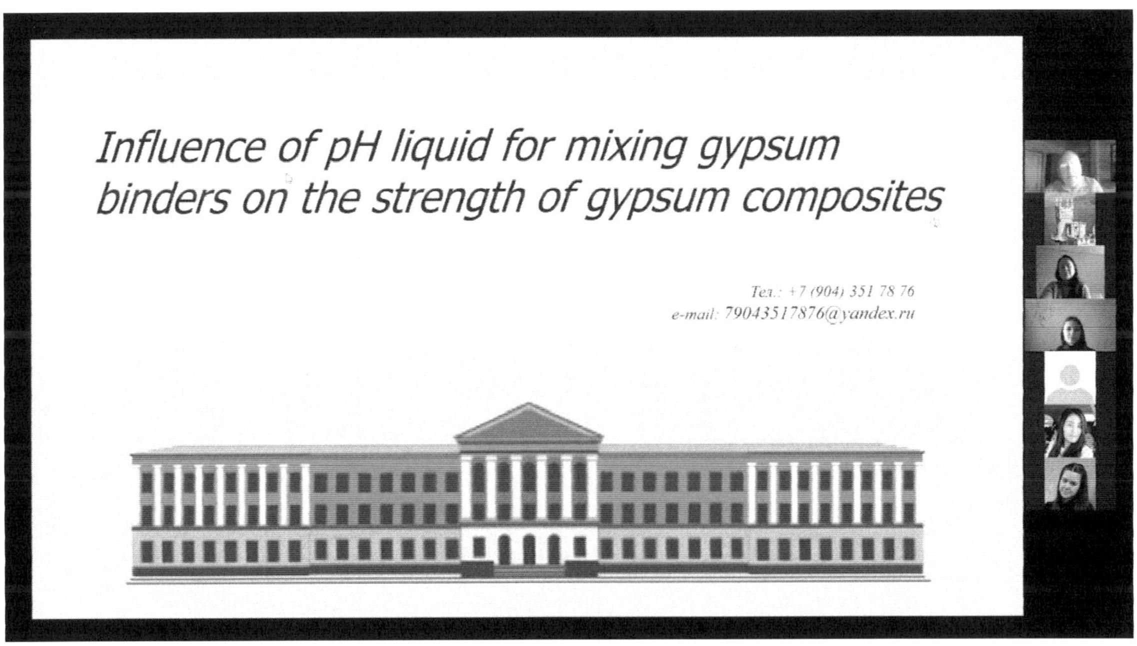

IOP Conf. Series: Materials Science and Engineering **889** (2020) 011001 doi:10.1088/1757-899X/889/1/011001

IOP Conf. Series: Materials Science and Engineering **889** (2020) 011002 doi:10.1088/1757-899X/889/1/011002

Peer review statement

All papers published in this volume of *IOP Conference Series: Materials Science and Engineering* have been peer reviewed through processes administered by the proceedings Editors. Reviews were conducted by expert referees to the professional and scientific standards expected of a proceedings journal published by IOP Publishing.

IOP Conf. Series: Materials Science and Engineering **889** (2020) 012001 doi:10.1088/1757-899X/889/1/012001

Study of diffusion oxide hardening on intermetallic materials

Elena Chekalova[1*] and Andrey Zhuravlev[1]

[1]Moscow Polytechnic University, Moscow, 107023 Avtozavodskaya St. 16, Russia

*alenka.2019@inbox.ru

Abstract. In the last decade, there has been a tendency to increase the use of titanium ortho-alloys on an intermetallic basis for the manufacture of critical parts. The technology of diffusion oxide hardening on intermetallic materials to increase hardness is proposed. The results of a metallographic study of the structure and phase composition of diffusion oxide hardening are presented. According to the results, the hardness of the VIT-1 intermetallic material in the initial state is 1250 MPa, and with hardening it is 1563 MPa, which is 25% higher than the base material.

1. Introduction

Recently, in industry, all attention has been focused on a new class of materials based on titanium aluminides with high specific strength characteristics, heat resistance, and heat resistance.

In these materials, the main alloying element is niobium, which increases strength, ductility, toughness and heat resistance. In addition to niobium, these materials also contain Mo, Ta, W, and Zr. This material is recommended for the manufacture of aircraft engine parts with an operating temperature of up to 700° C [1-6].

Despite a large number of studies on orthorhombic alloys, the mechanism of ortho-phase formation, and the study of the occurrence of phase transformations of ortho-alloys over a wide temperature range, the formation mechanism remains controversial. A set of service properties is formed not only by obtaining a specific phase composition, but also due to structural characteristics - morphological features of the structure of phase components, their distribution over the grain body, the presence of grain-boundary precipitates, etc.

Intermetallic compounds formed by two or more metals having a special crystal lattice, structure and properties are different from the properties of the starting components. Most intermetallic compounds have an ordered super structural lattice, the formation of which is associated with the interatomic interaction of components at temperatures below the melting point. In intermetallic compounds, there can be metallic, covalent and mixed types of atomic bonds in the crystal lattice. Intermetallic compounds with a metal bond of atoms are materials with properties similar to those of ductile metals that do not have high heat-resistant characteristics, and materials with mixed and covalent types of bonds are most interesting to increase heat-resistant characteristics [7-12].

Thus, these materials, which by their nature have a special crystal lattice, structure, and physicomechanical properties, are promising heat-resistant materials intended for use in rocket and space technology and in the field of aviation and engine building, but further research is also required to study the formation of strong directional bonds, structures with new physicochemical properties [13-16].

Based on this, in this work, the task was set to study the morphological changes of hardened intermetallic materials on physico-chemical properties.

The aim of the work is to study the diffusion oxide hardening of intermetallic materials on the structure, phase composition, and also on the study of hardness after oxide hardening.

2. Materials and methods

Discrete diffusion hardening was carried out on equipment, which consists of the UIV-1 device with a unipolar positive crown, from a work bed, fasteners, electromechanical units, electrical wiring, air duct, as well as a pressure regulator and flexible hoses (figure 1). Additionally, included is a monitor, computer, driver unit and system software. The UIV-1 device - is a nozzle - an ionizer combining the organization of a directed air flow and its activation by positive ions [17-20].

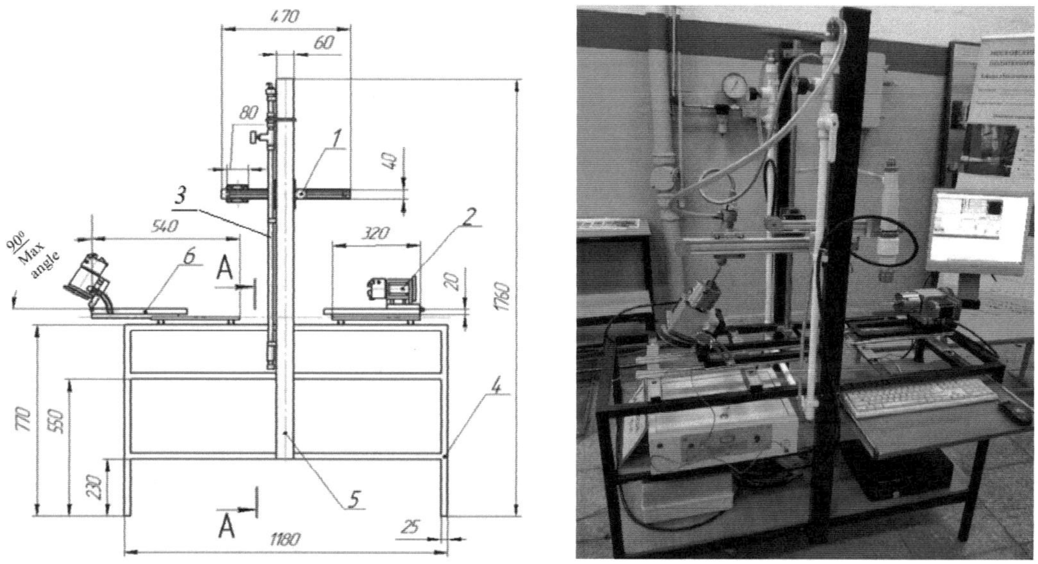

Figure 1. Scheme and photo of equipment with a sample mounting mechanism for hardening with the UIV installation: 1 - linear displacement module; 2 - control unit of the stepper motor; 3 - air duct; 4 - case; 5 - fixing rack; 6 – guide.

Samples are fixed in (2) the rotary axis with the control unit of the stepper motor (figure 1). Then set the predetermined frequency of its rotation or the specified feed linear displacement, according to a computer program. The processed sample can perform both rotational and reciprocating motion.

For research used samples from intermetallic materials, in particular VIT-1.

Metallographic studies of diffusion oxide hardening were carried out: the phase composition of the hardening was performed on a Dron 6 diffractometer in Fe-Kα radiation and the chemical composition of the hardening was determined using a JSM-5610 LV electron microscope. The presence of a low-vacuum operating mode of the microscope allows us to study non-conductive inorganic and organic objects without samples of preparation and application of conductive coatings, i.e. receive the image from a real surface. The presence of two types of detectors allows obtaining images in the modes of secondary and backward reflected electrons. The JSM-5610 LV scanning electron microscope is equipped with the EDX JED-2201 chemical microanalysis system, which allows simultaneous automatic qualitative and quantitative chemical analysis of up to 99 areas of interest in the image of the object under study.

Micro X-ray spectral analysis of the chemical composition allows one to determine the content of elements of diffusion oxide hardening with step-by-step scanning of the sample with an electron probe in the direction perpendicular to the side surface of the sample in the cross section of the microsection. The number of scan steps determines the thickness of the coating. X-ray diffraction analysis allows to determine the phase composition. Hardness was measured using a Solver Next microscope with a nano-sclerometric head (nanoindenter) in sclerometry mode.

IOP Conf. Series: Materials Science and Engineering **889** (2020) 012001 doi:10.1088/1757-899X/889/1/012001

3. Results and discussion

As a result of hardening of the sample from the VIT-1 intermetallic material, the presence of oxides is present on the surface, as evidenced by spectral chemical analysis (table 1).

Table 1. The chemical composition of the hardened sample.

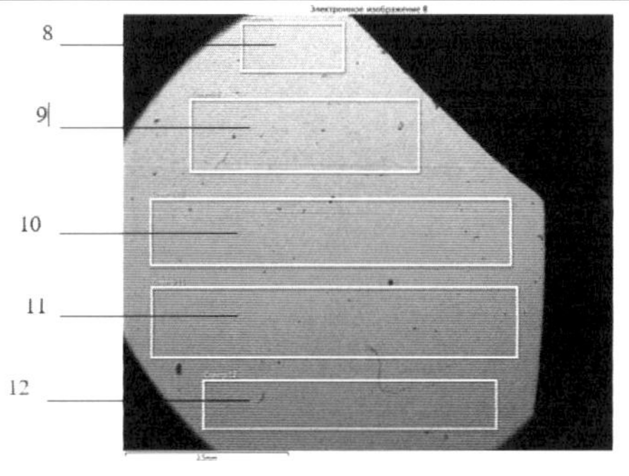

a

Spectrum	O	Al	Si	Ti	Zr	Nb	Ta	W	Total
spectrum 8	**7.83**	**9.72**	0.37	**40.65**	2.05	**37.05**	1.16	1.18	100.00
spectrum 9	**11.02**	**9.32**	0.44	**39.49**	1.97	**35.85**	0.93	0.99	100.00
spectrum 10	**11.47**	**9.18**	0.44	**39.31**	2.02	**35.79**	0.81	0.98	100.00
spectrum 11	**12.49**	**9.03**	0.32	**38.91**	1.85	**35.22**	1.05	1.13	100.00
spectrum 12	**12.95**	**8.90**	0.35	**38.69**	1.96	**35.03**	1.01	1.11	100.00
source base	0.05	**11.12**	0.19	**43.95**	2.28	**39.9**	1.16	0.91	100.00

b

The content of Si, W, Zr, Mo, Ta practically did not change, while it was found that oxygen is present only in the surface layer (12%) (table 1).

The increase in the oxygen content in the coating is explained by the ionization effect achieved by using a corona discharge.

Thus, it can be argued that the decrease in the content of Ti, Al and Nb in the surface layer is due to the oxidation and formation of oxides: titanium oxide (anatase) $TiO_2(\alpha)$, aluminum oxide (corundum) $Al_2O_3(\alpha)$ and the formation of intermetallic compounds $Ti_3Al(\alpha_2$- phase) и $Ti_2AlNb(\beta_2)$ having an ordered superstructural crystal lattice.

Intermetallic Ti_3Al (α_2- phase) has an ordered tetragonally curved face-centered structure. It should be noted that oxides have vacancies at the lattice sites, there is a cationic vacancy due to energy fluctuations, one of the atoms can receive enough energy from neighboring atoms to leave the lattice site, i.e. in place of one ion, two ions are formed at two other sites. Thus, any crystal located at a temperature other than zero will always contain a certain number of the indicated thermal defects.

The characteristic microstructure of diffusion oxide hardening on the samples is shown in figure 2.

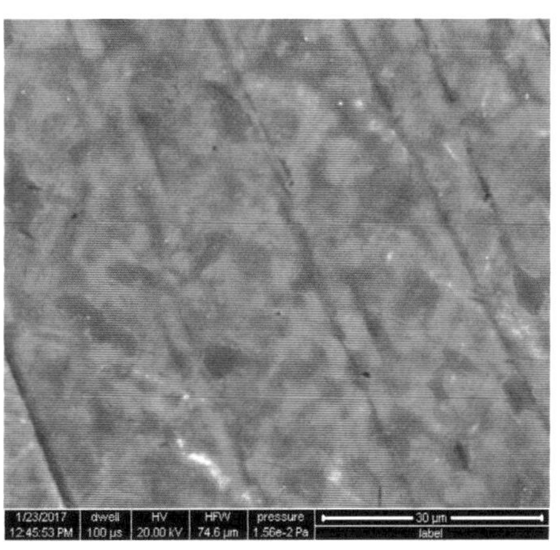

Figure 2. Microstructure of the surface of the sample with diffusion oxide hardening (x 2000).

The macrostructure showed that this hardening obtained by the diffusion method has a polycrystalline structure. In the mode of back-reflected electrons, the oxygen-saturated surface of the sample has a uniform color (see figure 2), which indicates uniform coverage.

In this case, the structure of the main alloy does not undergo noticeable changes in the surface layer, since the process occurs during ionization of the corona discharge, oxygen ions interact with the atoms of the alloy and form a thin oxide film.

X-ray diffraction analysis obtained the phase composition of the sample from the VIT-1 intermetallic compound (figure 3).

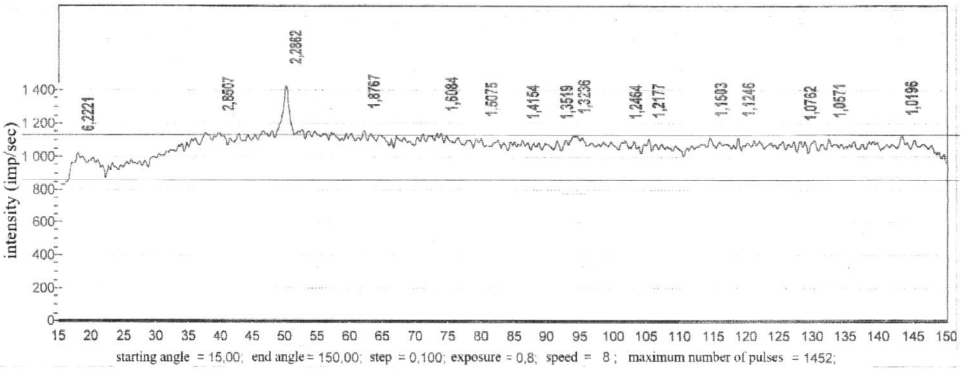

starting angle = 15,00; end angle = 150,00; step = 0,100; exposure = 0,8; speed = 8 ; maximum number of pulses = 1452;

№\(Alfa1)	angle	area	intensity	half latitude	interstrip	% max	% area
1	17,900	524,53	175	2,7602	6,2221	38,55	46,65
2	39,700	725,74	271	2,4742	2,8507	59,69	64,55
3	50,100	1124,39	454	2,2858	2,2862	100,00	100,00
4	62,100	602,59	254	2,1909	1,8767	55,95	53,59
5	74,000	462,82	228	1,8711	1,6084	50,22	41,16
6	79,900	240,33	196	1,1292	1,5075	43,17	21,37
7	86,300	316,78	179	1,6313	1,4154	39,43	28,17
8	91,454	196,43	146	1,2439	1,3519	32,16	17,47
9	94,000	538,87	206	2,4172	1,3236	45,37	47,93
10	101,900	285,30	155	1,6985	1,2464	34,14	25,37
11	105,300	246,34	181	1,2532	1,2177	39,87	21,91
12	114,600	267,30	157	1,5691	1,1503	34,58	23,77
13	118,794	164,38	158	0,9596	1,1246	34,80	14,62
14	128,168	171,56	127	1,2472	1,0762	27,97	15,26
15	130,220	154,63	133	1,0733	1,0671	29,30	13,75
16	132,600	303,48	129	2,1676	1,0571	28,41	26,99
17	143,376	194,43	170	1,0574	1,0196	37,44	17,29

a

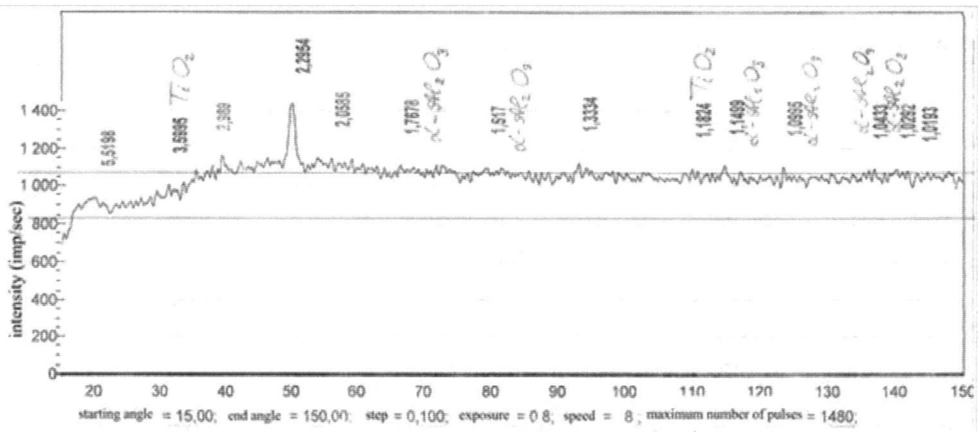

starting angle = 15,00; end angle = 150,00; step = 0,100; exposure = 0 8; speed = 8 ; maximum number of pulses = 1480;

№\(Alfa1)	angle	area	intensity	half latitude	interstrip	% max	% area
1	20,200	365,91	202	1,6687	5,5198	34,27	26,26
2	31,200	442,41	243	1,6997	3,5995	43,09	31,74
3	37,791	392,13	295	1,2257	2,9890	52,96	28,14
4	39,400	492,35	374	1,2158	2,8715	67,15	35,33
5	49,886	1393,67	557	2,3102	2,2954	100,00	100,00
6	56,100	967,22	326	2,7338	2,0585	58,53	69,40
7	66,400	485,34	293	1,5486	1,7678	52,06	34,82
8	79,300	746,47	236	2,9179	1,5170	42,37	53,56
9	93,100	353,62	220	1,4858	1,3334	39,50	25,37
10	109,903	192,04	174	1,0185	1,1874	31,24	13,78
11	114,658	314,64	156	1,8570	1,1499	28,01	22,58
12	117,100	249,66	145	1,8846	1,1347	26,03	17,93
13	123,378	143,90	175	0,7578	1,0995	31,42	10,37
14	136,200	198,02	104	1,7613	1,0433	18,67	14,21
15	140,265	294,16	90	3,0136	1,0292	16,16	21,11
16	143,472	52,62	(101)	0,4788	1,0193	18,13	3,78

b

Figure 3. Phase composition of the sample: a- before hardening; b- after hardening X-ray analysis confirmed the presence of oxides: TiO_2 (anatase) and α-Al_2O_3 (corundum) and the formation of intermetalides Ti_3Al(α_2-фаза) и Ti_2AlNb(β_2).

IOP Conf. Series: Materials Science and Engineering **889** (2020) 012001 doi:10.1088/1757-899X/889/1/012001

Metallographic studies were carried out on the colors of maturity; as a result, it was found that the composition of the hardened sample from the VIT1 intermetallic material includes the following oxides: titanium oxide (anatase) $TiO_2(\alpha)$; aluminum oxide (corundum) $Al_2O_3(\alpha)$ and intermetallic compounds $Ti_3Al(\alpha_2$- phase) and Ti2AlNb(β_2), as evidenced by color discoloration (figure 4).

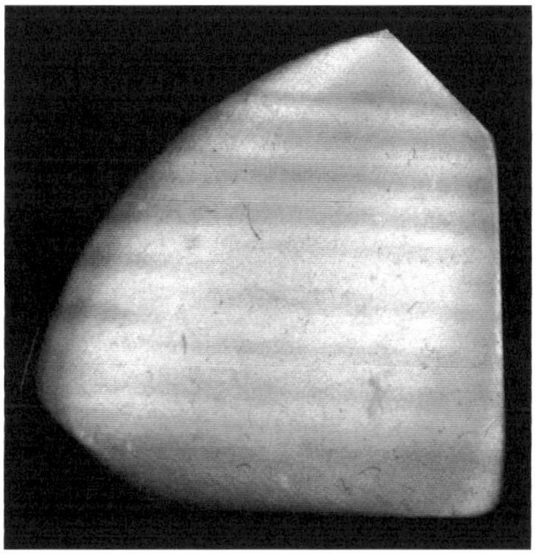

Figure 4. Discolouration color interference on the sample surface after diffusion oxide hardening.

Thus, it can be argued that the decrease in content – Ti, Al, Nb in the surface layer due to the oxidation process and the formation of oxides: $TiO_2(\alpha)$ (anatase) and $Al_2O_3(\alpha)$ (corundum) and the formation of intermetalides Ti_3Al (α_2- phase) and Ti_2AlNb (ortho-alloy β_2).

The oxygen content in the material affects the mechanical characteristics of the coating, in particular hardness. 4 scratches were made on a sample of VIT-1 intermetallic material. Each section was measured in several cross sections (blue stripes) (figure 5).

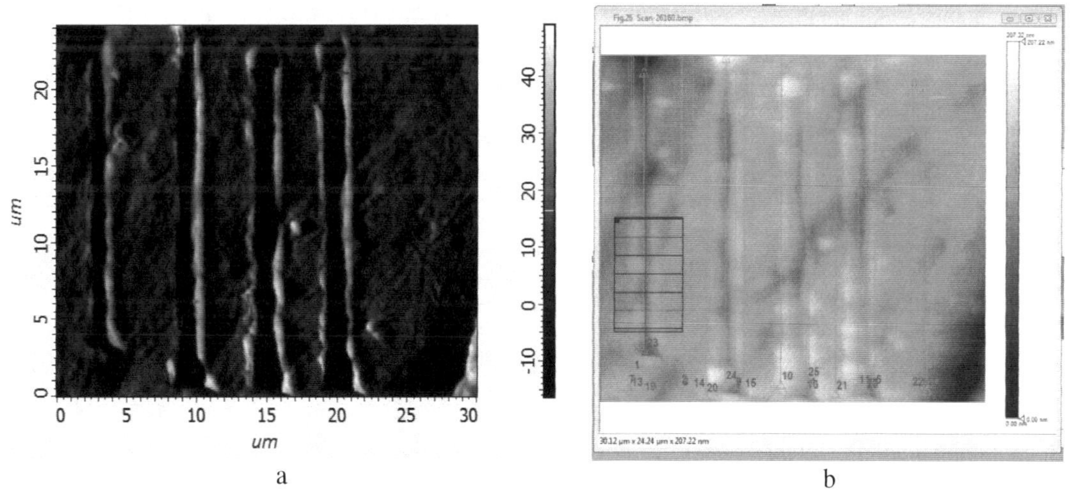

Figure 5. Scratch diagram of a coated sample: a) determination of hardness in sclerometry mode at a depth of 20 nm; b) cross sections of scratches in depth.

The measurement results are presented in figure 6.

According to the results, the hardness of the VIT-1 intermetallic material in the initial state is 1250 MPa, and with hardening it is 1563 MPa, which is 25% higher than the base material.

IOP Conf. Series: Materials Science and Engineering **889** (2020) 012001 doi:10.1088/1757-899X/889/1/012001

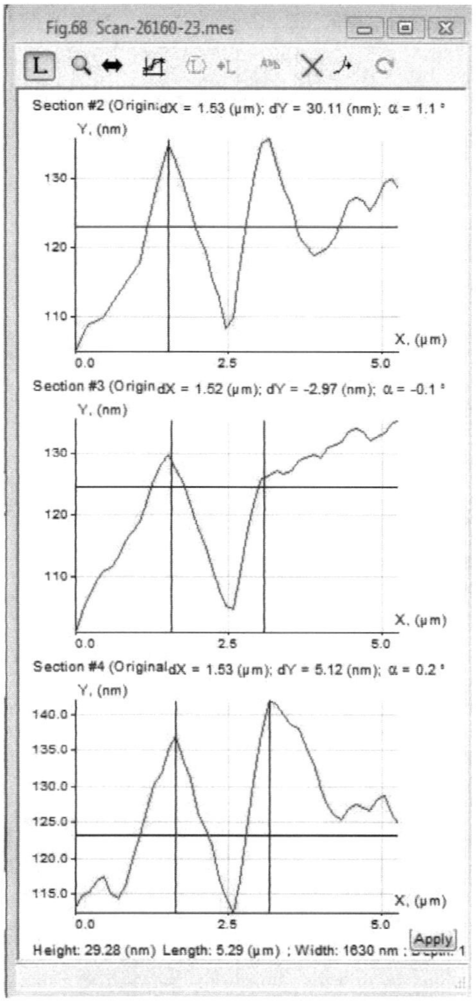

Figure 6. Hardness Results.

4. Conclusion

The proposed technology of diffusion oxide hardening on an intermetallic material can be different, since for each material the formation of the structure of the hardened material is individual in nature, depending on the chemical composition of the material. Diffuse oxide hardening prevents the propagation of cracks, and, therefore, contributes to an increase in resistance during operation.

5. References

[1] Kazantseva N V, Demakov S L and Popov A A 2007 Microstructure and plastic deformation of ortorombic titanium aluminides Ti_2AlNb. III. Formation of transformation twins upon the B2→O phase transformation *Phys. Met. Metallogr.* No 4 **103** 378-87

[2] Bendersky L A, Roytburd A and Boettinger W J 1994 Phase transformations in the $(Ti, Al)_3Nb$ section of the Ti-Al-Nb system -I. Microstructural predictions based on a subgroup relation between phases *Acta Metal. Mater.* No 7 **42** 2323-35

[3] Kazantseva N V, Grinberg B A Gulyaeva N.P. and et al. 2003 Microstructure and plastic deformation of ortorombic titanium aluminide Ti_2AlNb. II. Phase and structural transformation upon severe plastic deformation *Phys. Met. Metallogr* **96** 368-77

[4] Kazantseva N V, Demakov S L and Popov A A 2007 Microstructure and plastic deformation of ortorombic titanium aluminides Ti_2AlNb. IV. Formation of transformation twins upon the α_2→O phase transformation // *Phys. Met. Metallogr.* No 4 **103** 388 -94

[5] Cizeka J, Man O, Roupcovaa P, Lokeb K and Dlouhya I 2015 Oxidation performance of cold spray Ti–Al barrier coated γ-TiAl intermetallic substrates *Surf. Coat. Technol.* **268** 85-9
[6] Rai Amarendra K, Schmitt Michael P, Bhattacharya Rabi S., Zhu Dongming and Wolfeb Douglas E 2015 Thermal conductivity and stability of multilayered thermal barrier coatings under high temperature annealing conditions//*J. Eur. Ceram. Soc. No* 5 **35** 1605–12.
[7] Habazaki H, Tsunekawa S, Tsuji E and Nakayama T 2012 Formation and characterization of wear resistant PEO coatings formed on &-titanium alloy at different electrolyte temperatures *Appl. Surf. Sci.* **259** 711–18
[8] Cheng Y, Wu X, Xue Z, Matykina E, Skeldon P and Thompson G E 2013 Microstructure, corrosion and wear performance of plasma electrolytic oxidation coatings formed on Ti – 6Al – 4V alloy in silicate-hexametaphosphate electrolyte *Surf. Coat. Technol.* **217** 129–39
[9] Dunleavy C S, Golosnoy I O, Curran J A and Clyne T W 2009 Characterisation of discharge events during plasma electrolytic oxidation *Surf. Coat. Technol.* **203** 3410–19
[10] Yerokhin A, Leyland A and Matthews A 2002 Kinetic aspects of aluminium titanate layer formation on titanium alloys by plasma electrolytic oxidation // *Appl. Surf. Sci.* **200** 172–84
[11] Sun X T, Jiang Z H, Xin S G and Yao Z P 2005 Composition and mechanical properties of hard ceramic coating containing α-Al2O3 produced by microarc oxidation on Ti – 6Al – 4V alloy *Thin Solid Films* **471** 194–99
[12] Rakoch A G, Gladkova A A, Zayar Linn, Strekalina D M 2015 The evidence of cathodic micro-discharges during plasma electrolytic oxidation of light metallic alloys and micro-discharge intensity depending on pH of the electrolyte *Surf. Coat. Technol.* **269** 138–44
[13] Shelekhov E V, Sviridova T A Programs for X-ray Analysis of Polycrystals 2000 *Met. Sci. Heat Treat.* No 8 **42** 309–13
[14] Tillous K, Toll-Duchanoy T and Bauer-Grosse E 2009 Microstructure and phase composition of microarc oxidation surface layers formed on aluminium its alloys 2214 – T6 and 7050 – T74 *Surf. Coat. Technol.* No 19 **203** 2969–73
[15] Wenbin Xue, Zhiwei Deng, Ruyi Chen and et al. 2001 Microstructure and properties of ceramic coatings produced on 2024 aluminum alloy by microarc oxidation *J. Mater. Sci.* **36** 2615–19
[16] Pherson R 1973 Formation of metastable phases in flame and plasma-prepared alumina *J. Mater. Sci.* **8** 851–58
[17] Chekalova E A, Chekalov P D and Solomatina R D Published on April 20 2015 *The method of forming a wear-resistant coating on the surface of a metal part* (Patent 2548835 of the Russian Federation for the invention, IPC C23C8/36 Patent holder Chekalova E A)
[18] Chekalova E A 2014 *Increasing the durability of the cutting tool and heavily loaded parts by applying a diffusion mesh coating* (monograph, Moscow: Publishing House of the University of Mechanical Engineering) p 127
[19] Chekalova E A 2017 Study of wear-resistant coatings on VT3-1/E titanium alloy *Mat. Sci.* **9** 3-6
[20] Chekalova E A 2017 The study of diffusion local coating of the oxide type on the tool material *Mat. Sci.* **8** 24 - 9

IOP Conf. Series: Materials Science and Engineering **889** (2020) 012002 doi:10.1088/1757-899X/889/1/012002

The x-ray diffraction study of the composition of the powder materials from the waste steel X13

E.V. Ageeva[1], S.V. Khardikov[2]*, A. N. Novikov[3]

[1]Southwest state University, Kursk, 305040, Russian Federation.
[2]Southwest state University, Kursk, 305040, Russian Federation.
[3]OSU named after I. S. Turgenev, Orel, 302026, Russian Federation.

*hardikov1990@mail.ru

Abstract. To restore and protect the surfaces of parts using a layer of molten metal, various methods of surfacing are used, differing from each other by melting methods and compositions of the welding medium: electric arc, flame, plasma, laser, induction, etc. With this technology, metals of various chemical composition, including copper, bronze, cast iron, as well as nickel, cobalt and chromium alloys can be deposited on the working surfaces of steel structures. There are several ways to obtain such powder materials. The most promising method for obtaining powder materials is the electro-erosion dispersion method (EED), which is distinguished by its ecological purity of the process and relatively low energy expenditure. To study the structure of the particles of the obtained powders, their X-ray structural analysis was performed on a RigakuUltima IV X-ray diffractometer.

1. Introduction

Metal surfacing is used to restore the geometry of worn-out parts of machines and mechanisms, to form reinforcing metal layers on the surface of products and to create bimetallic structures. At its core, surfacing is one of the types of welding technologies, since it is based on the same physical and technological principles as traditional types of welding. To restore and protect the surfaces of parts using a layer of molten metal, various methods of surfacing are used, differing from each other by melting methods and compositions of the welding medium: electric arc, flame, plasma, laser, induction, etc. Using this technology, metals of various chemical composition, including copper, bronze, cast iron, as well as nickel, cobalt and chromium alloys can be deposited on the working planes of steel structures [1–6].

The technology of surfacing allows to achieve not only reliable adhesion of the applied metal to the base, but also to obtain the required physical and chemical characteristics of the deposited layer. The first is achieved by the qualitative preparation of the base product and the exact observance of technological regimes, and the second by the correct selection of welding materials. The essence of the surfacing consists in uniform deposition of narrow strips of molten metal on the surface of the part so that they merge into a continuous metal layer of a given thickness. The properties of the resulting coatings are directly dependent on the quality of the used materials. One of these materials is chromium steel. Powders with a high content of chromium can be used to obtain wear-resistant and heat-resistant coatings.

There are several ways to obtain such powder materials. The most promising method for obtaining powder materials is the electro-erosion dispersion method (EED), which is distinguished by the ecological purity of the process and relatively low energy expenditure [7–10].

For the development of technologies for the practical application of powder, obtained from waste, and evaluating the effectiveness of its use, complex theoretical and experimental studies are required.

2. Materialsandtechniques

To study the structure of the particles of the obtained powders, their X-ray structural analysis was performed on a RigakuUltima IV X-ray diffractometer.

Application area:

- phase analysis of samples;
- quantitative phase analysis of samples;
- determination of areas of coherent scattering and microstrain;
- texture analysis.

Features of the diffractometer series Ultima IV:

- goniometer radius - 185 mm on the output beam;
- slits of variable width, which allow to keep the irradiated surface of the sample unchanged;
- Θ / Θ vertical-type goniometer for all three configurations, adapted to install a wide range of additional optical components;
- high-speed X-ray detector D / teX Ultra, which has a high counting rate, high energy level of resolution and low noise.

Multi-purpose attachment for texture and residual stress analysis with turntables Multi purpose attachment MPA-IV χ (kai) - φ (phi) - Z stage includes:

- automanager of samples (10 ditch);
- software: qualitative and quantitative phase analysis, ICDD PDF-2 diffractogram database, crystallinity analysis, residual stress analysis, construction of direct and reverse pole figures, orientation distribution function.

Specifications radiation source:

– small-sized using a high-frequency converter, maximum power - 3 kW;
– tube voltage - 20–60 kV;
– tube current - 2–60 mA;
– tube anode material – Cu;
– focus size - 0.4 x 12 mm.

There is a vertical type goniometer with a fixed sample. The RigakuUltima IV X-ray diffractometer provides independent scanning of each Θs or Θd axis or in scanning mode with associated Θs / Θd axes.

Goniometer settings:

- radius - 185 mm;
- the range of the scan angle in the associated mode of the Θs / Θd axis is from –30 to +1620 (2Θ);
- deviation of the axis from –1.50 to +810, осиd axis from –950 to +1200;
- scanning step along the Θs or Θd axis - 0.0001–60;
- in the mode of coupled axes - 0.0002–120 (2Θ).

Scan speed in the associated axis Θs / Θd mode is 0.020 ~ 1000 (2Θ), independently of each axis is 0.010 ~ 500.

Two standard sets of Soller slits for working in focusing geometry and pseudo-parallel beam geometry. The device has fully automatic adjustment for goniometer, amplitude discriminator, counter, optical nodes and additional devices. The detector is a scintillation counter with linearity of 700,000 pulses (standard), a uniaxial semiconductor detector D / teX Ultra with a sensitivity exceeding the sensitivity of the scintillation counter by two orders of magnitude.

The powder material of the test sample is poured into a cuvette with a diameter of 20 mm and a depth of 0.5 mm and pressed so that the sample surface is parallel to the edges of the cuvette. The

IOP Conf. Series: Materials Science and Engineering **889** (2020) 012002 doi:10.1088/1757-899X/889/1/012002

cuvette is placed in the holder, as a result of which, on an adjusted goniometer, the sample surface is aligned with the focusing plane.

Survey parameters:
– range 10–95 degrees 2Θ;
–step - 0.020 degrees;
–speed - 1 degrees / min;
–operating voltage - 40 kV;
–current - 40 mA.

Diffraction pattern processing:
– PDXL program;
– smoothing - Savitsky-Golay method;
– background calculation - Sonneveld-Visser method;
– peak search - peak top method.

3. Conduct X-ray analysis

The properties of powder particles, on which their area of application depends, are largely determined by their structure.

The results of X-ray diffraction analysis of powders, obtained by the method of EED from chrome-containing wastes, are presented in Figure 1 and in Table 1.

On the experimental radiograph there are strong reflexes corresponding to:
- phase Fe_3O_4;
- phase Cr;
- phase Fe_3C.

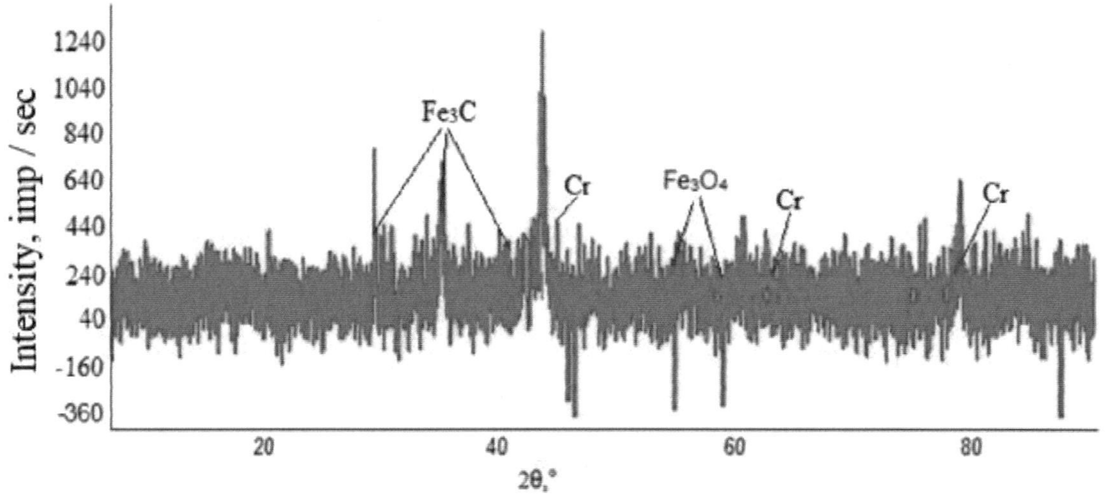

Figure1. The powder XRD pattern.

Table 1. The x-ray maxima of the powder in lighting kerosene

2θ, (degree)	Height (Hz)	Intensity W (degree)	Size (angstroms)
19.67(3)	113(31)	0.10(8)	4.509(8)
29.22(3)	188(40)	0.18(8)	3.054(3)
35.50(4)	343(54)	0.55(12)	2.527(3)
43.09(7)	146(35)	1.0(4)	2.098(3)
44.523(16)	773(81)	0.42(7)	2.0333(7)
57.387(16)	118(32)	0.36(17)	1.6044(4)
62.8(2)	73(25)	0.8(5)	1.479(5)
64.929(10)	181(39)	0.17(6)	1.4350(2)
82.37(13)	118(32)	0.7(4)	1.1698(15)

4. Conclusion

The phase composition of the powder, obtained by the method of EED, is determined by the nature of the WF: the presence of carbon, the dielectric constant and boiling point. Dispersion of waste alloys of ball-bearing steel in lighting kerosene led to carbon enrichment and the formation of the Fe_3C phase.

Thus, the phase composition of powders, obtained by the method of EED, is influenced by the complex of thermochemical properties of working fluids. This can determine the choice of WF to obtain a given composition of powders.

5. References

[1] Lazarenko B R and Lazarenko N I A. p. 70000 USSR, B 22f, 09/00 *A method for producing powders and a device for its implementation*.- No. 1371/321510; declared 04/01/1943; publ. 09/23/1964, Bull. 22, 2
[2] Karlsson J, Snis A, Engqvist H and Lausmaa J 2013 *J. of Mat. Proc. Tech.* **213** 2109-18
[3] Janitor M I2006 *Development of physico-chemical and technological fundamentals of the processing of tungsten-cobalt hard alloy by electroerosive dispersion*: *diss. ... cand. tech. Sciences,* Dvornik Maxim Ivanovich (Khabarovsk) p 116
[4] Ageev E V 2011 *Hard. tech. and coat.* **6** 8-14
[5] Latypov R A, Latypova G R, Ageev E V and Burak P I2013 *ISJ* **5** 80-86
[6] Ageev E V2011 *Electromet.* **10** 24–27
[7] AgeevE V, KirichekA V, AltuhovAYuand AgeevaEV2014J. *Nano- Electron. Phys.* **6** 3, 03001.
[8] AgeevaEV, AgeevEV, HoryakovaNMandMalukhovVS2014*J. Nano- Electron. Phys.* **6** 3, 03011.
[9] AgeevaEVAgeevEV KarpenkoVYuandOsmininaAS2014*J. Nano- Electron. Phys.* **6** 3, 03049.
[10] AgeevEV The patent 2449859, the Russian Federation, C2, B22F9/14. No 2010104316/02; appl. 08.02.2010; publ. 10.05.2012. - 4 p.

Acknowledgments

The work was supported by a scholarship of the President of the Russian Federation to young scientists and graduate students (SP-945.2019.1).

IOP Conf. Series: Materials Science and Engineering **889** (2020) 012003 doi:10.1088/1757-899X/889/1/012003

X-ray diffraction analysis of sintered products from electro-erosion materials, obtained from Cr17 alloy waste

A A Sysoev [1*], SA Grashkov[2] and NA Pivovar[3]

[1]Southwest State University, 305040 Kursk, Russian Federation
[2]Kursk state agricultural Academy named after I. I. Ivanov, 305004 Kursk, Russian Federation
[3] Kursk state agricultural Academy named after I. I. Ivanov, 305004 Kursk, Russian Federation

*evoking09@rambler.ru

Abstract. This article presents the phase composition study of sintered samples from electro-erosion materials of the Cr17 alloy, obtained in lighting kerosene. It was established that the main phases in the sintered sample are $FeCr_{0,29}Ni_{0,16}C_{0,06}$, Cr, Cr_7C_3.

1. Introduction

Nickel-free Cr17 stainless steel is a more cost-effective substitute for expensive nickel-containing corrosion-resistant and heat-resistant steels. Cr17 steel is ductile, durable and well welded, often used in annealed condition. The valves of hydraulic presses, conveyors for furnaces and drying lines, pipes, boilers are made from it.

The widespread use of Cr17 steel in various industries leads to a large accumulation of its waste requiring processing. Currently, there are many ways to recycle metal waste in order to reuse it. However, the disadvantages of the known methods are increased energy consumption, multi-operational technological process [1-7].

The most promising method for processing metal waste is the method of electroerosive dispersion (EED), which is distinguished by the environmental cleanliness of the process and relatively low energy costs.

Carrying out the planned measures will allow to solve the problem of waste disposal and their further use and thereby reduce the cost of production of the final product.

The aim of the work was to carry out X-ray diffraction analysis of sintered samples from electroerosive materials of Cr17 alloy, obtained in lighting kerosene.

2. Materials and Methods

To obtain powder from Cr17 steel wastes (GOST 5632-72), an EED device, developed by the authors, was used [7]. Waste was loaded into a reactor, filled with a working fluid - lighting kerosene, the process was carried out with the following electrical parameters: capacity of discharge capacitors 55 μF, voltage 120 ... 130 V, pulse repetition rate 95 ... 100 Hz.

The powder was consolidated by spark plasma sintering using the SPS 25-10 spark plasma sintering system (Thermal Technology, USA) according to the scheme shown in (Figure 1).

IOP Conf. Series: Materials Science and Engineering **889** (2020) 012003 doi:10.1088/1757-899X/889/1/012003

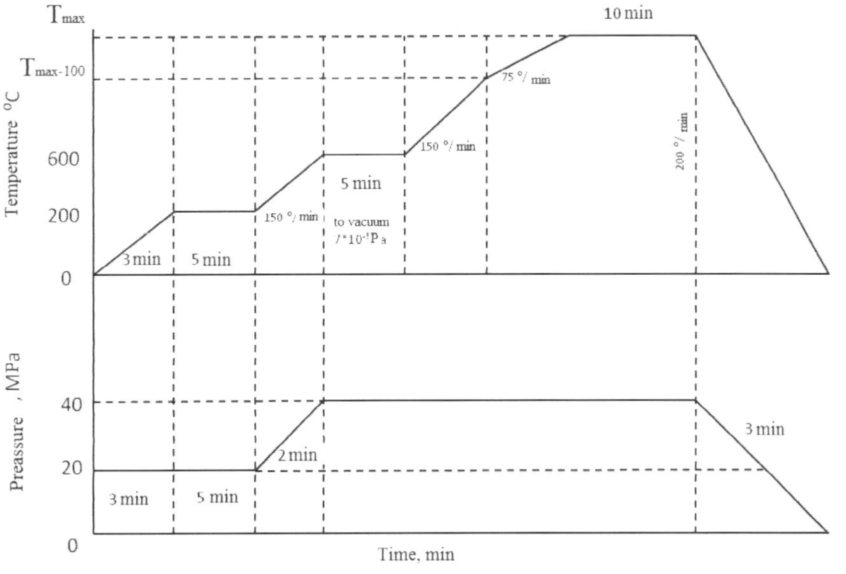

Figure 1. Consolidation of powders by spark plasma sintering (scheme).

The starting material was placed in a graphite matrix, placed under a press in a vacuum chamber. The electrodes integrated into the mechanical part of the press fed an electric current to the matrix and created spark discharges between the sintered particles of the material, providing intense interaction.

The advantages of spark plasma sintering technology are the uniform distribution of heat throughout the sample; high density and controlled porosity; lack of need for binding materials; uniform sintering of homogeneous and dissimilar materials; short cycle time; manufacturing of the part immediately in the final form and obtaining a profile close to the given [8-15].

The phase composition of the sample was studied by X-ray diffraction on a Rigaku Ultima IV diffractometer in Cu-K_{α} radiation (wavelength λ = 0,154178 nm) using Soller slits (Fig. 2).

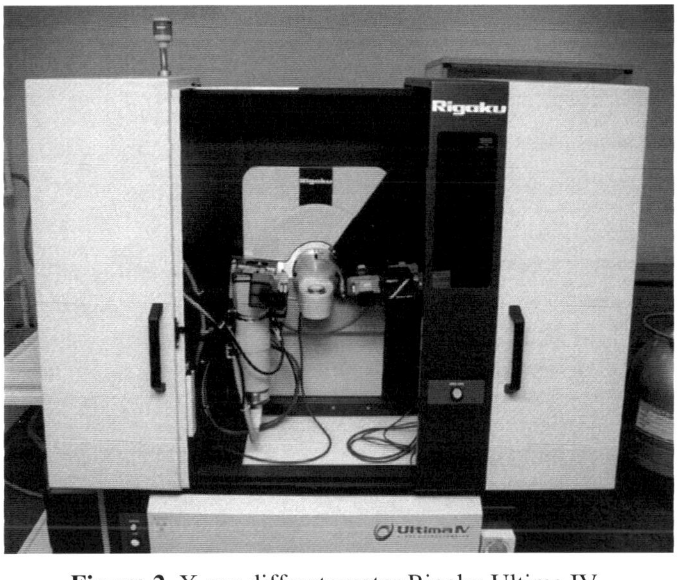

Figure 2. X-ray diffractometer Rigaku Ultima IV.

The diffraction spectrum for phase analysis is taken according to the θ-2Θ scanning scheme with focusing according to Bregt-Brentano in the range of angles of 5 ... 100 deg. 2 Θ. The shooting is carried out in dotted mode with a scanning step Δ (2θ) = 0.02 deg., speed 0.6 deg/min, operating voltage 45 kV, current 200 mA [16-20].

3. Results

To refine the profile of experimental X-ray diffraction patterns, the PDXL RIGAKU software package was used. The background was subtracted by the Sonneveld – Wisser method, the experimental profile was smoothed by the Savitsky – Golay method, and the components kα1 and kα2 were separated by the Rachinger method. To describe the diffraction maxima, a superposition of the Gauss function and the Lorentz function was used. The phase composition was determined using the ICCD PDF-2 database (2014). The diffraction pattern and phase composition of the test sample are shown in Figure 3 and Table 1.

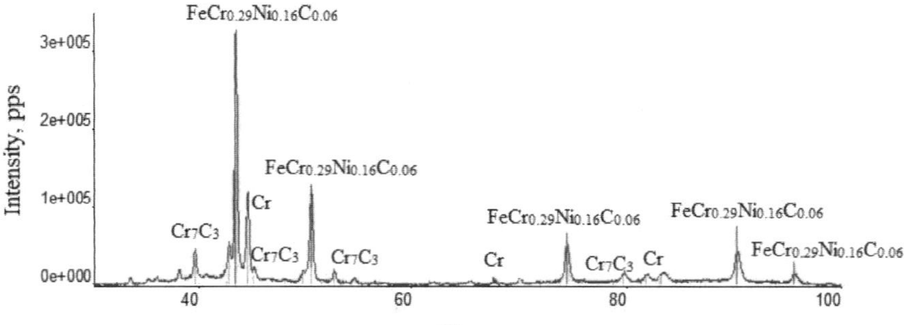

Figure 3. X-ray diffraction pattern of the test sample.

Table 1. The phase composition of the test sample.

Chemical formula	Type of crystal lattice	Lattice parameters
$FeCr_{0,29}Ni_{0,16}C_{0,06}$	225:Fm-3m cubic crystal lattice	a = 3.605153 Å b = 3.605153 Å c = 3.605153 Å
Cr	229:Im-3m cubic crystal lattice	a = 2.871221 Å b = 2.871221 Å c = 2.871221 Å
Cr_7C_3	51:Pmcm orthorhombic crystal lattice	a = 6.940538 Å, b = 12.029685 Å, c = 4.506928Å

4. Conclusion

Based on the performed x-ray microanalysis, it was found that the main phases in the sintered sample of electro-erosion materials of the Cr17 alloy obtained in illuminating kerosene are $FeCr_{0,29}Ni_{0,16}C_{0,06}$, Cr, Cr_7C_3. The study will determine the most relevant application area of the obtained samples and improve the quality of scientific and technological developments.

5. References

[1] Lazarenko B R and Lazarenko N I A. p. 70000 USSR, B 22f, 09/00 *A method for producing powders and a device for its implementation*.- No. 1371/321510; declared 04/01/1943; publ. 09/23/1964, Bull. 22, 2
[2] Gusev V N 1952 *Anodic-mechanical processing of metals* p 321
[3] Janitor M I 2006 *Development of physico-chemical and technological fundamentals of the processing of tungsten-cobalt hard alloy by electroerosive dispersion*: *diss. ... cand. tech. Sciences,* Dvornik Maxim Ivanovich (Khabarovsk) p 116
[4] Ageev E V 2011 *Hard. tech. and coat.* **6** 8-14

[5] Latypov R A, Latypova G R, Ageev E V and Burak P I 2013 *ISJ* **5** 80-6
[6] Ageev E V 2011 *Electromet.* **10** 24–7
[7] Ageev E V Patent 2449859, Russian Federation, C2, B22F9 / 14. *Installation for producing nanodispersed powders from conductive materials*, applicant and patent holder Southwestern State University. - No. 2010104316/02; application 02/08/2010; publ. 05/10/2012
[8] Ageev E V and Ageeva E V 2013 *Bull. of mech. eng.* **11** 51–7
[9] Ageev E V, Semenikhin B A and Latypov R A 2010 *Fund. Prikl. Probl. Tekhn. Tekhn.* **5** 39-42
[10] Oskolkova T N and Budovskikh E A 2013 *Metal. Sci. Heat Treat* **55** 96-9
[11] Karlsson J, Snis A, Engqvist H and Lausmaa J 2013 *J. of Mat. Proc. Tech.* **213** 2109-18
[12] Gu D D, Meiners W, Wissenbach K and Poprawe R 2012 *Int. Mat. Rev.* **57** 133-64
[13] Radek N 2009 *Maint. and Rel.* **4** 10-6
[14] Ribalko A V and Sahin O 2003 *Surf. & Coat. Tech.* **168** 129-35
[15] Chen Z and Zhou Y 2006 *Surf. & Coat. Tech* **201** 1503-10
[16] Galinov I V and Luban R B 1996 *Surf. & Coat. Tech* **79** 9-18
[17] Azarova E V, Levashov E A and Ralchenko V G 2010 *Metall.* **8** 50-5
[18] Pereverzev A and Ageev E 2019 *MATEC Web of Conf.* **298** 37
[19] Ageev E V, Khardikov S V, Vorobyev E A and Sysoev A A 2019 *MATEC Web of Conf* **298** 127
[20] Latypov R A, Ageeva E V and Latypova G R 2019 *MATEC Web of Conf* **298** 125

Acknowledgments
The reported study was funded by RFBR, project number 19-33-90053

Texture and anisotropy of the mechanical properties of 2 mm sheets of Mg-9Li-3Al-0.4Ce alloy

S Ya Betsofen[1], I A Grushin[1], A A Petrov[1] and K A Speransky[1]

[1]MAI-Moscow Aviation Institute (National Research University), 121552, Moscow, Russia

Email: s.betsofen@gmail.com

Annotation The phase composition, texture and anisotropy of the mechanical properties of 2 mm sheets of Mg-9Li-3Al-0.4Ce alloy were studied by X-ray diffraction and mechanical tests. It is shown that the alloy contains two solid solutions, the α phase with HCP lattice based on Mg and the β phase with BCC lattice based on Li. Intermetallic $Al_{12}Mg_{17}$ is also present in the alloy. Alloy sheets are characterized by a basal α-phase texture. At the same time, the most durable prismatic orientations are mainly located in the transverse rolling direction (TD), which leads to higher strength properties in this direction compared to the rolling direction (RD). The texture of the basicβ-phase in RD and TD does not differ and, accordingly, does not contribute to the anisotropy of strength properties.

1. Introduction

Magnesium alloys are widely used in aerospace engineering, electronics and transport due to their low density and high specific characteristics of strength and stiffness [1]. With all these advantages, magnesium alloys have three main disadvantages: flammability, low corrosion resistance and reduced ductility, due to the strong anisotropy of the mechanical properties [2, 3]. The first two problems are largely solved by applying protective coatings, and the anisotropy problem is solved by rational doping, primarily REE (cerium, neodymium, dysprosium, samarium, yttrium, gadolinium) [4]. The ductility of magnesium can also be increased by alloying with lithium, which not only plasticizes magnesium, but also significantly reduces the density of alloys (density of lithium is 0.58 g/cm^3 and magnesium 1.74 g/cm^3)[5-7]. One of the problems is the search for optimal compositions of Mg-Li alloys [8 - 11]. The demand for these ultralight alloys is very high, however, there is very little prospecting because of the difficulty of smelting ingots from alloys of the Mg-Li system. In this work, we studied the texture and anisotropy of mechanical properties of the Mg-9Li-1Al alloy.

2. Materials and methods

We studied 2 mm thick sheets of the Mg-9Li-3Al-0.4Ce alloy obtained by rolling at temperatures of 300–350 ° C. X-ray diffraction analysis was carried out on a DRON-7 diffractometer in filtered CuK_α radiation. The texture was determined by the method of inverse pole figures (IPF), while in order to obtain IPF for the RDand TD, type-setting samples were used. The pole densities (P_{hkl}) for each of the m-poles (hkl) at the IPF were estimated by the ratio of the integrated intensities of the reflections of the textured (I_{hkl}) and textureless reference (R_{hkl}) with Morris normalization:

$$P_{hkl} = \frac{I_{hkl} / R_{hkl}}{\sum_{i=1}^{m} \left(A_{hkl} \frac{I_{hkl}}{R_{hkl}} \right)_i} \quad (1)$$

3. Results and discussion

The analysis of the X-Ray patterns of the alloy (figure 1) confirmed the presence of two solid solutions of the α phase with HCP lattice based on Mg and the β phase with BCC lattice based on Li; weak reflections of the intermetallic $Al_{12}Mg_{17}$ are also present on the X-ray diffraction pattern. The microstructure of alloy after rolling (figure 2) is a deformed, elongated in the direction of rolling grain β-phase located in the matrix of the α-phase. An analysis of the IPF for the three directions of the RD, TD, and ND showed (figure 3) that the α-solid solution is characterized by a basal texture with a more pronounced pole density of prismatic orientations in the TD compared with the RD. The BCC β-solid solution is characterized by the texture (110). The texture of the α phase corresponds to the texture of rolling magnesium, while the texture of the BCC solid solution differs from the texture of rolling the BCC metals, and fully corresponds to the texture of the phase transformation of the HCP α phase with the basal texture to the BCC phase with the texture (110) in accordance with the orientation relations Burgers: $(0001)_{HCP}$ // $(110)_{BCC}$ [12].

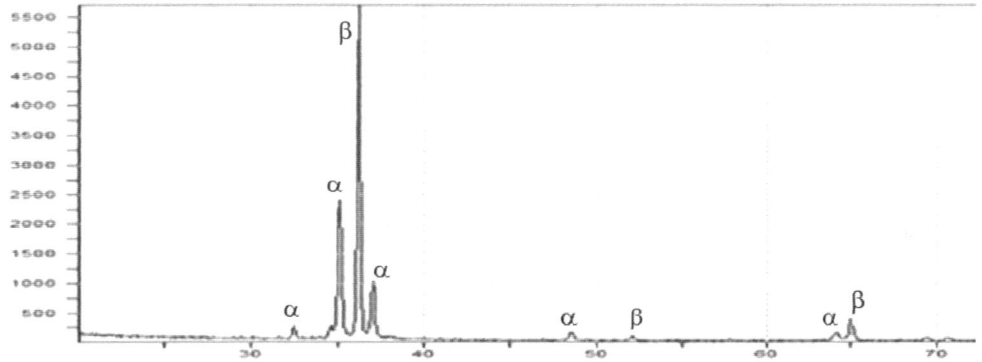

Figure 1. Diffraction pattern of sheets of Mg-9Li-1Al-0.4Ce alloy in CuK_α radiation.

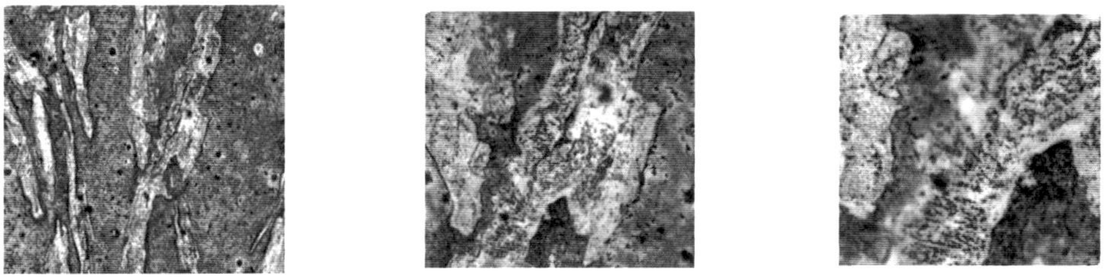

x100 x280 x600

Figure 2. Microstructure of a 2 mm Mg-9Li-1Al-0.4Ce alloy sheet.

Anisotropy of yield strengths and relative elongations were found in the Mg-9Li-3Al-0.4Ce alloy sheets (table 1), characterized by a higher yield strength (10%) in TD and higher ductility in the rolling direction (43%). To interpret the relationship between texture and anisotropy of strength properties, Taylor factors for basal slip were calculated by averaging orientation factors for three orthogonal sheet directions. The values of the Taylor factors (M) for basal slip are more convenient to use compared to the Schmid factors (Φ) when interpreting the anisotropy of the strength properties, since the strength is inversely proportional to the Schmid factor, but proportional to the Taylor factor:

$$\sigma_t=\tau_\kappa/\Phi= \tau_\kappa M \qquad (2)$$

where: σ_t - yield stress; τ_κ - critical resolved shear stress (CRSS).

Averaging was performed for all orientations on the IPF for RD and TD sheets, taking into account the pole density of reflections (P_{hkl}) as a weighting factor for averaging, and then the average value was determined as the inverse of the average value:

$$\overline{M}_{\mathrm{RD(TD)}} = 1/\overline{\Phi}_{\mathrm{RD(TD)}}\,;\quad \overline{\Phi}_{\mathrm{RD(TD)}} = \frac{1}{n}\sum_{n}(P_{hkl}^{\mathrm{RD(TD)}}\Phi_{hkl}^{basal}) \qquad (3)$$

Calculations showed that the Taylor factor in TD is higher than in RD by 36%. Considering that the amount of HCP phase is not more than 50%, it can be considered that the anisotropy of the yield strength of alloy sheets (10%) is completely determined by the texture of the HCP phase.

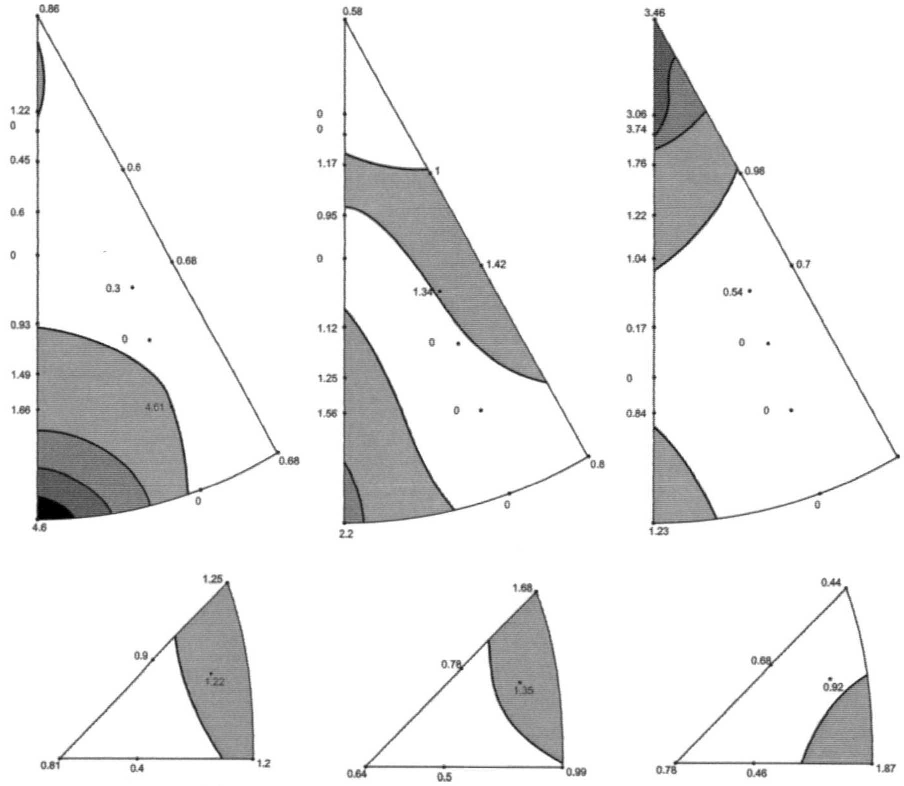

TD RD ND

Figure 3.IPF for RD, TD, and ND in a 2 mm sheet of Mg-9Li-3Al-0.4Ce alloy.

Table 1. The results of mechanical tensile tests for the alloy Mg - 9Li - 1Al-0,4Ce

	RD	**TD**
$\sigma_{0,2}$, [MPa]	176	195
σ_B, [MPa]	215	220
δ, %	28.5	16.7
ψ, %	42.3	26.6

4. Conclusions

1. The phase composition, texture, and anisotropy of the mechanical properties of 2 mm Mg-9Li-3Al-0.4Ce alloy sheets were studied.

2. It is shown that the alloy is two-phase and contains an α phase based on Mg with an HCP lattice and a β phase based on Li with a BCC lattice.

3. The basal α-phase texture and the (110) β-phase texture are present in the alloy sheets, while the texture of the α-phase corresponds to the rolling texture of magnesium, while the texture of the β-phase corresponds to the texture of the α→β phase transformation in accordance with the orientation Burgers relationships.

IOP Conf. Series: Materials Science and Engineering **889** (2020) 012004 doi:10.1088/1757-899X/889/1/012004

5. References

[1] You S, HuangY, Kainer K and Hort N 2017 *Journal of Magnesium and Alloys* **5** 239 53
[2] Betsofen S, Osintsev O, Grushin I, Petrov A and Speranskii K 2019 *Russian Metallurgy (Metally).* **4** 361–73
[3] Lin Y, Wu H, Zhou G, Chiu C and Lee S 2008 *Materials & Design.* 29 2061-5
[4] Betsofen S, Osintsev O, Grushin I, Petrov A and Speranskii K 2019 *Russian Metallurgy (Metally).* **4** 346–60
[5] Haferkamp H, Boehem R, Holzkamp U, Jachik C, Kaese V and Niemeyer M 2001 *Mater. Trans.* **42** 1160–6
[6] Wu R, Qu and. Zhang M 2010 *Rev. Adv. Mater. Sci.* **24** 14–34
[7] Al-Samman T 2009 *Acta Materialia* **57** 2229-42
[8] Wang T, Zhang M, Niu Z and Liu B 2006 *J. Rare Earth.* **24** 797– 800
[9] Wu R, Deng Y and Zhang M. J *2009 Mater. Sci.* **44** 4132–9
[10] Li R, Pan F, Jiang B, Dong H and Yang Q 2013 *Materials Science and Engineering: A.* **562** 33-8
[11] Zhu T, Sun J, Cui C, Zhang T, Wu R, Betsofen S, Leng Z, Zhang J and Zhang M 2014 *Materials Science and Engineering A.* **600** 1–7
[12] Betsofen S, Volkova E and Shaforostov A. 2011 *Russian Metallurgy (Metally)* **1** 66–71

Oxide films resistance to hydrogen penetration in VT6 titanium alloy

O N Gvozdeva[1*], A V Shalin[1] and A S Stepushin[1]

[1]MAI, 125993, 4 Volokolamskoeshosse, Moscow, Russia

*gon7133@mail.ru

Abstract. The ability of hydrogen to penetrate through the oxide film formed during air-atmosphere annealing at elevated temperatures on VT6 titanium alloy samples was studied in this paper. It was investigated that oxide film prevents the penetration of up to 0.4 wt.% hydrogen. It was shown that it is possible to change the depth of hydrogen penetration and the surface layer structure by varying the amount of added hydrogen. It was determined that the formation of a finely dispersed surface structure after the thermal hydrogen treatment allows to increase the surface hardness by 8-10 HRC, compared to the center of material.

1. Introduction

Armor weight reduction while maintaining a high level of its working properties has been a major field of research in the area of vehicles anti-impact protection [1-3]. This is because the most popular IV+ protection class armor must be applied on such vehicles while preserving its possibility of safe movement on public roads [4]. One of the ways to solve the existing problem is usage of much lighter titanium alloys instead of steels [6, 7]. There are hundreds of industrially manufactured Russian and international titanium alloys known at this moment [8]. Therefore, to choose the most acceptable of these alloys, and to develop the processing technology aimed to increase their dynamic stability are relevant. Such materials must contain a hard surface layers and a ductile bulk, which is necessary for the deformation of the damaging element and the greater impact energy absorption [6, 9-11]. This could be achieved by creating layered composite materials [12,13], materials with multilayer «sandwich-type» structures [14,15] or gradient structured materials [16-18].

Gradient structures, which consistof gradual transition from one type of structure to another between the surface and the center of material, could be created using local heating with high-energy heat sources or by intense plastic deformation [16, 17]. In particular, a reversible hydrogen addition can be used to create gradient structures in titanium alloys [19]. However, when it comes to armored materials, it is necessary to modify the structure from one side only, i.e. to create a «one-way» or «linear» gradient structure. In this case, top priority task is to «isolate» the remaining sides of the semi-finished product from hydrogen penetration.

This work is a continuation of the research conducted by the authors in this direction [20-22]. In [20, 21], the temperature-time parameters of air atmosphere heat treatment were determined, which made it possible to form an oxide film on VT6 samples with high «insulating» properties against hydrogen penetration. It was shown [20-22], that it is possible to create a «one-way» or «linear» gradient structure in VT6 alloy samples using thermal hydrogen treatment.

Therefore, the aim of this work was to study the oxide film resistance to hydrogen penetration, as well as to determine the influence of hydrogen addition to VT6 alloy samples structure formation and its properties.

2. Materials and procedures

Investigations were performed on the 15×13×13mm samples cut from 13mm thick industrial manufactured hot-rolled plate made of VT6 titanium alloy.

Samples oxidation was performed in air atmosphere using SNOL-2.2.5.1.8/10-I3 furnaces. Vacuum annealing was perfomed in Vega-3M vacuum furnace. Addition of hydrogen up to 0.3-0.5 wt. % was performed in Sieverts device at β-area temperatures.

The amount of added hydrogen was determined by weight gain with an accuracy of 0.0001 g. The amount of residual hydrogen was determined by the spectral method on an ISP-51 spectrograph device with an electronic analytical attachment MORS-1/2048/PCI.

Microstructures were investigated using a Cals Zeiss Axio-Observer.A1m graphic microscope at up to x1000 magnification. Light-field mode in air atmosphere was used. The hydrogen penetration depth was determined on metallographic specimens by the change of microhardness. The Vickers microhardness was measured on a MicroMet 5101 device with a diamond tetrahedral pyramid and a load of 50 g. The results were processed using MicroHardness software. Rockwell hardness was measured on a MacroMet 5100T device with a diamond cone and a load of 1500 MPa.

3. Results and discussion

The initial structure of the VT6 alloy plate is well worked and consists of β-phase and partially deformed lamellar β-grains, which corresponds to type 3 of 9-typestandart scale [23] (fig. 1a). All samples were annealed at the β-area temperature, followed by slow cooling to room temperature to form a macrolaminal structure. The resulting structure is also uniform in cross section and is represented by large β-grains surrounded by a α-border with lamellar α-phaseinside the β-graincollected in a colonies and corresponds to type 9of 9-type standard scale [23] (Figure 1b).

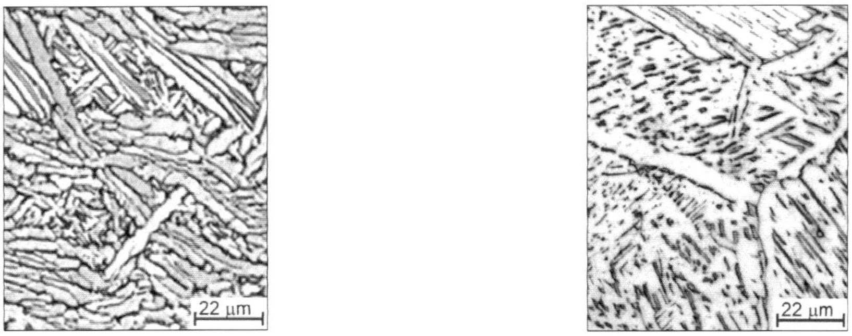

(a) (b)

Figure 1.Microstructure of VT6 alloy plate in initial state (a) and after β-area temperature annealing (b).

The influence of one-sided hydrogen addition on the patterns of structure formation in surface layers was studied at the initial stages of work. All samples were preliminarily oxidated in an air atmosphere at a temperature of 90°C for 4 hours to form a protective oxide film. Then, before the beginning of the hydrogen addition, oxide and alpha layers were removed from one side of the sample by mechanical treatment.

It is known [24, 25] that the diffusion mobility of hydrogen atoms is much higher than that of the main alloying substitution elements. Therefore, it is possible to control the completeness of the β→α-transformation in the titanium alloys over the sample cross section by changing the amount of added hydrogen and the duration of hydrogen annealing. Besides, it was shown in previous studies [26], that in order to convert a large lamellar (α+β)-structure into a finely dispersed one, it is necessary to make sure that β→α- transformation is complete during hydrogen addition and the alloy structure is practically «β-phase only» after room temperature cooling with just a small amount of α″-martesite up to 10%. Taking into account that addition of hydrogen lowers the polymorphic transition temperature, transformation of alloy structure to single β-phase state will occur at lower temperature. Besides, hydrogen mainly located in surface layers during one-sided hydrogen penetration and therefore less amount of hydrogen is sufficient to transform surface structure.

Hydrogen addition was performed at a temperature of 800°C with the amount of added hydrogen up to 0.3-0.6 wt. % in this work. To ensure an inhomogeneous distribution of hydrogen over the sample cross section, accelerated cooling was carried out immediately after the completion of the hydrogen absorption process.

Metallographic analysis of oxide-free samples side hydrogenated to 0.3% and 0.5% wt. showed that its structure is represented by α″-martensite and β-phase (Figure 2a).Since almost all of hydrogen is concentrated in the surface layer due to specially created annealing conditions, the structure conversion does not occur over the entire sample cross section, but only to a certain depth. Amount of α″-martensite decreases with distance from the surface layers increases, as well as the amount of α-phase increases (Figure 2b).

At a distance of 3500 μm from the surface, the structure no longer differs from its annealed state and is represented by large β-grains surrounded by a α-border with lamellar α-phase inside the β-grain (Figure 2c).

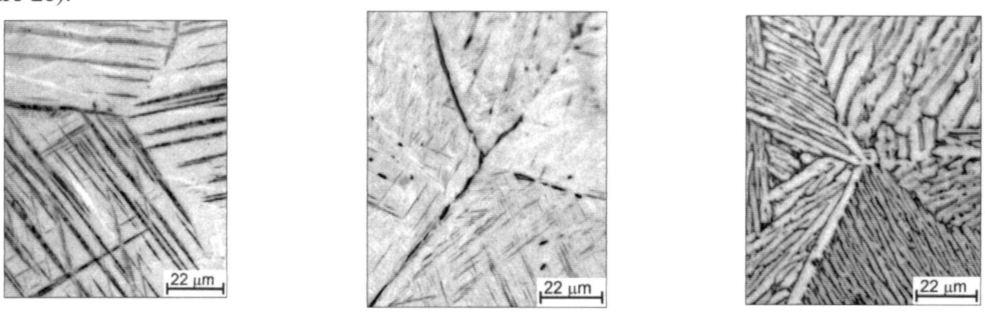

(a) (b) (c)

Figure 2. Microstructure of VT6 alloy samples after hydrogen addition up to 0.3 wt.% at the depth of 500 μm (a), 1500 μm and 3500 μm (c) from the oxide-free surface.

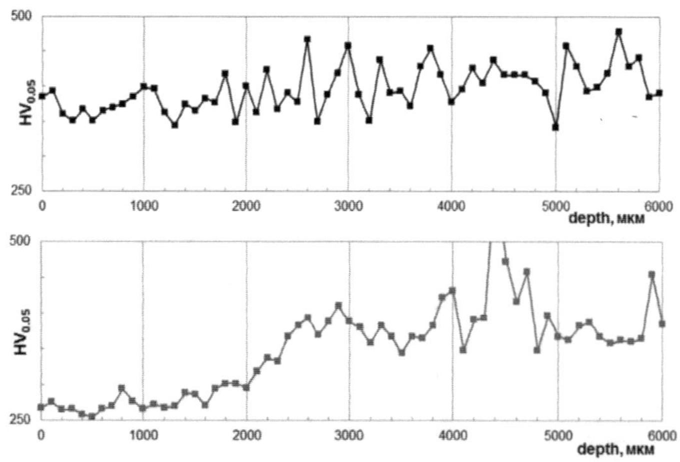

(a)

(b)

Figure3. Microhardness distribution in VT6 alloy samples cross section after 900°C oxidation and hydrogen addition up to 0.3 (a) and 0.5wt% (b) from the direction of oxide-free surface.

The microhardnesswas measured in the transverse direction. On each measurement line, 50 prints were set in increments of 100 μm. The depth where the microhardness began to change stepwise was taken as the boundary of the transformed structure layer.Analysis of microhardness distribution over sample cross section showed that transformed structure layer has uniform hardness of 350 $HV_{0.05}$, while the hardness of bulk area varies from 340 to 450 $HV_{0.05}$ due to its large lamellar morphology (Figure 3a). Thus, the thickness of the transformed layer is 1800 μm for samples with 0.3 wt. % of hydrogen and 2800 μm for the ones with 0.4 wt. % of hydrogen

Analysis of structure transformation and microhardness distribution over samples cross section from the oxide film «isolated» sides (Figure 4a) showed that at the distance of 100μm from surface the sample structure are no longer differs from its annealed state (before hydrogen treatment) (Figure 1b). This indicates a weak penetration of hydrogen through the oxide film.

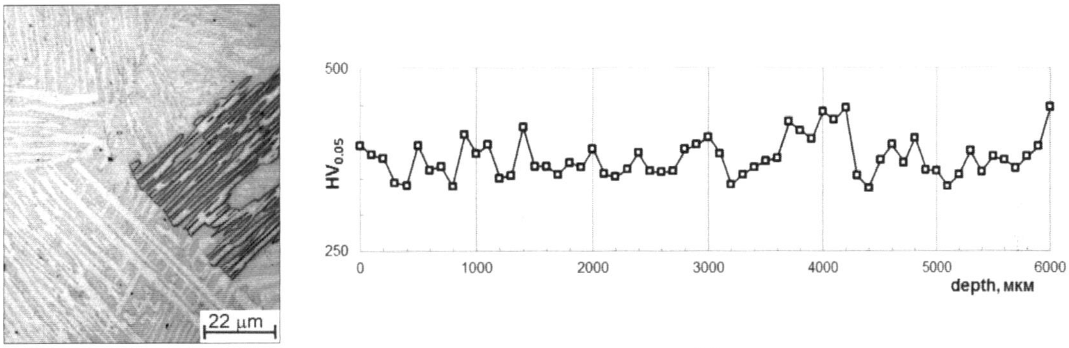

(a) (b)

Figure 4. Microstructure of VT6 alloy samples after hydrogen addition up to 0.3wt.%at the depth of 100 μm from oxide film «isolated»sides (a) and its microhardnessdistribution (b).

Similar results were obtained on samples with 0.5 and 0.6wt % of hydrogen added. The character of structure changes and microhardness distribution from the direction of oxide-free sample side is preserved. The depth of transformed layer with increasing amount of added hydrogen from 0.5 to 0.6wt % gradually increases from 3000 to 4000μm, respectively.

However, as the hydrogen content increases, the formation of α″-martensite is observed in the structure (Figure 5a) from «isolated» by oxide film sides (Figure 5a). Moreover, the almost unchanged microhardness value of 300 $HV_{0.05}$ is observed on a depth up to 2500 μm (Figure 5b). This indicates a partial loss of the oxide film "protective" properties. In addition, this may be the reason that the actual depth of the transformed structure layer from the direction of oxide-free sample side could be much larger during the hydrogen addition of 0.4 wt%.

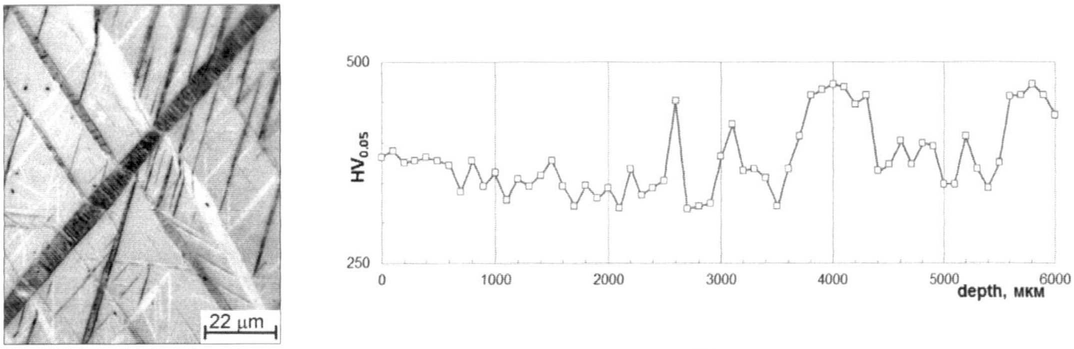

(a) (b)

Figure 5. Microstructure of VT6 alloy samples after hydrogen addition up to 0.5wt.%at the depth of 500 μm from oxide film «isolated» sides (a) and its microhardness distribution (b).

At the final stage of work, vacuum annealing was carried out at a temperature of 625 °C to remove hydrogen to its safe amount and to form the final «linear» gradient structure. Preliminary hydrogenated to 0.3% and 0.4%wt samples were annealed.

The studies showed that in the process of gas removal during low-temperature vacuum annealing, the β→α-transformation develops and a dispersed structure forms in the surface layers of oxide-free sample side (Figure 6a). Structure becomes less disperse as distance from the surface increases (Figure 6b), and at a distance of 3500 μm from surface the structure no longer differs from its annealed state (Figure 6c).

Figure 6. Microstructure over cross section of preliminary hydrogenated up to 0.3 wt.% VT6 alloy samples after subsequent low-temperature vacuum annealing at a distance of 500μm (a), 2000μm (b) and 3000 μm (c) from oxide-free surface.

Analysis of microhardness distribution over sample cross section showed that the depth of transformed layer together with transition zone is 2300 and 2800μm after preliminary hydrogenation of 0.3% and 0.4 wt. %, respectively (Figure 7 and Figure 8). Macrohardness of surface layer with finely dispersed (α+β)-structure is about 40 HRC, and the inner layer with unchanged large lamellar (α+β)-structure is 32HRC.

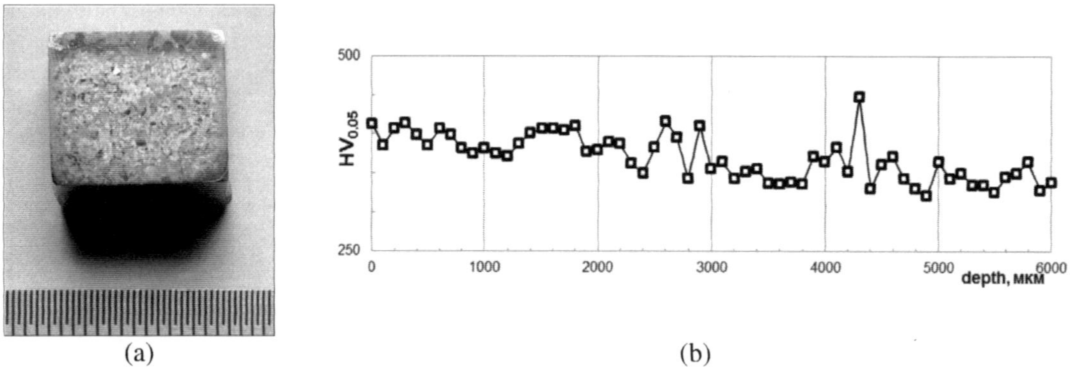

Figure 7. Macrostructure (a) and microhardness(b) distribution of preliminary hydrogenated up to 0.3 wt.%VT6 alloy samples after subsequent vacuum annealing at 625°C from the direction of oxide-free surface.

(a)

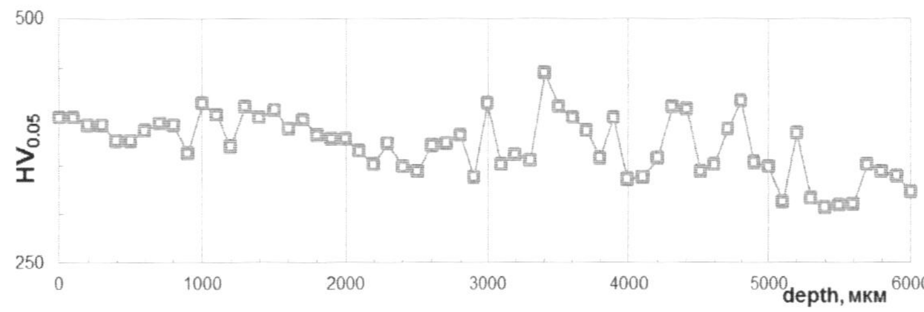

(b)

Figure 8. Macrostructure (a) and microhardness (b) distribution of preliminary hydrogenated up to 0.4 wt.%VT6 alloy samples after subsequent vacuum annealing at 625°C from the direction of oxide-free surface.

4. Conclusions

Studies have shown that the thermal hydrogen treatment of VT6 alloy samples with 5 of 6 sides «isolated» by oxide film allows to create a «one-way» or «linear» gradient structure that changes from finely dispersed on one side to a large lamellar on the opposite. It is shown that with an increase in amount of added hydrogen from 0.3 to 0.6 wt%, an increase in the depth of the transformed layer is observed from 1800 to 4000 μm, respectively. It was established that the oxide film formed during oxidation at 900°C effectively protects from hydrogen penetration up to 0.4wt.% only. It is also shown, that a dispersed structure formation in the surface layers allows to increase the hardness to 40 HRC. Thus, a «one-way» or «linear» gradient structure formed using thermal hydrogen treatment in VT6 alloy can provide increased impact strength and slowed crack proliferation rate in its samples.

5. References

[1] Kupryunin D G and Chistyakov E N 2015*World and Safety***2** pp 6–10
[2] Zuzov V N and Shash N 2018 *News of Higher Educational Institutions. Engineering***4** pp 42-8
[3] Gavrilov E V, Kulakov N A and Lyubin A N 2018 *Russian Patent № 2 667 981*
[4] GOST R 50963-96 2003 (Moscow: Government Standards Publisher) p 30
[5] Rossikhin K G and Selifonov V V 2010 *"Automobile and tractor engineering in Russia: development priorities and training" Conference, Dedicated to the 145th Anniversary of MSTU «MAMI»*pp 269-77
[6] Kupriyun D G, Gavse A L and Chusov S Yu 2017 *Defense equipment issues***121-122** pp 114-22
[7] Gavze A L, Chusov S Yu, Yankov V P Tetyukhin V V, Zhloba A V, Berestov A V and Bondaryuk N N 2013 Titan1 pp 46 – 8
[8] KhorevA I 2014 Mechanical Engineer Technology 1 pp 5-10
[9] Ilyin A A, Skvortsova S V, Spektor V S, Kudelina I M and Oreshko E I 2011 Titan1(31) pp 26-9
[10] Illarionov A G, Zhloba A V, Leder M O, Stepanov S I, Berestov A V and Popov A A 2018Titan2pp 16-21
[11] Reichman S H 2007 US Patent №2303230
[12] Kryukov D B and Kozlov D V 2016 Penza State University Bulletin 2 (14) pp 103–8
[13] Agarkov G A and Bliznik M G 2018 Russian Patent № 2 647 963
[14] Borovleva I N 2018 Russian Patent № 2 653 903
[15] Khovrich M V, Saharov S A and Kamaev E A 2007 Russian Patent №2 296 288
[16] Golenkov V A, Radchenko S Yu, Dorofeev O V and Dorokhov D O 2009 Fundamental and applied problems of engineering and technology 3 (275) pp 42-6

[17] Raab G I, Kodirov I S, Alyoshin G N, Raab A G and Cenev N K 2019 Magnitogorsk State Technical University BulletinV 17 I 1 pp 64-75
[18] Pettersson A, Magnusson P, Lundberg P and Nygren M. Titanium 2005 International Journal of Impact Engineering pp 387-99
[19] lyin A A, Skvortsova S V, Spektor V S, Kudelina I M and Mamontova N A 2011 Technology of Light Alloys2 pp 37-41
[20] Skvortsova S V, Gvozdeva O N, Shalin A V and Stepushin A S 2019 Titan 3 pp 25-31
[21] Gvozdeva O N, Shalin A V, Makarov F V and Stepushin A S 2019 18th International Conference «Aviation and Cosmonautics-2019». Abstracts p 246
[22] Skvortsova S V, Gvozdeva O N, Shalin A V and Stepushin A S 2019 Fifth interdisciplinary scientific forum with international participation "New materials and promising technologies." Collection of materialsV 1 pp 596-600
[23] GOST 26492-85 (Moscow: Government Standards Publisher) p 32
[24] Ilyin A A, Skvortsova S V, Mamonov A M and Kollerov M Yu 2006 Physico-Chemical Mechanics of Materials V 42 I 3 pp 33-9
[25] Kolachev B A, Ilyin A A, Nosov V K and Mamonov A M 2007 Technology of Light Alloys 3 pp 10-26
[26] Mamonov A M, Safaryan A I, Agarkova E O, Zhilyakova M A2018 Metal Science and Heat TreatmentV 60 I 1-2pp 80-8

Acknowledgements
The work was performed on the equipment of the collective resource center «Aerospace materials and technologies» of the Moscow Aviation Institute.

IOP Conf. Series: Materials Science and Engineering **889** (2020) 012006 doi:10.1088/1757-899X/889/1/012006

Improving the reliability of metal structures of buildings and structures during reconstruction

A Mezentseva[1*], Z Gel'manova[1] and A Konakbayeva[1]

[1]Karaganda state industrial university, 30 Republic Avenue, Temirtau, Republic of Kazakhstan

*avmez@mail.ru

Abstract. Recently, the direction of prestressing metal structures, the regulation of forces in which is made by various methods without the use of high-strength steel is one of the hot spots of research. In this paper, an unconventional method for creating pre -stress in tubular elements using an expanding substance is applied. As used by a non-explosive destructive mixture (NDM).

1. Introduction
Strengthening and repairing load-bearing structures with unconventionally stressed elements is one of the most pressing problems in construction. The Republic of Kazakhstan has a large number of buildings and structures that have a service life of more than 50 years . Aging, fatigue, and corrosion of building structures and structures are the most poorly studied. This work will help to reduce the gap to a certain extent and at the same time develop science and practice in improving the load-bearing capacity of bearing elements of buildings and structures.

2. An experiment
Compositions composed of a combination of different substances were tested as an expanding material. The best results are obtained for an expanding material consisting of calcium oxide (OST 21-27-78), a retarder of the hydration reaction of calcium oxide (OST 81-79-74), and also possible additives in certain conditions of nitrate and potash [1-5].

It was found that the aqueous solution of the expanding material increases in volume as a result of the reaction by about 23 times relative to the initial one. The working mixture is a powdered material, usually gray in color, non-flammable, non-explosive, with weak alkaline properties with a volume density of 1.251. 35 t / m^3 and water absorption up to 30%. The volume mass of the working mixture solution is 2.16 t / m^3. Developed pressure up to 30 MPa. According to the conclusion of the Republican sanitary and epidemiological station of the Ministry of health of the Republic of Kazakhstan, the components of the working mixture meet the hydrohygenic requirements for materials and products for industrial purposes.

When preparing a working mixture of powder, the amount of water for mixing should be from 28% to 30% of the powder weight. Mechanical stress in the working mixture occurs as a result of increasing its volume due to the hydration reaction in confined conditions. The hydration reaction is provided by the main component of the powder – calcium oxide. The content of calcium oxide reaches from 70% to 90% of the total weight of the powder. Additional components included in the powder are used to maintain certain modes of expansion of the working mixture.

Experiments were performed on a special stand to determine the values of stresses that occur in an expanding material (NDM solution). The General appearance of the stand included a frame with a DOSM-3-5 dynamometer and sliding cylinders of different diameters and heights [5-14, 19]. Sliding cylinders were filled with a solution of expanding material, closed from the bottom and top with a lid and tightly. With a preliminary voltage, they were installed to a dynamometer, according to the indications of which they judged the resulting stresses. To do this, it was necessary to divide the results recorded by the dynamometer by the area of the inner surface of the sliding cylinders containing the expanding material [15-18, 20-21]. A total of 60 preliminary experiments were conducted under the same conditions.

3. Results of experimental studies

Table 1 shows the results of experiments to determine the stresses in sliding cylinders of different diameters and heights. For 24 hours under the same conditions of the experiments , a voltage approximately equal to 17 MPa occurs, which is approximately 0.7 of the maximum voltage in the expanding material.

Table 1. Results of stress determination experiments changes in the working mix (NDM).

Sketchslidingcylinders'	Parameterscylinders'				Results		
	D(cm)	H(cm)	F(cm^3)	n	σ_t=N/F(MPa)	T(hour)	σ_t/σ_{0max}
	2.5	2.5	19.6	7	17.0±1.5	24	0.68
	3	6	28.4	7	16.8±1.1	24	0.67
	3.6	4	45.2	7	18.7±0.9	12	0.35
	4	3	37.7	7	18.1±0.8	12	0.34

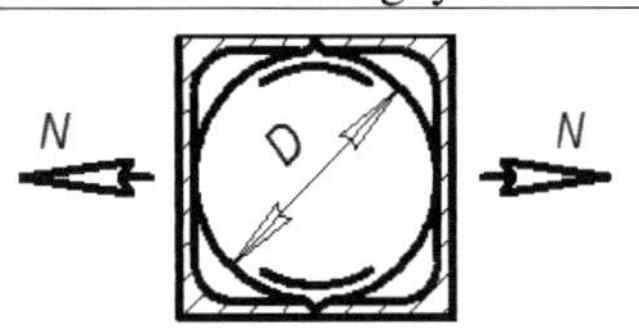

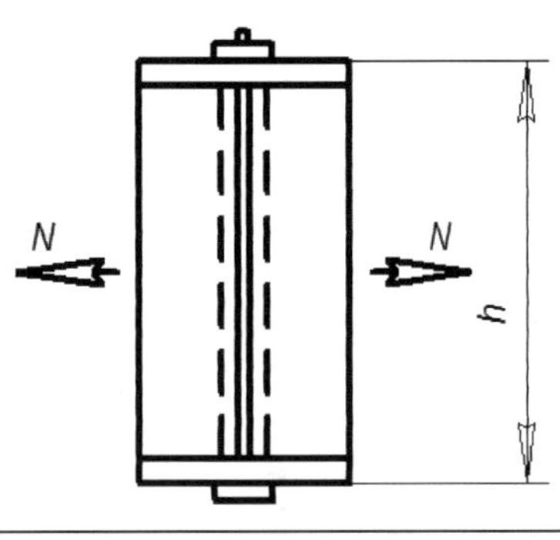

The student's distribution coefficient for seven repetitions of experiments (n) and the confidence probability P_S=0.95 is equal to t_S=2.5.

The curve of the spread of the measured values of forces (stresses) from the expansion of the LDC is subject to the normal distribution law (Gauss curve). Processing of the measurement results showed that in order to obtain reliable data with a probability of P=0.95 and an accuracy of K=5%, it is necessary to conduct 7 repetitions of each experiment. This mixture works effectively at a positive temperature. Figure 1 shows the experimentally obtained dependence of the efficiency of increasing the voltage in the working mixture in a closed volume over time at a positive hardening temperature, which shows that the increase in the voltage in the working mixture occurs within two months, and for the first day the working voltage of the mixture reaches about 70% of the maximum. The remaining 30% increase in voltage occurs gradually.

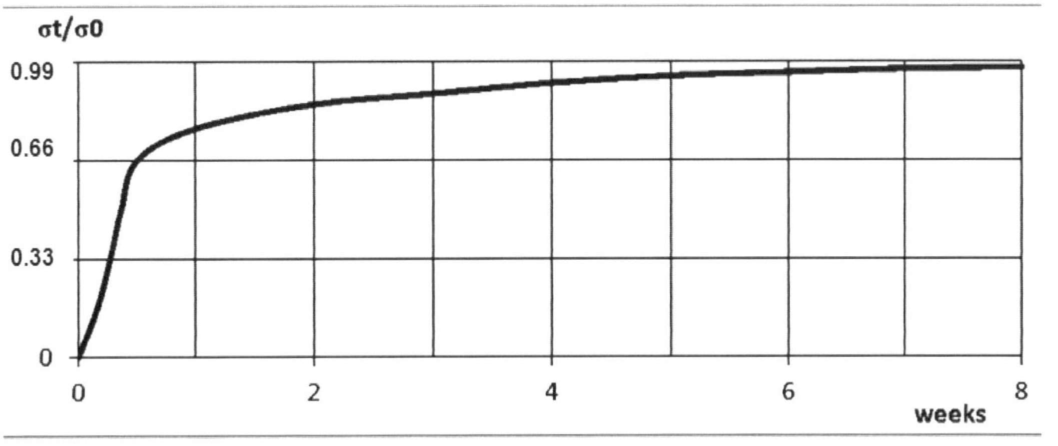

Figure 1. Graph of the intensity of stress growth over time expansion of the working mixture.

The first two hours characterize the hardening of the working mixture placed in a closed volume. Then it begins to gradually expand and increase the stresses (figure 2). The maximum rate of expansion of the mixture falls on the first 610 hours of its hardening. In the future, the rate of expansion of the mixture gradually decreases.

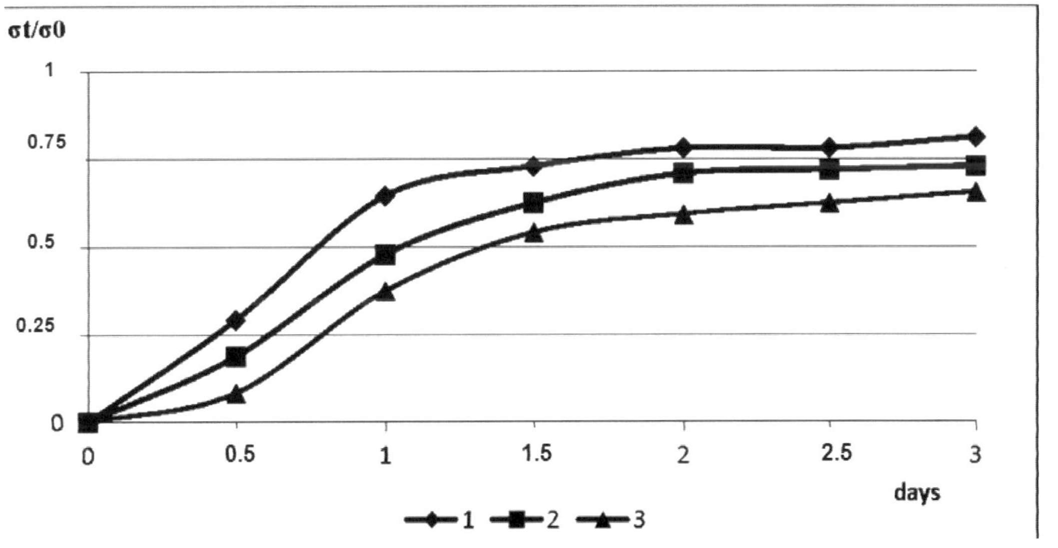

Figure 2. Growth of stresses during expansion of the working mixture; 1 – clean working mix; 2 – working mix with 20% sand added by weight; 3 – working mix with 40% sand added by weight.

When the working mixture solidifies in hot water (t=5060°c), the intensity of stress increases faster (figure 2). Therefore, it is advisable to use hot water at a generally low positive temperature from +1°C to +5°C.

Inert additives can be used to change the stress in the expanding mixture. As an inert additive for the preparation of an expanding mixture, it is advisable to use dry fine sand. The addition of dry sand in an amount of 20% to 40% to the material slightly affects the amount of its expansion and stress reduction (figure 3).

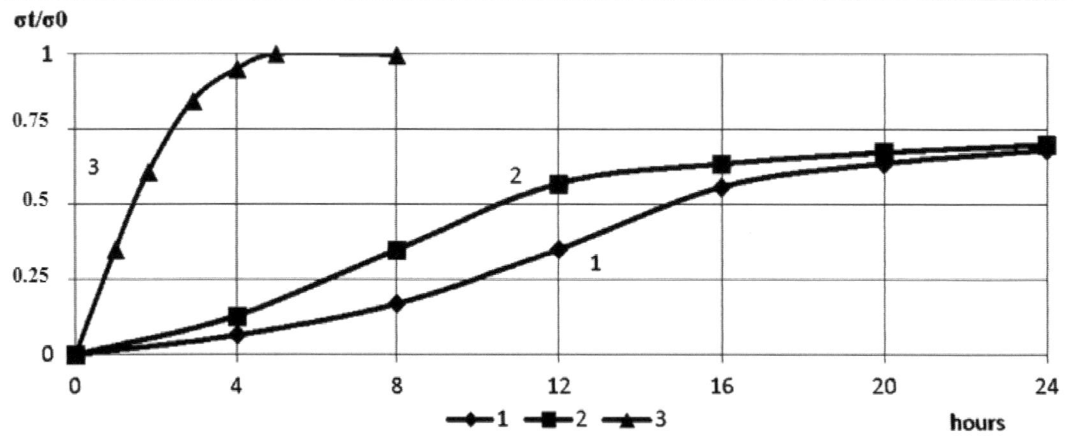

Figure 3. Growth of stresses during the expansion of the working mixture on the first day;1 – closing the mixture on water at room temperature ; 2 – closing the mixture on water at t=50-60°C and hardening the mixture at 20°C; 3 – heating the mixture to intensify the hydration process.

4. Conclusions
In the present work, research has been carried out and the characteristics of the work of tubular elements that are strained using a non-explosive expanding mixture have been identified. The possibility of creating a pre- stress using non-traditional methods when strengthening the load-bearing capacity of building structures is proved. The advantages of the non-traditional method of creating pre- voltage under study are that it is less labor- intensive, the amplification effect is achieved quickly, and it does not require bulky equipment.

5. References

[1] SNiP of the Republic of Kazakhstan (Building Codes) 2002 *Steel structures. Design standards* 5.04-23-2002

[2] Hilo E R and Popovich B S 1985 *Strengthening of building structures* (Lviv: Higher school: Lviv University Publishing House)

[3] OST (Industry Standards) 21-27-78 (Republic of Kazakhstan)

[4] OST (Industry Standards) 81-79-74 (Republic of Kazakhstan)

[5] Plotnikov V M, Kazbanov O A, Smolkina T P, Petrov S A and Kuznetsov S B *Method of strengthening building structures* (Preliminary patent of the Republic of Kazakhstan (19) KZ(13) A(11) 9843)

[6] Kikin A I, Sanzharovsky R S, Trull V A 1974 *Structures made of pipes filled with concrete* (Moscow: Stroizdat)

[7] Onufriev N M 1965 *Strengthening of reinforced concrete structures of buildings and structures.* (Leningrad-Moscow: Stroizdat)

[8] Mingalyov A G and Piotrowicz A A 2016 *Scientific, technical and economic cooperation of the APR countries in the XXI century* **1** 400-3

[9] Speransky B A 1970 *Lattice metal prestressed structures* (Moscow: Moscow Publishing house of literature on construction)

[10] Emelyanov R V and Ilyakhin S V 2018 *Proc. of the Int. sci. and practical conf.* **7** 145-6

[11] Belsky M R 1975 *Strengthening of metal structures under load* (Kiev: Ed. Budivelnik)

[12] Gaidarov Yu V 1971 *Prestressed metal structures (New types and applications)* (Leningrad: Ed. literature on construction)

[13] Lashenko M N 1987 *Improving the reliability of metal structures of buildings and structures during reconstruction* (Leningrad: Stroyizdat)

[14] Rebrov I S 1988 *Strengthening rods of metal structures: Design and calculation* (Leningrad: Stroyizdat)
[15] Brodk, Jan 1975 *Tubular steel structures* (Leningrad: Stroyizdat, translated from Polish)
[16] Kikin A I and et.al. 1974 *Constructions made of steel pipes filled with concrete* (Moscow: Stroizdat)
[17] Flyagin A S, Vorsin V A and Ufimtsev V M 2018 *Proc. of VSUET* **2(80)** 290- 96
[18] Baturhodjayev E D 2013 *Proc. of the XVII Int. Academician M. A. Usova Symposium of students and young scientists, dedicated to 150 anniversary of birthday of academician V. A. Obruchev and the 130th anniversary of academician M. A. Usov, and the founders of the Siberian geological school* (National research Tomsk Polytechnic University, Problems of Geology and exploitation of mineral resources) 254-5
[19] Val V N, Gorokhov E V and Uvarov V Yu 1987 *Strengthening of steel frames of single-storey industrial buildings during their reconstruction* (Moscow: Stroizdat)
[20] Tolmachev K X 1960 *Rgulation of the stresses in the metal span structures of bridges* (Moscow: Autotransmit)
[21] Mezentseva A V, Bozshalova L T and Belozerova D S 2018 *Bulletin of the Perm national research Polytechnic University. Construction and architecture* **2** 110-6

IOP Conf. Series: Materials Science and Engineering **889** (2020) 012007 doi:10.1088/1757-899X/889/1/012007

Increasing the operational reliability of car variators due to creating regular surface microrelief by laser ablation

Elmar Yagyaev[1*], Leonid Shron[2] and Dmitriy Meniuk[2]

[1]Crimean Engineering and Pedagogical University, per. Uchebnyj, 8, Simferopol, 295015, Russian Federation
[2]Sevastopol State University, Universitetskaya st., 33, Sevastopol, 299053, Russian Federation

*elmar1875@gmail.com

Abstract. The paper analyzes the car variator reliability issue. The variator performance is ensured by the surface layer quality, manufacturing accuracy, and the pulley and steel V-belt surface microrelief. The decisive factor in ensuring the variator reliability is the pulley finishing technique. Creating regular microreliefs on the finished pulley surface by laser ablation has been proposed, which act as micro-channels to improve the drainage of oil from the contact pattern. Creating oil channels on the pulley surface by laser ablation will ensure their arrangement and allow obtaining the desired pattern, direction, depth, and length.

1. Introduction

The progress and development of the motor vehicle industry are constantly ongoing; earlier, the car manufacturer, engine type, and vehicle class have been the key factors when choosing a car. Currently, when buying a car, one of the decisive factors is the type of gearbox installed.

There are four transmission types used worldwide:

- Manual (MGB);
- Automatic (AGB);
- Semi-automatic (SAGB);
- Variable-speed (Variator, CVT).

Most buyers choose cars with an automatic, semi-automatic, or variable-speed gearbox. This is because they show better behavior and greater comfort compared to a manual gearbox. Both automatic or semi-automatic and variable-speed gearboxes have advantages and disadvantages. A significant advantage of variators is the continuously variable gear ratio, which allows leveraging the engine traction capabilities and its optimal operation, and great fuel saving.

The globally accepted abbreviation for variator is CVT (English: continuously variable transmission). The main disadvantage of variators is a rather low operational reliability. Let us analyze the reason for the variator reliability decrease and propose solutions to improve it.

Variable-speed gearbox CVT consists of two adjustable cone pulleys and a steel V-belt (Figure. 1) [1]. Since the cones of each pair of the pulleys move relative to each other, their conical shape causes the belt to go up and down between the pulley grooves. The active gear ratio varies depending on the belt position in each pulley. The CVT gearbox pulleys create a continuous exchange of gear ratios by constantly changing the belt position between them. This provides a limited range of gear ratios according to the pulley diameters, and all gear ratio values are available within this range.

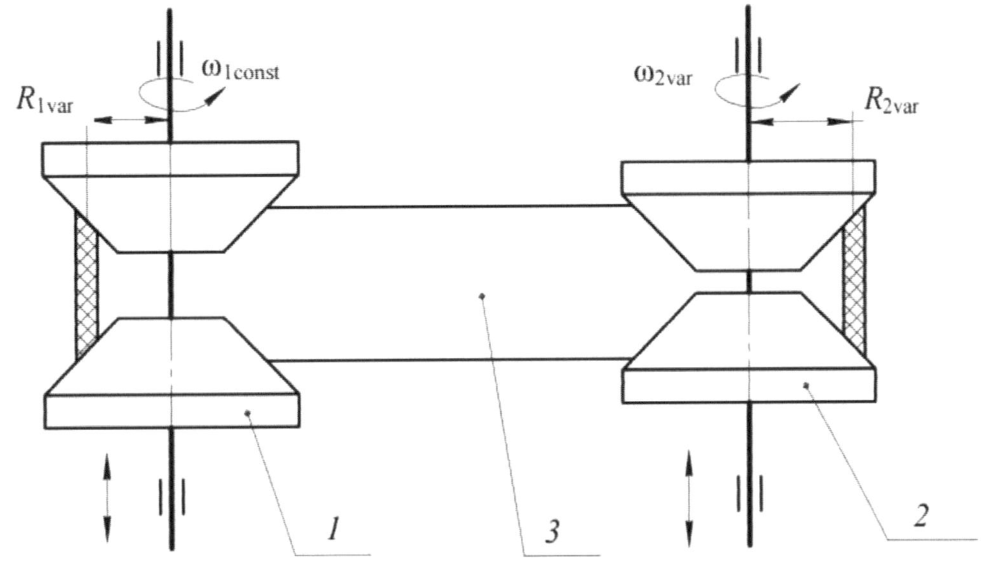

Figure 1. Diagram of a Variator with Adjustable Cone Pulleys: 1 – drive adjustable pulley; 2 - slave adjustable pulley; 3 - steel V-belt.

Torque is transmitted due to friction between the pulleys and the side surface of asteel V-belt. This design ensures high strength, durability, flexibility, and low noise during operation (Figure. 2) [2-3].

Figure 2. CVT Assembly: Two Pulleys and a Belt.

The variable-speed gearbox reliability depends on the life of the main CVT unit – the pulley-belt friction pair. In the course of operation, torque is transmitted due to friction between the pulley-belt contact pair. The special CVTF (Continuously Variable Transmission Fluid) affects the provision of the required coefficient of friction.

The development of a steel V-belt (Figure. 3) consisting of 10-14 layers of 0.2 mm thick steel strips made of maraging steel 18Ni16.5Co5Mo and numerous segments strung on them has ensured transmitting torque up to 180 Nm and allowed widely using variators in cars [4].

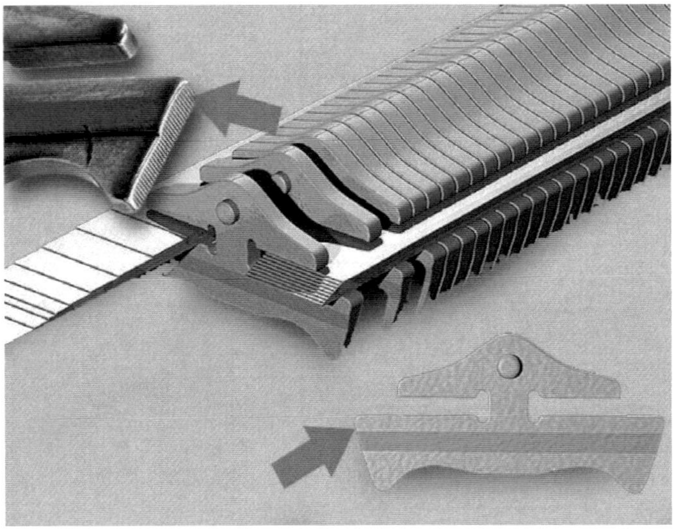

Figure 3.Steel V-belt [4].

The paper's objective is the possibility of creating regular microrelief on the pulley-belt friction pair surfaces by laser ablation to increase the variator reliability.

2. Materials and methods

In the variator, torque is transmitted from the drive pulley to the slave one due to friction between the steel belt and the pulleys. To create the required friction force between the belt and the pulley, the belt is tensioned with a preliminary tension F_0. (Figure. 4). At rest and when idle, each belt branch is tensioned with the same force F_0.

When a car starts moving at an operating load T_1, the tension is redistributed between the belt branches. The leading branch is additionally loaded to F_1 value, and the slave branch is released to F_2 value [1].

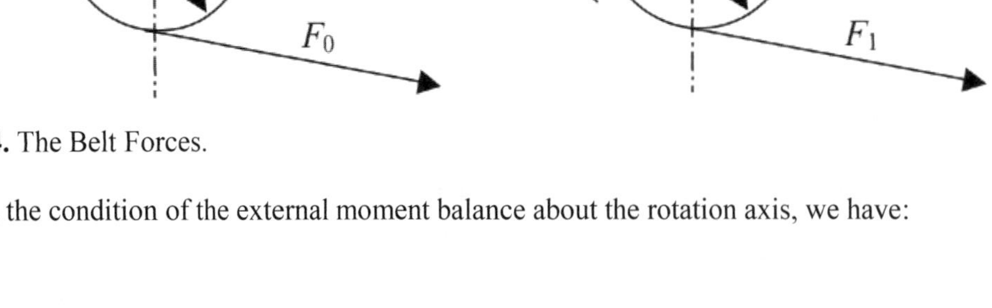

Figure 4. The Belt Forces.

From the condition of the external moment balance about the rotation axis, we have:

$$-T_1 + \frac{F_1 D_1}{2} - \frac{F_2 D_2}{2} = 0. \tag{1}$$

Since $T_1 = \frac{F_t D_1}{2}$, then from equation (1) we obtain

$$F_1 - F_2 = F_t. \tag{2}$$

The total belt length remains unchanged during operation since the leading branch elongation is compensated by an equal contraction of the slave branch. It means that whatever extent the slave branch tension increases to that extent the driven branch tension reduces, i.e.

$$F_1 = F_0 + \Delta F \text{ , } F_2 = F_0 - \Delta F \text{ .} \tag{3}$$

When adding the equations, we obtain:

$$F_1 + F_2 = 2F_0. \tag{4}$$

When solving equations (1) and (4) together, we obtain

$$F_1 = F_0 + 0{,}5F_t, \tag{5}$$

$$F_2 = F_0 - 0{,}5F_t. \tag{6}$$

Equations (5) and (6) describe the changes in the leading and slave branch tension depending on the load Ft but do not consider the friction force between the belt and the pulley. Since the creep occurs throughout the entire wrap, then the dependence between F_1 and F_2 is expressed by the Euler's formula.

$$F_1 = F_2 e^{f\alpha}. \tag{7}$$

Equation (7) describes a relation between the branch tension forces for a gear operating at the load F_t, the friction coefficient *f,* and the wrap angle *α*. This allows determining the minimum belt pretension F_0 required, at which the transmission of a given load F_t is still possible.

If $F_0 \lhd \frac{F_t}{2}\left(\frac{e^{f\alpha}+1}{e^{f\alpha}-1}\right)$, then the belt will creep.

The analysis of the above dependences shows that an increase in the f and α values allows increasing the torque transmitted and consequently improves the variator reliability.

These conclusions will be treated as the main way to improve the variator reliability using the principle of increasing the pulley-belt friction. Reduced wear of the contact surfaces depends on the precision of manufacture and the pulley and metal V-belt surface microreliefs.

The studies on the pulley-belt friction pair reliability are most often aimed at improving the V-belt design [4]. Using belts of improved design or those made of new materials does not completely solve the issue.

When manufacturing and repairing pulleys, special requirements for the contact area of the pulley-belt friction pair are not being considered. The processing type and the pulley surface microrelief may contribute to both increasing and reducing the friction coefficient.

3. Result and discussion

The technology proposed is based on creating a guaranteed microrelief on the pulley-belt friction pair surface ensuring the required frictional properties and wear resistance of parts [5, 6]. The performance and reliability of the pulley-belt friction pair can be improved by creating regular microreliefs on the finished surface, which act as micropumps [7].

Grooves placed at a negative angle of inclination γ clockwise from the rotational velocity vector will act as micropumps draining liquid out of the cavity. The pulley-belt friction pair operation can be controlled by changing the depth and density of grooves.

Processes for obtaining partially regular microreliefs and completely regular relief have been considered in [7]. After grinding, the surface microrelief is a chaotic pattern of micro bulges and micro dents of various configurations and depths.

After finishing by grinding, on the pulley surface, annular grooves have formed (Figure. 5a) that increase the oil-absorption power of the surface and lead to deterioration of the contact pattern of the belt-pulley friction pair. During operation, it contributes to belt slip and scoring (Figure. 5b). Conventional techniques available do not allow arranging the micro bulges and micro dents to obtain the required oil channel pattern.

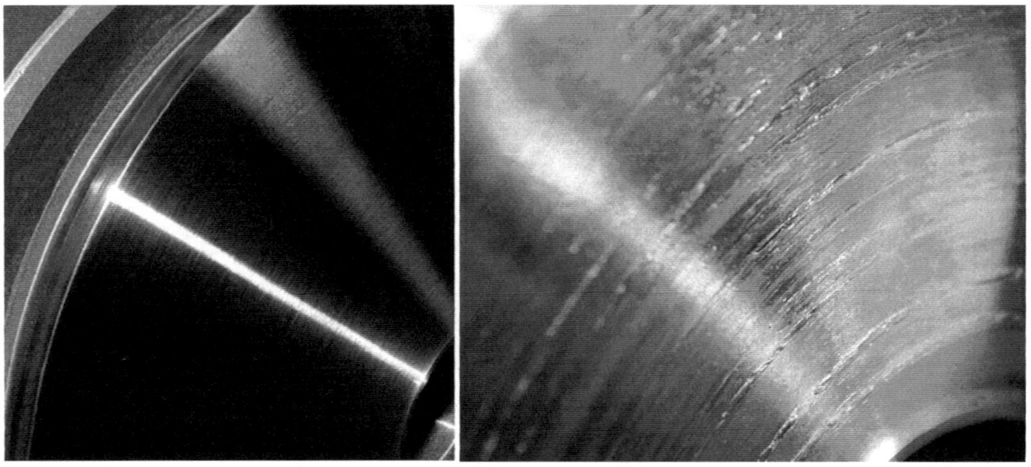

Figure 5. CVT Pulley a) after grinding, b) with scoring during operation.

A model of the existing and proposed microrelief structure of the pulley-belt friction pair surface is shown in Figure. 6.

In recent years, the development of nano-, pico-, and femtosecond diode-pumped solid-state lasers has led to a quantum leap in the creation of new cutting techniques [8–14]. The use of short-pulsed lasers has ensured such a power density in the processed area at which the substance evaporates, and laser ablation occurs.

Various laser cutting, drilling, and milling processes have been developed based on diode-pumped lasers generating nanosecond pulses.

The power density of the laser radiation incident on the processed surface with a reflection coefficient R(λ) at the given area and duration of exposure can be approximately estimated by the below equation [9]:

$$W = \frac{(1-R)E_i}{S \cdot t_i}, \tag{8}$$

where$R$ is the reflection coefficient of the material processed at the radiation wavelength, $E_i$is the pulse energy in Joules, S is the radiation incident surface area in cm^2, and t_i is the pulse duration in seconds. Depending on the processing type and the properties of the material processed, to obtain the required power density on the sample surface, laser radiation of a certain power with a given pulse duration should be applied.

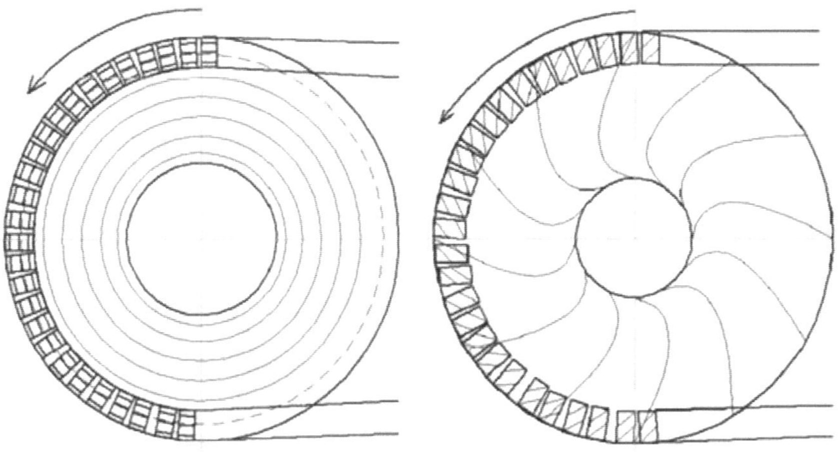

Figure 6. The Pulley-Belt Friction Pair Microrelief Model.
a) existing surface structure b) proposed surface structure.

According to the zone map of typical laser processing, the heat strengthening is performed within the interval W = 10^5-5x10^5 W/cm^2. Within the range W = 5x10^5-5x10^6 W/cm^2, the temperature reaches a level suitable for surfacing and welding of metals. At a laser power density of W > 5x106 W/cm^2, laser ablation allows drilling, cutting, and milling various materials. Precision laser processing is based on drilling, cutting, and milling.

The laser ablation unit based on diode-pumped laser generating nanosecond pulses is shown in Figure 7. This unit is equipped with a laser, a precision rotary positioning stage, an optical radiation focusing system, and an automated processing control and monitoring system.

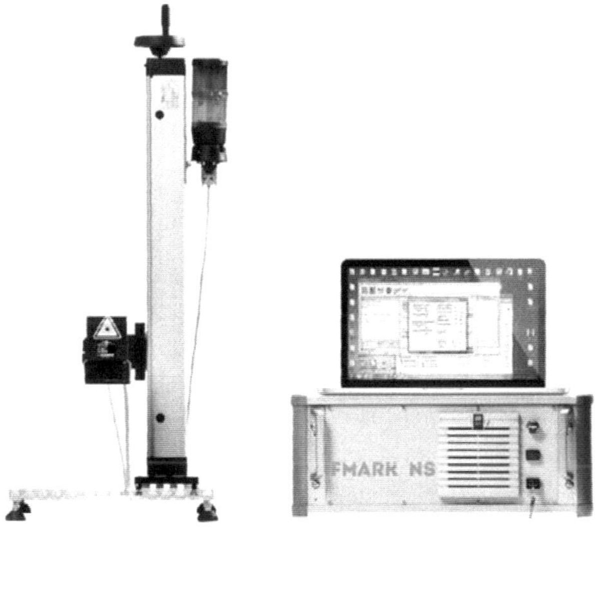

Figure 7. Experimental Laser Ablation Unit, Ytterbium Fibre Laser IPG Photonics, 1.064 μm Operating Wavelength.

4. Conclusion

(1) The research results have shown that existing conventional techniques do not allow obtaining an arranged microrelief of oil channels with the pattern required.

(2) After finishing by grinding, on the pulley surface, annular grooves have formed that increase the oil-absorption power of the surface and lead to deterioration of the contact pattern of the belt-pulley friction pair.

(3) The use of laser ablation technique allows obtaining the required microrelief on the variator pulley surface, ensures high-quality processing and high accuracy of the geometric pattern of the structures processed.

(4) The variator reliability can be improved through the formation of regular microreliefs by laser ablation.

5. References

[1] Dulevich A F,Osoko S A, Nikonchuk A N 2007 *Machine Parts Design and Calculation of Belt Drives and Variators* Minsk BSTU p.120

[2] Van der Sluis, Brandsma F A, van Lith J, van der Meer K, van der Velde Aand Pennings B 2002 Stress Reduction in Push Belt Rings Using Residual Stresses *Proceedings of the CVT2002 Congress* VDI-Berichte 1709 Munich pp.383–402.

[3] Aladagli I 2015 *Advanced CVT Modeling and Control Eindhoven*: Technische Universiteit Eindhoven p.151.

[4] PenningsB, TranMD, DerksM, BrandsmaA and van SchaikM (2005)New CVT Push-Belt Design Featuring a New Maraging Steel to Cover All Front Wheel Drive Powertrains*Conference: SIA Engine and Transmission Congress – Progress for the next ten years?* Paris.

[5] Schneider Yu G 1982*Operational Characteristics of Parts with Regular Microrelief* 2nd ed. L.: Engineering p.248.

[6] Provolotsky A E,Lapshin S P 2004Creating Developed Surface Microreliefs Cutting and Tools in Process Systems *Coll. Sc. Tech. Proc. Kharkiv NTU KhPI* **66** 153–62.

[7] Shron B L,Yagyaev E E andNovoselov Yu K2015Improving the Shaft-Cuff Connection Reliability through Creating Regular Microreliefs on the Contact Surfaces by Laser Ablation *Crimean Engineering Science Notes of Pedagogical University* **2** (50) 105–9.

[8] Pyachin S A,Pugachevsky M A 2013*New Techniques to Produce Functional Nanomaterials: Laser Ablation, Spark Exposure* Khabarovsk p.38.

[9] Grigoryants A G, Shiganov IN and Misyurov AI 2006*Laser Processing Processes* Textbook for universities Publishing House of MSTU. N.E. Bauman p.665.

[10] Vorkunov RYu andBriukhanov VV 2013Femtosecond Ablation of Copper and Silver Nanoparticles in Water*Bulletin of KSTU* **31** 23–9.

[11] TangM, ShimV, PanZ Y, ChooY S and HongM H2011Laser Ablation of Metal Substrates for Super-hydrophobic Effect *JLMN-Journal of Laser MicroNanoengineering* **6** (1) 6–9.

[12] Barmina EV 2010Generation of Nanostructures in Laser Ablation of Metals in Liquids:New Results*Quantum Electronics* **40** (11) 1012–20.

[13] Cheng G J, Pirzada D and Zhou M 2007Microstructure and mechanical property characterizations of metal foil after microscale laser dynamic forming*J. Appl. Phys.* **101** 063108.

[14] Vaks ED, Milenky MN and Saprykin LG 2013*The Practice of Precision Laser Processing* Moscow Technosphere p.696.

IOP Conf. Series: Materials Science and Engineering **889** (2020) 012008 doi:10.1088/1757-899X/889/1/012008

Constructive-technological peculiarities of the machinery for implementing the technology of the surface modification by high-entropic materials with SME under the conditions of high-voltage pulse electromechanical influences

R L Plomodialo[1*], Zh M Blednova[1] and D V Dmitrenko[1]

[1] Kuban State Technological University, 2 Moskovskaya Street, Krasnodar, 350072, Russia

*sutk_kubstu@mail.ru

Abstract. The paper shows the reasonability of using high-entropic materials for a number of multifunctional technical applications on the basis of analysis, systematization and generalization of the market's demands for operating the machine building products by composition construction of the surface layers using TEPT materials. For the implementation of the technology of surface layers formation from high-entropic materials with SME a multifunctional technological complex was designed and manufactured to develop the technology of surface modification by high-entropic materials with SME under the conditions of high-voltage pulse electromechanical influences.

1. Introduction

During last fifteen years high-entropic alloys (HEAs) attract special attention of material scientists and the whole engineering community as the research into these alloys revealed their attractive properties, such as high solidity and durability, high fatigue resistance and fracture strength, resistance to high-temperature oxidation, high corrosion resistance, and unique electric and magnetic properties. All these properties allow considering the materials as functional structurally sensitive ones. Due to high cost of HEAs because of expensive elements (such as Nb, W, Cr, V, Ni, Ti and Co) inclusion into their composition, films and coatings of high-performance HEAs on the inexpensive metallic bases arouse a great interest [1]. Different technologies are used for obtaining high-entropic coatings: magnetron [2], laser [3], plasma [4], gas-thermal [5]. The most needed methods are high-speed methods: high-velocity oxygen fuel (HVOF) [6] which is, essentially, a combined method of processing, comprising spraying and intensive plastic deformation, which provides the increase of adhesion strength.

The recent researches have shown that among a great variety of HEAs a special place is occupied by less-studied HEAs with thermoelastic phase transformations (TEPT) and materials with shape memory effect (SME) [7]. The peculiarity of these alloys consists in the ability to reveal structural-adaptive properties during the operation activity under definite conditions. The use of materials with TEPT expands the capabilities of the surface modification in increasing durability and survivability of the product. Materials with TEPT can be incorporated into composites, including hybrid ones, in different forms: layers, wires, fibers, and powder inclusions. The efficiency of their usage in composite materials consists in their ability to absorb the energy of the force influence via superelastic deformation or recovery stress, decreasing the impact of the influence on the composite structure.

Analysis, systematization and generalization of the market's demands for operating the machine building products by composition construction of the surface layers, using materials with TEPT, have shown that the alloys with shape memory are mostly used in engineering supplements, such as deep-water drives, valves, underwater installations, connectors, packings, self-tightening fastening details and sandblast control. The phase transformations property allows materials with SME to withstand erosion stresses, this is important for the minimization of erosion effect, which borehole devices often experience. The superelasticity of the materials with SME provides good energy dissipation, and can overcome different defects and limitations, which usual methods of passive damping face. The unique cryogenic properties of HEAs make them to be of great demand for cryogenic application, such as jackets of rockets, pipelines and equipment with liquid oxygen or nitrogen. The usage sphere of high-entropic compositions with TEPT is expanding for multifunctional application.

The technologies of surface layers and compositions formation from 2, 3 and 4-component materials with TEPT are elaborated (Patents of the Russian Federation № 2502829, 2535432, 2569871, 2563910) and described in details in some works [8-11]. The formation of coatings from materials with SME for high-entropic materials (comprising 5 and more components) causes considerable difficulties [12, 13].

For obtaining a qualitative composition coating from high-entropic materials it is necessary to seek to provide maximum adhesion strength of the coating with the host-material. To research the process of surface modification by high-entropic materials with SME under the conditions of high-voltage pulse electromechanical influences we need a technological complex which meets the following requirements:

- a chamber for improving adhesion strength with a protective atmosphere (to exclude the formation of oxides that decrease the possibility of thermoelastic phase transformations);
- the possibility to research the improvement of adhesion strength by different methods (by simultaneous surface plastic deformation and high-energy heating by high-voltage pulse);
- universality, the possibility to vary force, electric and electromechanical sources of influence.

2. Materials and methods

Nowadays the use of HEAs as coatings is limited due to insufficient study. Based on the analysis, systematization and experience of the previous researches, we have chosen the compositions on the basis of Ti, Ni, Zr, Hf, Co, Cu for HEAs with TEPT $Co_{17.6}Cu_{18.8}Ti_{9.6}Zr_{18.3}Hf_{35.7}$, $Ni_{17.8}Co_{17.7}Ti_{9,8}Zr_{18,6}Hf_{36}$, $Ni_{17.6}Cu_{18,8}Ti_{9.6}Zr_{18,3}Hf_{35,7}$, $Ni_{17.8}Co_{17.7}Ti_{9,8}Zr_{18,6}Hf_{36,1}$. Various structures of the high-entropic compositions were chosen in the way to discover the influence of nickel and cobalt on the structural-phase state and the functional properties of the coatings [14]. All the coatings include Ti, Ni, Zr, Hf, Co, Cu in different combinations, which we have already researched in the compositions of three- and four-component coatings (TiNiCu, TiNiCo, TiNiZr, TiNiHf, TiNiHfCu).

To research the process of surface modification by high-entropic materials with SME under the conditions of high-voltage pulse electromechanical influences we developed a multifunctional technological complex. This complex allows regulating the applied pressure and simultaneous transmitting of high-voltage pulse discharges through HEAs, preliminary having been formed by high-velocity oxygen fuel (HVOF). To conduct this research a high-voltage pulse machinery was designed and manufactured. The machinery has a switched condensing energy storage (100…1400 μF), allowing to form high-voltage discharges with a regulated initial voltage amplitude in the range from 1 to 5 kV.

The registration of current values and pulse width was done by measuring voltage drop on the shunt rating 750 μΩ by the dual-link oscillographLeCroy with the conversion frequency 300 MHz.

3. The constructive-technological peculiarities of the machinery

To carry out the research of the process of surface modification by high-entropic materials with SME under the conditions of high-voltage pulse electromechanical influences a multifunctional technological complex was offered and implemented (see figure 1). It was used for obtaining a qualitative compositional coating from high-entropic materials. The complex consists of a multifunctional vacuum chamber 1, a system of pumping out the air 2, a system of gas supply 3, a high-voltage pulse generator 4.

To preserve the initial coating structure it is necessary to study the impact of the duration of electric pulses thermal influence. To decrease the heating duration a high-speed heating by high-voltage pulse generator is employed. A high-voltage electropulse consolidation is one of the most high-speed methods of obtaining materials and products from electroconductive compositions [15-17].

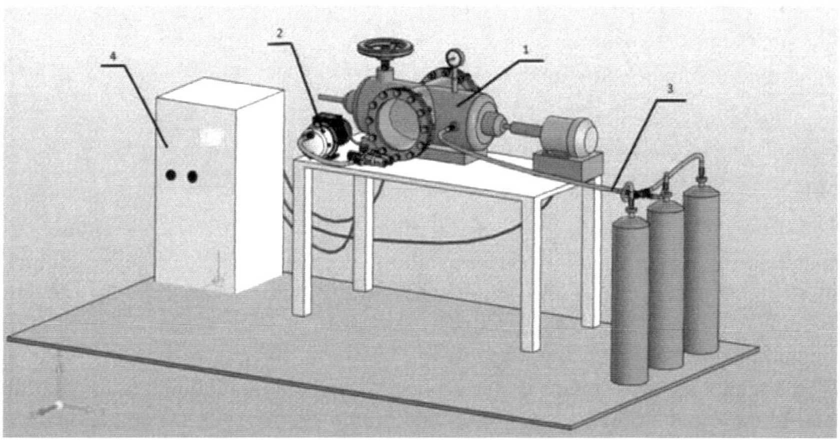

Figure 1. A technological complex for developing the process of surface modification by high-entropic materials with SME under the conditions of high-voltage pulse electromechanical influences.

During the research a multifunctional vacuum chamber was designed and manufactured. It consists of the chamber's case 1, an observation window 2, a case of bearings 3, fastening shafts of the workpiece 4, a changeable module for electrical and electromechanical source of influence 5, devices for surface plastic deformation 6, and a pressure-and-vacuum gage 7 (figure 2).

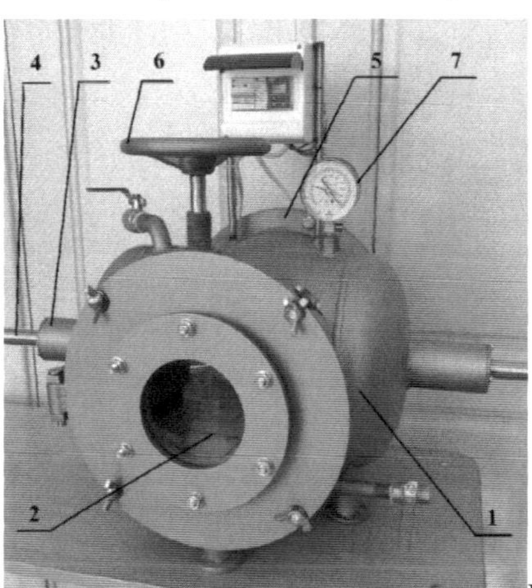

Figure 2. A multifunctional vacuum chamber.

To research the technology of the surface modification by high-entropic materials with SME under the conditions of high-voltage pulse electromechanical influences a changeable module with electrical vacuum lead-ins for a high-voltage pulse generator is set on the reverse side of the chamber. The scheme of the research of adhesion strength improvement by simultaneous influence of mechanical pressure and high-voltage current pulses is presented in figure 3.

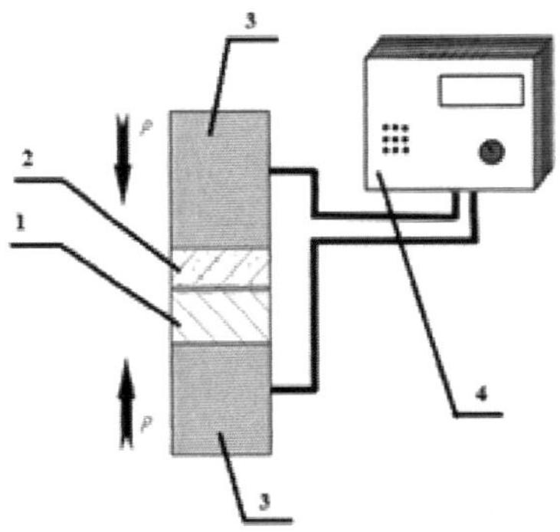

Figure 3. The scheme of the research of adhesion strength improvement by simultaneous influence of mechanical pressure and current high-voltage pulses.

Under the influence of high-voltage pulse the main thermal energy release occurs on the contact of the host-material with the coating. The temperature of the interparticle contacts in the process of current pulse influence may reach the value up to the one corresponding to the condition of the electrical thermal explosion of the interparticle contact [15]. The average temperature in the host-material volume during the high-voltage current pulse influence is considerably lower than the temperature between the host-material and the coating, this is conditioned by the presence of hightened resistance on their interface. The peculiarities of the thermal processes, going on in the material, cause the necessity of the force influnce on the workpiece with the coating.

In the developed machinery the condensing storage C is charged by a high-voltage constant current source up to the required voltage through the current-limiting charging resistance R_{ch}. The discharge of the condensers occurs with the help of a commutating device on the load resistance R_l, which in itself is HEAs, preliminary formed by HVOF on the host-material. For the emergency system of the condensing storage discharge the possibility of connecting discharge resistance R_d is foreseen. In the circuit of the condensing storage discharge shunt R_{sh} is also placed, on which the voltage drop is measured by the oscillograph.

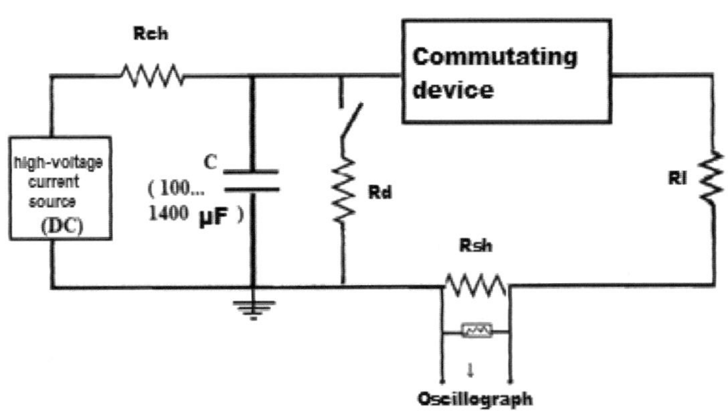

Figure 4. The scheme of the high-voltage pulse machinery.

IOP Conf. Series: Materials Science and Engineering **889** (2020) 012008 doi:10.1088/1757-899X/889/1/012008

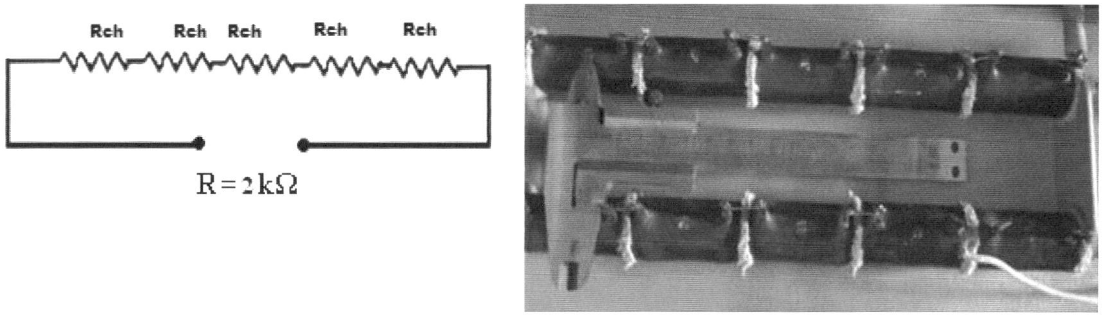

Figure 5. The scheme and construction of the current-limiting charging resistance.

Figure 6. The construction of the condensing storage.

The time of the thermal energy lead-in into the coating and the host-material is determined by the duration of the high-voltage current pulse. Figure 7 presents the current pulse amplitudes and the time of the high-voltage when influencing HEAs, preliminary formed by HVOF, in different modes of the condensers' storage charge of the high-voltage pulse generator.

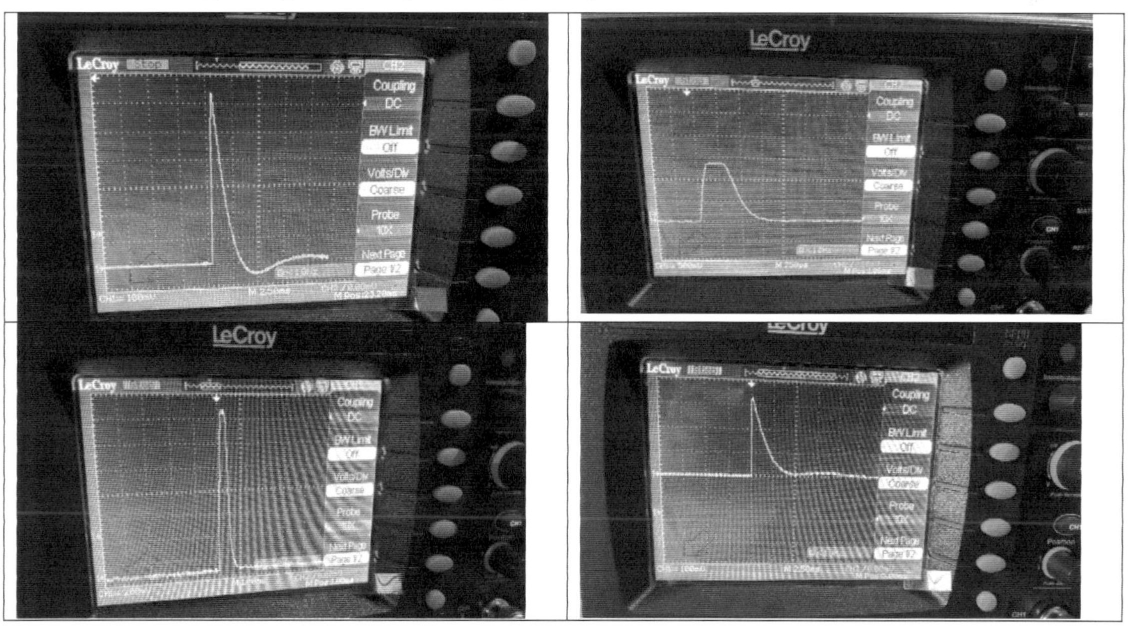

Figure 7. Registration of the current values and pulse width.

4. Conclusions

Based on the received results, we may conclude that for obtaining high operational characteristics of the coatings during surface modification by high-entropic materials with SME under the conditions of high-voltage pulse electromechanical influences it is necessary to identify optimum values of the process' technological parameters (the amplitude and the duration of current pulses and mechanical pressure, influencing the coating).

The construction has been designed and the technological complex has been manufactured to develop the process of the surface modification. The use of the multifunctional vacuum chamber together with the sources of power and loading system allows flexible modifying of different variants of electromechanical influence upon the workpieces with the applied compositional high-entropic coatings with SME.

Acknowledgments

The research was carried out with the financial support of the Russian Scientific Fund (grant № 19-19-00331).

References

[1] LiuPing and LiawPeterK2018 Microstructures and properties of high-entropy alloy films and coatings: a review, *Mat. Res. Letters*, **6:4** 199-229 DOI: 10.1080/21663831.2018.1434248

[2] Braeckman B R, Boydens F, Hidalgo H and et al. 2015 High entropy alloy thin films deposited by magnetron sputtering of powder targets *Thin Solid Films* **580** 71–6

[3] Qiu XW, Zhang YP and Liu C G2014 Effect of Ti content on structure and properties of Al2CrFeNiCoCuTix high-entropy alloy coatings *J Alloy Compd.* **585** 282–6

[4] Tian L-H, Xiong W, Liu C and et al. 2016 Microstructure and wear behavior of atmospheric plasma-sprayed AlCoCrFeNiTi high-entropy alloy coating. *J Mater EngPerfor.* **25** 5513–21

[5] Wang LM, Chen CC, Yeh J W and et al. 2011 The microstructure and strengthening mechanism of thermal spray coating NixCo0.6Fe0.2CrySizAlTi0.2 high-entropy alloys *Mater Chem Phys*. 126 880–5

[6] Löbel M, Lindner T, Kohrt C and Lampke T 2017 Processing of AlCoCrFeNiTi high entropy alloy by atmospheric plasma spraying, *IOP Conf. Series: Materials Science and Engineering* **181** 012015 doi:10.1088/1757-899X/181/1/012015

[7] BlednovaZh M and RusinovP O 2019 Formation of surface layers from highly entropic materials with shape memory effect*AIP Conf. Proc.* **2167**, 020035-1–020035-4https://doi.org/10.1063/1.5131902

[8] Blednova Zh M, Rusinov P O and Balaev EYu 2014 Methodological approaches and technical solutions on formation of nanostructured TiNi layers by high-velocity oxygen fuel *The proceedings of Samara scientific centre of RAS* **15,** 4(2) 484-8

[9] Blednova Zh M and Rusinov P O 2017 Compositional construction of surface layers from the materials with shape memory effect: monograph of FGEE HE "CubSTU" (Krasnodar: Publishing House of CubSTU) p 280

[10] Blednova Zh M, Balaev EYu, Dmitrenko DV and Plomodialo R L 2018 The ways of improving adhesion strength of coatings from the materials with shape memory effect, formed by high-velocity oxygen fuel*The proceedings of the International scientific-practical conference dedicated to the 100th anniversary of FGEE HE "Cuban State Technological University" Mechanics, equipment, materials and technologies* 107-16

[11] BlednovaZ, RusinovP, BalaevE and Dmitrenko D2020 Improving product performance by forming surface compositions from shape memory effect materials with a gradient of properties and phase transformation temperatures. *Mat Design Process Comm.*1–7 https://doi.org/10.1002/mdp2.132

[12] Löbel M, Lindner T, Kohrt C and Lampke T 2017 Processing of AlCoCrFeNiTi high entropy alloy by atmospheric plasma spraying *IOP Conf. Series: Materials Science and Engineering***181** 012015 doi:10.1088/1757-899X/181/1

[13] TangZhi, YuancChe-Wei TsaiTao, Jien-WeiYeh, D.LundinCarl andK.LiawPeter 2015 Fatiguc behavior of a wrought Al0.5CoCrCuFeNi two-phase high-entropy alloy *ActaMaterialia* **99**,15 247-58. doi.org/10.1016/j.actamat.2015.07.004
[14] BlednovaZh M and RusinovP O2019 Formation of Multifunctional Surface Compositions with Hierarchical Structure from High Entropical Materials with Memory Effect of the Form. *AIP Conference Proceedings* **2167** 020035 https://doi.org/10.1063/1.5131902.
[15] GrigoryevEG and OlevskyE A2012 Thermal processes during high-voltage electric discharge consolidation of powder materials *ScriptaMaterialia* **66** (9)662-5
[16] GrigoryevE G, OlevskyE A and YudinA V and YurlovaM S 2015 Wave mode high voltage consolidation of powder materials*Computational Materials Science* **100 A** 8–14
[17] Olevsky E A and GrigoryevE G 2017*Proceedings of 5th International Scientific Workshop on Advanced Technologies of Materials Field-Assisted Consolidation-2016,IOP Conf. Series: Materials Science and Engineering* **218** 011001

IOP Conf. Series: Materials Science and Engineering **889** (2020) 012009 doi:10.1088/1757-899X/889/1/012009

Improving the adhesive strength of coatings from multicomponent materials with thermoelastic phase transformations by external high-energy influences at various processing stages

Zh M Blednova[1], G V Kurapov[1*] and E Yu O Balaev[1]

[1]Institute of Mechanical Engineering and Car Service, Kuban State Technological University, Krasnodar, Russia

*kurapov_georgii@mail.ru

Abstract. The paper presents the results of a comprehensive study of the influence of external high-energy ultrasonic and electropulse influences on the adhesive strength of a coating from the material with thermoelastic phase transformations (TEPT) at various processing stages: preparing the base; preparing the sprayed material at the finishing stage after coating. It has been shown that as a result of complex high-energy influences at the processing finishing stage, one observes a distinct fusion of the coating from the material with TEPT (Ni-33%, Ti-49%, Zr -49%) with a base and the absence of the interface, confirmed by electron microscopic studies. The fusion zone is 10–20 μm with the coating thickness of 1 mm; there are practically no coating elements in the base material at the depth of 20 μm. The experimental studies have confirmed an increase in adhesive strength by 1.2-1.5 times up to 130-150 MPa.

1. Introduction

In modern industrial production under intense competition conditions, both on the Russian and global markets, the requirements to products are increased: by the metal structure, chemical composition, mechanical properties, geometric dimensions, the occurrence of defects accounting the quality, production time and products cost [1]. Under these conditions, taking into account the determinative role of surface layers in the damage and destruction accumulation, improving performance characteristics, increasing reliability and resource can be successfully resolved at the final processing stage on the basis of the technologies grounded on the principles of layer-by-layer synthesis using combined (hybrid), functionally oriented macro-, micro-, and nano-technologies [2].

A promising area in the implementation of layer-by-layer synthesis technologies is the functionalization and intellectualization of products using materials with thermoelastic phase transformations, including highly entropic ones with shape memory effect (SME) [3-7]. The increasingly expanding use of alloys with TEPT in engineering is connected with their exceptionally wide functional and mechanical capabilities: unique effects of thermomechanical memory, high strength and damping properties, thermomechanical reliability and durability, wear-proof and corrosive resistance [7, 8]. At present, a number of technologies have been developed for the formation of surface layers from two-component (TiNi, NiAl), three- and four-component (TiNiHf, TiNiHfCu, TiNiZr, TiNiMo, TiNiNb) [9] materials with SME using laser and argon-arc surfacing, plasma spraying, explosion surfacing; durable and reliable surface layers of the required thickness and dispersion have been obtained [8-10]. These methods differ in the type of heating the sprayed coating material and in the method of particle acceleration. The common thing about all these methods is heating the sprayed material up to a high-plastic state or melting, accelerating particles or drops by a gas stream and their subsequent interaction with the treated surface. Among the specified variety of methods of forming surface layers, the most universal and effective ones are high-speed methods (ultrasonic gas-flame, ultrasonic plasma, gas-dynamic spraying), characterized by a high speed of particles flight (more than 500 m/s) and a decrease in porosity (less than 2-5%) [11].

One of the problems of the functional surface layers formation from materials with TEPT on high-loaded products is the coating strength adhesion to the base, which ensures the reliability and durability of the surface-modified product. The analysis of domestic and foreign information sources describing the current state of theoretical studies on adhesion of a sprayed particle and a substrate showed that most works rely on the theoretical model of describing the particle adhesion to the substrate surface. Despite this, there is currently no unified theory describing the process due to the impossibility of accounting and describing the whole variety of factors influencing the adhesion process.

The adhesive strength of coatings made from materials with TEPT depends on a number of factors, the main of which are: preparing and cleaning the surface of the base, chemical compatibility of the base and the applied layer, the difference in their linear expansion coefficients [12]. The ways to increase adhesion are diverse (applying intermediate layers, having a high potential for interaction both with the substrate and the coating; optimization of the base temperature, granulometric composition, activation characteristics and thermalphysic properties of the sprayed material, operating modes, including particle flight speed and combustible gas composition) and depend on the requirements for the coating, and the product's operating conditions. Various external high-energy influences are used to improve the adhesion: acoustic stimulation, ultrasonic influence, thermal treatment initiating diffusion processes in the contact zone of the coating with the base [13, 14]. However, for the materials with SME, heating the coating to the temperatures ensuring the occurrence of diffusion processes is not always possible and is connected with the complexity of ensuring the necessary structural-phase state of the functional layers.

The aim of this study is to elaborate methods for increasing the adhesion strength of coatings from the materials with thermoelastic phase transformations by external high-energy influences at various stages of processing and to develop recommendations for increasing the adhesion strength of products in specified operating conditions

To achieve this goal we investigated:

- the influence of ultrasonic influences at the surface preparation stage on the adhesive strength of coatings from multicomponent materials with TEPT;
- the optimization of the granulometric composition of the sprayed material with TEPT at the stage of mechanical activation;
- the impact of ultrasonic influences on the adhesive strength of the coatings from multicomponent materials with TEPT at the finishing processing stage;
- the impact of complex ultrasonic, electropulse and thermal force influences on the adhesive strength of the coatings from multicomponent materials with TEPT at the finishing processing stage.

Another section of your paper

2. Materials and research methods. Peculiarities of the forming surface layers technology from the materials with TEPT

When developing the forming surface layers technology from the materials with TEPT according to raised standards to adhesion, Steel 45, widely employed in mechanical engineering, was used as a base, and nickel, having an unlimited solubility with iron and chemical affinity with the material of the functional layer, was used as a transition layer. To form the functional layer we used a three-component material with TEPT Ti33Ni49Zr18, possessing a high-temperature memory effect (with the following temperatures of phase transformations Mf = 459K, Ms = 522K, As = 488K, Af = 571K, transformation sequence B2 ↔B19'), which is in a martensitic state at the room temperature.

The surface layers formation from the materials with TEPT was carried out under conditions of high-energy influences by using a complex technology, including: preparing the base surface, dispersing and mechanical activating the applied multicomponent material with TEPT, applying adhesive and functional layers in the protective atmosphere, and subsequent thermal treatment using ultrasonic and electropulse thermal force influences, providing the formation of a nanoscale structure.

The preparation of the base includes mechanical treatment to create a surface with a developed microstructure with the following activation by surface plastic deformation with ultrasonic treatment by bead-blasting and chemical treatment consisting in degreasing the surface and etching with a mixture of hydrochloric and nitric acids. Preparation of the applied material consists in the mechanical activation (MA) of the material, which provides the necessary granulometric composition, energy state, and reactivity of the material. The mechanical activation was done in the modernized Hephaestus-2 ball mill AGO-2U [15], in which the mechanical action is done by a series of successive mechanical impulses (impacts) from actuating media transmitting portions of mechanical energy, reinforced by the ultrasonic action, to the processed material. The used planetary ball mill has the following operating parameters: the drum volume is 150 cm3, the ball acceleration is 40g, the ball to load ratio is (10-15): 1, the ball diameter is 6-8 mm. The maximum loading of the drum with grinding balls is 250 g, and with the processed material is 100 g. The rotation frequency of the carrier is 630, 890, 1090 rpm; the drum rotation frequency is 1290, 1820, 2220 rpm; the centrifugal acceleration of the grinding media is 300, 600, 1000 m/s2. During the scrutiny the influence of MA time on the powders' morphology and specific surface, on the phase composition change of the powder mixture, and on the distribution of elements and their compounds in MA powders was studied.

The coating was done by high-speed gas-flame spraying (HGS), having all the advantages of high-speed coating methods, which are essentially a combined process of spraying and plastic deformation of particles at a contact temperature with the base due to the high speed of collision. The coatings made from the materials with SME were sprayed using the upgraded GLC-720 apparatus in the protective argon atmosphere. HGS provides the adhesion of coatings from the materials with SME up to 120 MPa without an additional processing [16], the detonating formation of coatings provides the adhesion strength of the coating to the base up to 160 - 240 MPa. The experience shows that for high-loaded products this adhesion level does not provide the necessary level of reliability. During the usage of products with coatings from the materials with SME, the coatings undergo both a temperature change, leading to a martensitic-austenitic transformation, and a change in the coating material structure due to atom shear. This leads to stresses occurrence at the coating – base interface and, during the repeated phase transformations, can cause delamination, which is typical for coatings with a low value adhesive strength, provided only by the adhesion mechanical component. It is known that gas-flame spraying with reflow provides the adhesion to the base of 300-400 MPa [17]. To increase the reliability of products coated with SME materials it is necessary to develop additional measures or look for alternative ways to improve the adhesion.

A device for the implementation of ultrasonic processing in one cycle with the coating by the high-speed gas-flame spraying method (Figure 1a), and the subsequent electropulse treatment with the thermal force of the coated substrate is presented in Figure 1b.

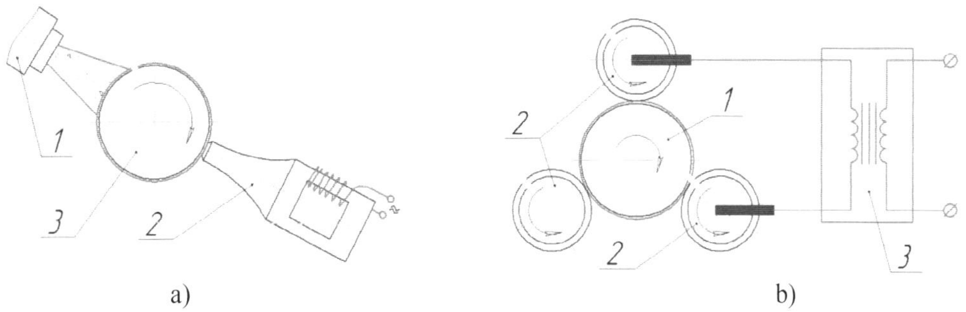

Figure 1. The scheme of the formation of surface layers with improved adhesive strength (1 – a burner for a high-speed gas-flame spraying; 2 – an ultrasonic source; 3 – a detail) - a); a fastening scheme for a detail of surface layers with improved adhesive strength (1- a detail; 2- three-roller SPD mechanism; 3- a welding transformer) - b).

The electropulse treatment with a thermal force influence is a combined treatment based on electrical, thermal and mechanical influence [18]. It is implemented by passing the pulsed electric current of high density in direct contact with the treated surface, using a conducting deforming element in the form of a roller. The depth of the thermal action zone is limited by the plastic deformation zone and depends on the pulses frequency, current density, time of thermal pulse influence, or the detail rotation speed, and the speed of the deforming element movement. The direction of motion of the electrodeforming tool is determined by the direction of the maximum stresses action in the product during usage. The optimization of the indicated technological parameters allows providing a sufficient reflow zone along the depth of the coating and the base to ensure the adhesion of the coating to the base at the value level of 0.4 - 0.5 of the ultimate base strength. For products such as shafts, operating under the conditions of cyclic loading, it is reasonable to implement the deforming element movement along a helical line.

The pin method was used to assess the adhesion strength of the coating to the base. The tensile test was done on the machine Instron 8801 according to State Standard 28844-90 "Gas-thermal, reinforcing and restoring coatings". Electron-microscopic studies were carried out on the microscope JSM-7500F. The X-ray spectral analysis was done on the diffractometer DRON-7M using Cu-Kα radiation, the stereoscopic studies were done on the microscope MSP-1.

3. Discussion of the results

To form a strong physicochemical bond at the coating – base interface the base surface was activated, including surface plastic deformation by rolling and ultrasonic treatment (UST) by a strengthening spherical element. The mechanical SPD influence and the physical UST influence contribute to the destruction of oxide layers on the base surface and to a sharp increase in the number of defects in the surface layer. Due to cold work hardening and the increase in the number of defects, coming to the base surface, SPD increase the number of setting areas during coating spray and create compressive residual stresses in the base material, which can partially decrease due to a base temperature increase under certain coating application conditions. The combined action of mechanical and physical activation allows increasing the density of dislocations, increasing the number of vacant sites, which are active centers on the base surface, enhancing the flow of physical and chemical processes at the "coating-base" interface and contribute to bulk diffusion of the materials. Thus, ultrasonic treatment is one way to activate the base surface before coating. The ultrasonic treatment machinery is shown in Figure 2. After the base surface activation the samples underwent a chemical treatment, including surface degreasing and etching with a mixture of hydrochloric and nitric acids.

Figure 2. The machinery for ultrasonic treatment: 1 – an ultrasonic generator; 2 - coated sample; 3 – a waveguide with and indenter; 4 – a magnetostrictive transducer; 5 - a water cooling system; 6 – wires.

The preparation of the applied material with SME consisted in mechanical activation (MA) with ultrasonic influence, done in a water-cooled ball mill Hephaestus-2 AGO-2U. The investigation of the granulometric composition evolution, phase transformations, and structure evolution in multicomponent materials with an electron-phase transformation of Ni-33%, Ti-49%, Zr-18% at different stages of the MA process are shown in Figure 3. The industrially produced powders of titanium-nickel (PN55T45) with a fraction up to 20 microns, and zirconium (PZrK-1) with a fraction up to 40 microns were used as the initial materials. The initial powders were prepared by removing excess moisture by drying at the temperature of 100-110 ° C during an hour. After that the components were dosed by weight to create a mixture with the necessary ratio of elements: Ni-33%, Ti-49%, Zr-18%. Figure 3 shows the particle size distribution based on the image analysis using the ImageJ program as well as the elemental composition of the powder at various stages of mechanical activation.

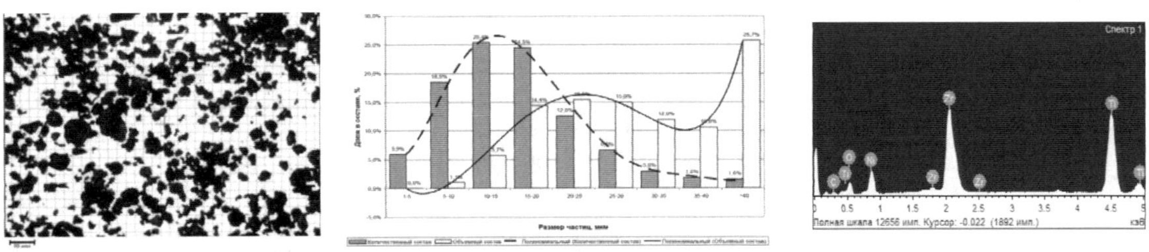

The average particle size of 16.1 microns a)

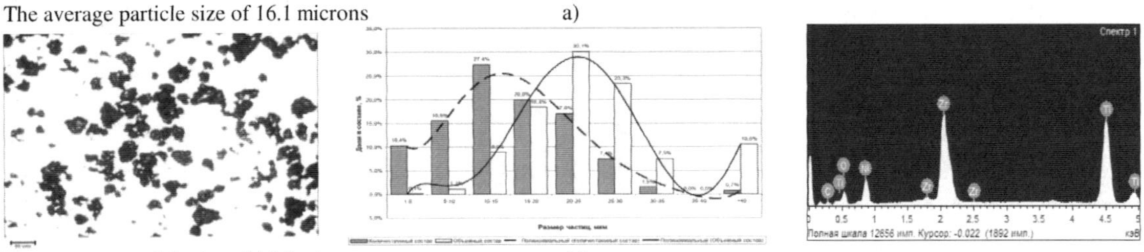

The average particle size of 15.2 microns b)

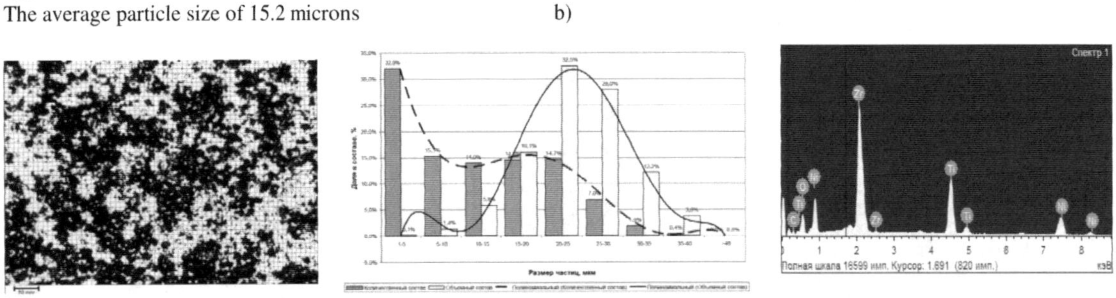

The average particle size of 12.2 microns c)

Figure 3. The mechanical activation of the composite mixture Ni-33%, Ti-49%, Zr-18% during: a) 60 min; b) 90 min; c) 120 min.

The mechanical activation of the mixture of powders Ni-33%, Ti-49%, Zr-18%: MA during 60 min does not significantly reduce the fractional composition of the mixture (up to 16.1 μm). A decrease in the average particle size to 12.2 μm occurs after MA during 120 min, and then the particle size stabilizes and further MA does not give a significant reduction in particle size; with an increasing mill drum rotational speed, the stabilization occurs earlier.

The analysis of the average particle diameter changes at MA of Ni-33%, Ti-49%, Zr-18% and at the carrier rotation frequency of 800 rpm-1 showed that an average particle size of 10 μm is achieved over a large period of time, approximately 150 minutes.

The electron microscopic analysis combined with the energy dispersive X-ray spectral analysis allows determining the possible formation of phases during the mechanical activation and mechanosynthesis. At MA and mechanosynthesis not only particle grinding, but also the formation of compounds occurs. So, after a two-hour MA of the powder mixture, conglomerates are formed from particles smaller than 1 μm in size, and the chemical composition corresponds to the already defined set of chemical compounds Zr2Ni, NiZr, TiNi, Ti2 (NiZr). Zr atoms dissolve well in titanium; a continuous series of solid solutions is formed without the formation of intermediate phases. Only insignificant amounts of these elements dissolve in nickel, and a number of intermediate intermetallic compounds are formed (Figure 4).

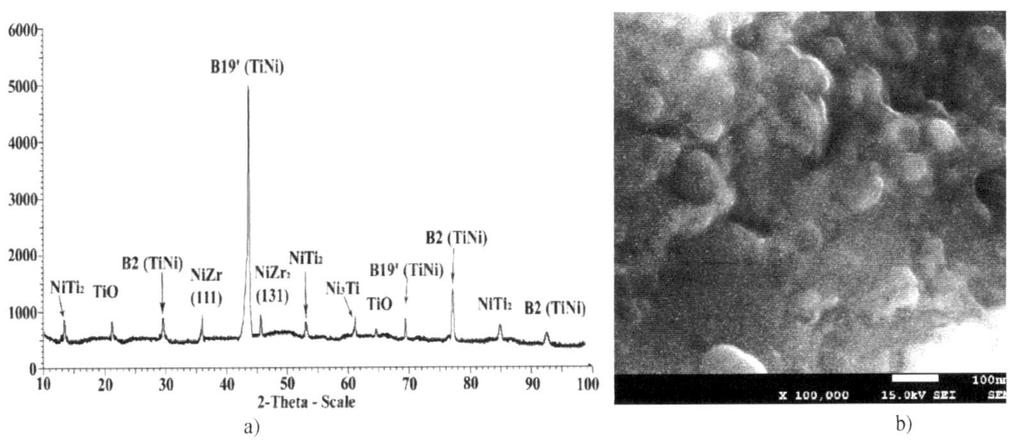

Figure 4. The diffraction pattern of the coating Ni-33%, Ti-49%, Zr-18%, obtained by high-speed gas-flame spraying -a); the microstructure of the coating Ni-33%, Ti-49%, Zr-18% × 100,000 - b)

Based on the study of the MA process of multicomponent materials with SME, it was found that: an increase in the MA time leads to obtaining a finer powder; homogenization is observed, a particle size is reduced, and due to cold work hardening, the strength characteristics improve; new powder material compositions are formed that in chemical composition correspond to the material necessary to obtain coatings, this is confirmed by chemical analysis of the surface layers formed by HGS MA powder. The analysis of the experimental results showed that with increasing fineness of grinding, the number of active centers increases. However, this tendency is observed up to a certain specific surface area value, after which the process slows down significantly. An intensive increase in specific surface TiNiZr area value, when dispersed in a ball planetary mill, is observed during 1.5 h [15]. According to the results of the study, the modes of mechanical activation and HGS are recommended [7], which ensure the formation of a nanoscale structure (Figure 4, b).

After the formation of the TiNiZr coating at the finishing stage, in order to increase the adhesion and the formation of the functional properties of the material with TEPT, the samples underwent ultrasonic, electropulse and force influences. The electropulse influence of the current is accompanied by a significant heating of the coating in the defects' area, and at the coating-base interface, and rapid heat removal to the bulk of the base, and the force influencing the electrodeforming tool (roller electrode) compacts the coatings and, thus, reduces the imperfection. As a result of such thermal deformation influence the structural and stress-strain state of the surface layer changes, leading to the residual stresses occurrence. To relieve residual stresses annealing was performed at the temperature of 650 ° C in the argon atmosphere. The electron microscopic analysis of the exposure zone showed a pronounced fusion of the coating with the base and the absence of the interface that occurs after sputtering (Figure 5); this is confirmed by the results of the X-ray spectral analysis. The elemental analysis, performed in three zones – in the coating after HGS, at the base – coating interface after finishing processing and in the main material, showed that the fusion zone or contact spot is 10–20 μm with the coating thickness of 1 mm (Figure 5). In the base material at the distance of more than 20 microns, there are practically no alloying elements of the coating.

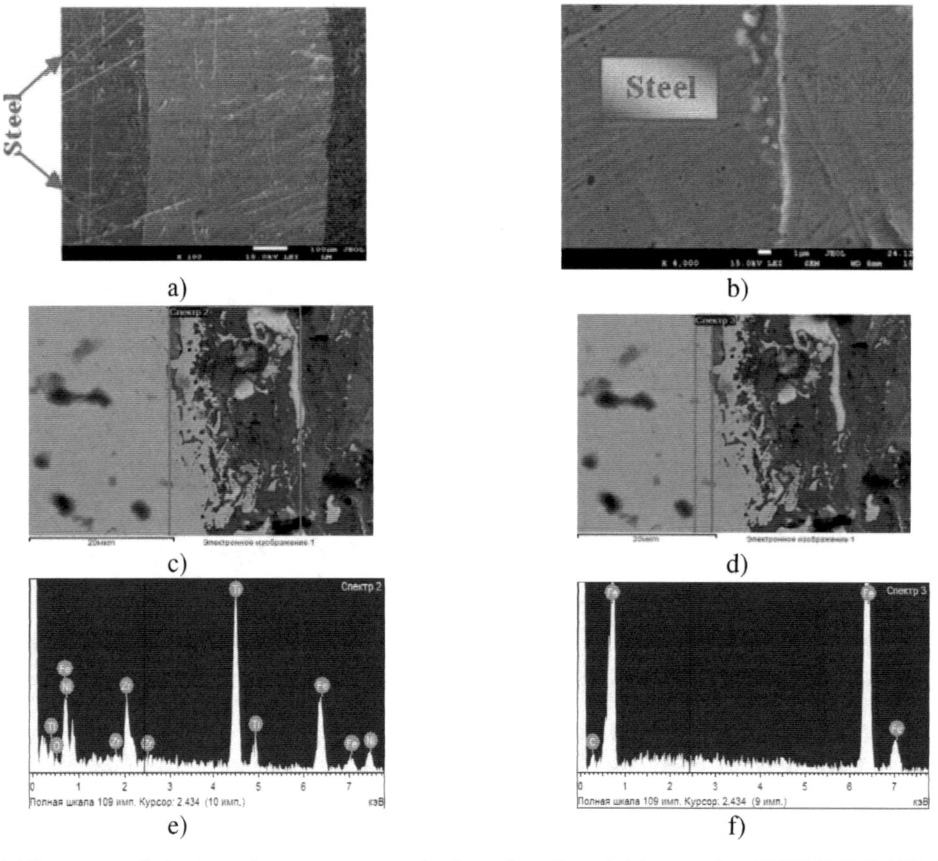

Figure 5. The zone of electropulse exposure at the interface "steel 45 - coating Ti33Ni49Zr18" formed by high-speed gas-flame spraying: a), b) - x 50; c), d) - x 10000.

To assess the effectiveness of the described technological operations using UST at various stages of the coating formation from materials with SME Ni-33%, Ti-49%, Zr-18%, the adhesion strength tests with fracture surface analysis - x 50 were performed. Figure 6 shows the workplace for profilometric studies of the samples surface. A stereoscopic pancratic microscope MSP-1 (1) is installed at the workplace, this allows shooting the front part of the sample. The profilometry results are displayed on the computer using the AMSu program (3). The surface roughness of the samples under study was measured using the «Abris PM7» profilograph-profilometer (2).

IOP Conf. Series: Materials Science and Engineering **889** (2020) 012009 doi:10.1088/1757-899X/889/1/012009

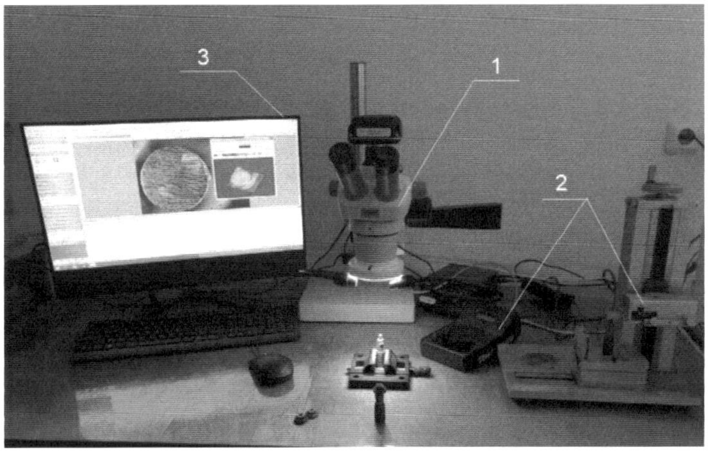

Figure 6. The workplace for studying profilometry of the sample surface: 1 - microscope stereoscopic pancratic MSP-1; 2 - profilograph-profilometer "Abris PM7"; 3 –computer with a working software

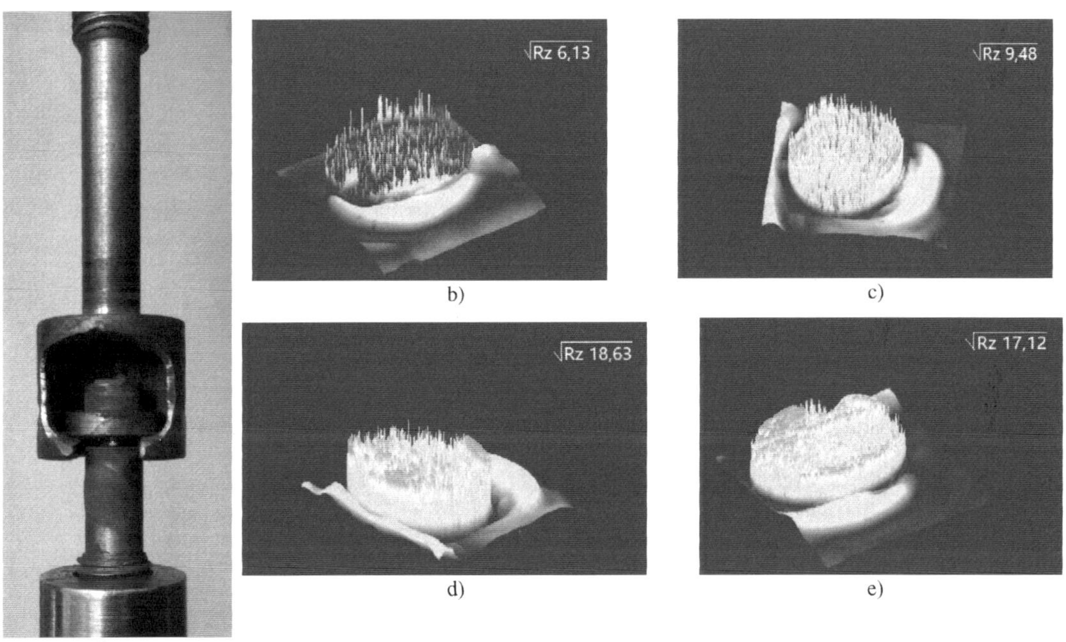

Figure 7. The moment of the sample destruction when being tested for adhesive strength by the pin method - a); profilograms of destroyed samples: before coating - b); after coating - c); after coating of the material with an SME Ni-33%, Ti-49%, Zr-18%, using ultrasonic influences at different stages of processing - d, e)

Figure 7a shows a sample for testing adhesion by the pin method at the time of destruction and the results of profilographic studies of the samples before and after applying coatings with SME Ni-33%, Ti-49%, Zr-18% by high-speed gas-flame spraying using UST at various stages of processing.

The analysis of the obtained profilograms indicates that an ordinary viscous fracture with the initiation of fracture on the surface is characteristic of a sample of steel 45 without coating. This section is marked by a brighter color on the profilogram (Fig. 7, b). A specimen with a HGS coating without UST has a mixed fracture, the coating is brittle-destroyed, the middle part of the sample is characterized by a more even distribution of protrusions and depressions, the presence of coating material is observed on the protrusions (Fig. 7c); this indicates an increased value of the adhesive strength. The highest adhesive strength is achieved by the implementation of the technological operations with UST described above. Fig. 6d, e shows that less than 50% of the fracture surface occurred at the base-coating interface; this allows concluding that adhesion is increased, but the cohesion strength is insufficient. Since the reliability of the base-coating compound is ensured by the mechanical and chemical components, the result obtained indicates the prevailing influence of the mechanical component on the adhesive strength during high-speed gas-flame spraying. UST, being a powerful stimulating factor for the intensification of diffusion mass transfer, increases the chemical component of adhesion. The significant spread in the values of adhesive strength is due to the need for careful testing of UST modes.

The considered sequence of technological operations during the formation of coatings from a multicomponent material with thermoelastic phase transformations Ni-33%, Ti-49%, Zr-18%, HGS using external high-energy influences at various stages of processing, including ultrasonic, electropulse and thermal force influence, is an effective way to improve the adhesion strength of coatings and, as a result, the reliability of coated products.

4. Conclusions

A comprehensive study of the impact of external high-energy ultrasonic and electropulse influences on the adhesive strength of a coating from the material with thermoelastic phase transformations at various processing stages: preparing the base; preparing the sprayed material at the finishing stage after coating.

The ultrasonic treatment of the base before coating can be characterized as a method of mechanical activating the base surface before coating.

The analysis of the granulometric, elemental, and phase composition of the applied material in the process of mechanical activation made it possible to determine the optimal parameters of mechanical activation that ensure the formation of a nanoscale coating structure.

As a result of complex ultrasonic, electropulse and thermal force influences, at the finishing stage of processing, we may observe a distinct fusion of the coating from the material with TEPT Ni-33%, Ti-49%, Zr-18% with the basis and the absence of the interface, confirmed by electron-microscopic studies. The fusion zone is 10–20 μm with the coating thickness of 1 mm; there are practically no coating elements in the base material at the depth of 20 μm. The experimental studies have confirmed an improvement in adhesive strength of 1.2-1.5 times up to 130-150 MPa.

5. References

[1] Glezer A M 2012 The principles of creating multifunctional structural materials of a new generation *The Successes of the Physical Sciences* **182** (5) 559 - 67

[2] Kheifets M L 2001 *The formation of the materials properties in the layered synthesis of details* (Novopolotsk: PSU) p 156

[3] Blednova Zh M and Rusinov P O 2015 Intellectualization of Surface layers, Working Under Cy-clic Loading and Reversing Friction *Ap. Mech. and Mat.* (Trans Tech Publications, Switzerland) **798** 440-6

[4] Exarchos D A, Dalla P T, Tragazikis I K, Dassios K G, Zafeiropoulos N E, Karabela M M, Crescenzo C De, Karatza D, Musmarra D, Chianese S and Matikas T E 2019 Development and Characterization of High Performance Shape Memory Alloy Coatings for Structural Aerospace Appli-cations *Mat.*, **12** (21), 3510; https://doi.org/10.3390/ma12213510

[5] Firstov G S, Kosorukova T A, Koval Y N and Odnosum V 2015 High entropy shape memory alloys *Mater. Today: Proceedings Materials Today: Proceedings 2S* 499-504

[6] Blednova Zh M and Rusinov P O 2019 Formation of Surface Layers from Highly Entropic Materials with Shape Memory Effect AIP Conference Proceedings () *AIP Conference Proceedings* **2167** 020035

[7] Blednova Zh M, Rusinov P O and Balaev E Y 2020 Improving product performance by forming

surface compositions from SME materials with a gradient of properties and phase transformation temperatures *Material Design & Processing Communications* 1–7

[8] Blednova Zh M, Makhutov N A and Chaevsky M I 2009 *Surface modification by shape memory materials* (Krasnodar: Publishing House-South) p 354

[9] Rusinov P O and Blednova Zh M 2015 *Surface Engineering Using Multicomponent materials with the effect of shape memory (chapter 3)* (Collective monograph, Results of science, Moscow: RAS) p 231

[10] Belyaev S, Rubanik V, Resnina N, Rubanik V Jr and Lomakin I 2014 Functional properties of Ti50 Ni50 – Ti49. 3Ni50. 7'shape memory composite produced by explosion welding *Smart Materials and Structures* **23** (8), 085029

[11] Carmen De Crescenzo, Despina Karatza, Dino Musmarra and Simeone Chianese 2018 Ni-Ti Shape Memory Alloy Coatings for Structural Applications: Optimization of HVOF Spraying Parameters. *Adv. in Mat. Sci. and Eng.* **1** 1-10 DOI: 10.1155/2018/7867302

[12] Revun S A and Balakirev B V 2002 *Peculiarities of the adhesive bonds formation during gas-thermal spraying of coatings* (Moscow: Physics and Chemistry of Materials Processing) **2** 55-62

[13] Stolyarov V V 2012 The role of external influences in nanostructured titanium alloys. *Proc. of the RAS. Physical Ser.* **76** (1) 108-13

[14] Rubanik V V, Rubanik V V, Lesota A V, Sibirev A V Belyaev S P and Resnina N N 2018 Realization of the shape memory effect under ultrasonic and thermal influence in titanium nickelide *Proc. of the 60th international conference "Actual problems of strength"* (Vitebsk: Ed. VSTU, May 14-18, 2018.) p 493

[15] Blednova Zh M, Rusinov P O, Bezmogorychnaya M E 2017 Influence of Particle size Distribution, Energy condition and Reactivity of the Sprayed Material on the Formation of Surface layers 'Nanostate Made of Shape Memory materials. *Key Engineering Materials.* **730** 8-14

[16] Klimenov V A, Kovalevskaya Zh G, Uvarkin P V, Nekhoroshchkov O N, Ivanov Yu F and Kukarenko V A 2004 Ultrasonic surface modification and its impact on the properties of coatings *Moscow Physical Mesomechanics* **7** (P 2) 157-260

[17] Khmelevskaya V B and Moseyko E S 2012 Study of plasma spray coating with ultrasonic treatment for hardening ship propeller shafts. *Metalworking. New materials and production technologies* **1** (67) 40-3

[18] Misochenko A A, Tsarenko Yu V, Rubanik V V and Stolyarov V V 2017 The combination of the effects of pulsed current and ultrasound in an alloy with shape memory in tension. *Mat. of the international scientific conference "Ultrasound: problems, developments, prospects"* (Ufa, September 25-29, 2017) 41-3

Acknowledgments
This work was financially supported by the Russian Science Foundation (Agreement No. 19-19-00331).

IOP Conf. Series: Materials Science and Engineering **889** (2020) 012010 doi:10.1088/1757-899X/889/1/012010

Development and experimental studies of new types of sheet metal stamping equipment

A Yu Botashev[1], R A Bayramukov[1], N U Bisilov[1] and R S Malsugenov[1]

[1]North Caucasian State Academy, 36, Stavropolskaya str., Cherkessk, 369001, Russia

E-mail: melov.mel@mail.ru

Abstract. In this paper, developed devices for gas sheet stamping are presented. The devices allow to significantly reduce the cost of products. They can be used both for food manufacturing, petrochemical industry and other branches of industry. Prototypes of products obtained experimentally are presented.

1. Introduction

A large proportion of the products of sheet rolling production is used in mechanical engineering for the production of thin-walled machine parts. The best quality of these parts is provided when making their sheet stamping. Traditional stamping equipment is only effective in mass production. However, many industries, in particular food, agricultural, chemical and energy engineering, aircraft engineering, rocket engineering, are characterized by low-volume production [1]. In the conditions of small-scale production, pulsed stamping methods are more effective, the varieties of which are electrohydraulic [2], Explosive Forming [3], hydrodynamic [4, 5], gas detonation forming [6] and other methods[7-10]. In the methods described above, the stamping process takes place in the cold state of the blank part. Many alloys of aluminum, titanium, molybdenum, tungsten have a low plasticity; therefore, when stamping parts of complicated shape, it is necessary to heat the blank part to the temperature range of warm or hot treatment.

2. Problem statement

The aim of this work is to develop new types of devices for sheet metal stamping with heating a workpiece as applied to conditions of small-scale production. To achieve this goal, we have developed new types of sheet metal stamping equipment, a device for gas stamping with two-sided heating of a workpiece, a gas molding device, a gas stamping device with a piston-type pressure multiplier [11].

3. Problem solving

The diagram of gas stamping device with two-way heating of a blank is shown in figure 1. The device comprises a matrix 1 and a combustion chamber 2, which are provided with inlet valves 4, 8, exhaust valves 6, 9 and spark plugs 5, 7. The stamped blank 11 is clamped between combustion chamber 2 and annular piston 13. Combustion chamber 2 and cavity of matrix 1 are filled with a combustible mixture through inlet valves 4 and 8, for instance, natural gas and compressed air. Combustible mixtures are ignited via spark plugs 5 and 7. In the process, as a result of the chemical interaction between combustible gas and the air, high-temperature and high-pressure gas is formed in both cavities, (pressure increases 7...8 times, temperature reaches 2000...2100 °C). Under two-way action of the gas, the blank 11 is heated intensively for a short period of time. After sample temperature has reached a specified value, valve 9 is opened and gas is released from the cavity 1. As a result, under the action of the gas pressure from the combustion chamber 2, blank part 11 deforms and fills the cavity of matrix 1.

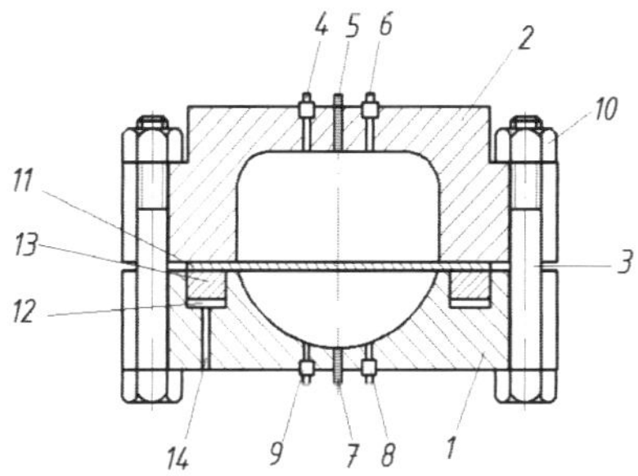

Figure 1. Schematic representation of the experimental setup of the gas detonation forming process: 1 – matrix; 2 - combustion chamber, 3 – bolt; 4, 8 – inlet valves; 5,7 – Igniter plug; 6,9 – exhaust valves; 10 nut; 11 workpiece; 12 – cavity; 13 – annular piston; 14 –channel.

Theoretical and experimental studies [6] have shown that this method of stamping ensures intensive heating of thin-walled blanks to a specified temperature at a relatively low pressure of a combustible gas mixture. In particular, the pressure of the combustible gas mixture of 1.0 cp 1.2 MPa is sufficient for heating a steel blank of a thickness of 1 mm up to 900 °C,he heating time of the workpiece is 0.5 ... 0.6 s.The required value of the pressure of the fuel mixture for the implementation of the stamping process can be determined by the following relation:

$$\boldsymbol{P_c} = \frac{1}{\lambda_v}\frac{1}{1-\tau_{\text{н}}}\frac{2\delta}{R_{min}}\boldsymbol{\sigma_s}\left(\mathbf{1}+\frac{V_{\text{м}}}{V_k}\right)^{k} \tag{1}$$

where, $V_{\text{м}}$, $V_{\text{к}}$ are the volume of the matrix and combustion chamber; k is the adiabatic exponent; σ_s is the yield point of the blank part of material; $\lambda_v$is the degree of pressure increase when the fuel mixture burns at a constant volume; δ is the thickness of the blank part; $\tau_{н}$ is the blank part heating time; $\boldsymbol{R_{min}}$is the minimum radius of curvature of the blank part surface.

For practical testing of this stamping method, an experimental setup for gas stamping was created (figure 2).

The setup is very compact, its overall dimensions are 600x600x1250 mm.Installation does not require a special foundation. The device comprises a system for supplying and monitoring of energy carrier, for igniting said fuel and for igniting the flanged part of a stamped blank. The energy source is a mixture of compressed air and propane butane or other combustible gas, which is formed directly in the combustion chamber and cavity of the matrix.

Clamping of the workpiece is carried out under the pressure of liquid. The apparatus allows for stamping parts of a sheet workpiece with a diameter of up to 400 mm. The shape of the part is determined by the shape of the matrix. The apparatus was used to carry out experimental studies of stamping of several types of part.

Figure 3 shows a honeycomb construction panel. Stamping of honeycomb construction panels out of aluminum sheet 2 mm thick was performed using a grid matrix with square cells measuring 90x90 mm.

The fuel mixture pressure of 0.4 ... 0.5 MPa ensured obtaining good quality parts.

In general, the results of the conducted experimental studies have shown that the method of gas stamping with double-sided heating of the blank part is applicable for stamping parts of both simple and complicated configuration with small curvature radii i.e. it has a wide range of technological capabilities. Therefore, this method can be effectively used in small-scale and pilot production , especially for stamping hard-to-deform alloys parts.

However, this device is not effective for stamping parts of small height, for instance, cap-type parts. In this case, due to the small height of the cavity of the matrix, the combustion process of the fuel mixture in it is unstable. Taking this fact into account, a method of gas molding with backpressure was developed to stamp thin-walled parts having a small height,schematic representation of the experimental setup shown in Fig. 4.A distinctive feature of this device is that cylinder 11 with piston 10 is fixed to body 2. Cavity 9 of cylinder 11 is communicated with combustion chamber 5 and cavity 12 is connected by pipeline 13 to cavity 18 of matrix 17

Figure 2. Spherical bottom.

Figure 3. Honey comb panels.

To carry out the molding process, combustible gas and compressed air are supplied to the combustion chamber 5, and a combustible mixture is formed in it, and to the cavity 18 of the matrix 17, only compressed air is supplied, which also, through the pipe 13, enters the cavity of the cylinder 11. When spark plugs 7 ignite a combustible mixture, the pressure in the combustion chamber 5 begins to increase. Under this pressure, the piston 10 moves, displacing the air from the cavity 12 into the cavity 18 of the matrix 17 while doing so. Due to this, an increase in pressure in the combustion chamber 5 is accompanied by an increase in pressure in the cavity 18. This prevents the deformation of the workpiece 16 and prevents it from contacting the surface of the matrix 17. This allows for intense heating of the workpiece. When its temperature reaches a predetermined value, the valve 22 opens, and compressed air is discharged from the cavity of the matrix 17. Meanwhile, affected by combustion products, the workpiece 16 is deformed in the direction of the matrix 17, as a result of which the manufactured part is molded.

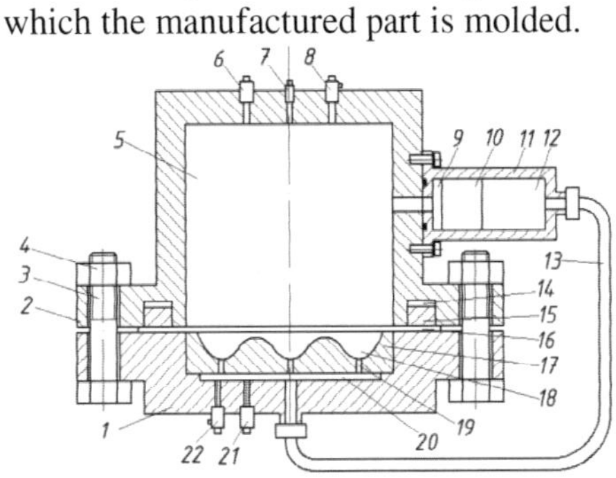

Figure 4. Diagram of the device for sheet gas molding: 1 - matrix holder; 2 - housing; 3 - bolt; 4 - nut, 5 - combustion chamber; 6.21 - intake valves; 7 - spark plug; 8.22 - exhaust valves; 9,12, 14,20 - cavities; 10 - piston; 11 - cylinder; 13 - pipeline; 15 - annular piston, 16 - blank; 17 -matrix; 18 - matrix cavity; 19 – channel.

According to the scheme shown in figure 4, a setup for backpressure gas molding was created (figure 5).

Figure 5. Experimental installation for gas molding with backpressure.

Using this setup, backpressure gas molding studies were conducted. During the studies, aluminum sheet workpieces 0.5 mm, 1 mm and 2 mm in thickness and steel sheet workpieces 3 0.5 mm and 1 mm in thickness were molded. As a result of the studies conducted, the technology of molding several types of parts ensuring the production of parts of good quality has been developed. Photographs of the two types of parts are shown in figure 6. Based on the results of the studies, it can be concluded that the device for backpressure gas molding can be used in small-scale production for the manufacture of thin-walled parts of various configurations.

a
Spherical bottom with central cavity

б
Heat exchanger panel with spiral channel

Figure 6. Stampedproducts.

The considered devices allow for stamping thin-walled parts made of steel up to 1.5 mm thick, and of non-ferrous alloys — up to 2.5 ... 3 mm thick. For stamping parts of greater thickness, it is necessary to provide a significantly greater gas pressure on the surface of the in-process workpiece. In this regard, we have developed a gas stamping device with a piston pressure multiplier [7]. The scheme of this device is shown in figure 7.

For significant increase in the gas pressure on the surface of the workpiece, the device is equipped with an operating cylinder 6 comprising a piston 7, which is located between the combustion chamber 8 and the matrix 2. Moreover, the combustion chamber 8 comprises the prechamber 19, equipped with spark plugs 20.

To complete the stamping process, through the inlet valves 9, 12, 14, combustible gas and compressed air are supplied to the operating cylinder 6, the combustion chamber 8 and the cavity 3 of the matrix, as a result of which combustible gas mixtures are formed in them. At the same time, the mixtures in the combustion chamber 8 and the working cylinder 6 have the same pressure, and in the cavity 3, its pressure is 2 ... 3 times less. Combustible mixtures in the working cylinder 6 and in the cavity 3 are ignited simultaneously. As a result, when the mixtures are burned, the pressure in the two cavities increases sharply, but the pressure difference between the two still remains, that is, the pressure in the cavity of 3 is 2 by 3 times lower.

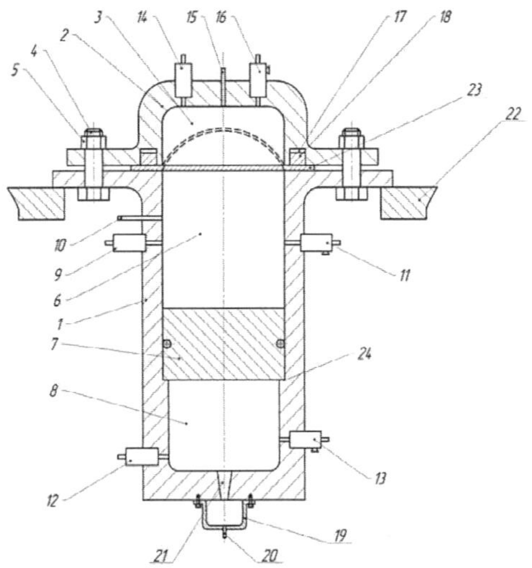

Figure 7. Diagram of a gas stamping device with a piston pressure multiplier.

Under the action of the difference of pressures and the blank is drawn into a cavity 3. Since the gas in the cavity 3 is compressed, the deformation ceases when a certain bending of the blank is applied. This position of the blank is shown in figure 7 in the bar line. Then, with a candle of 20, a combustible mixture is lit in the forechamber 19. The resulting combustion products in the form of a torch are transmitted to the combustion chamber 8, which causes the combustible mixture therein to burn intensively. The pressure in the combustion chamber increases sharply, which causes intense displacement and acceleration of the piston. At the same time, under the action of the piston 7, the pressure and temperature of the gas in the operating cylinder increase rapidly, which ensures further heating of the workpiece until the temperature range of hot processing is reached. The exhaust valve 16 is then opened and the gas from the cavity 3 is released. Meanwhile, under the pressure of the gas in the cylinder 6, the workpiece is deformed and fills the entire cavity 3 — a forming process of the manufactured part is completed.

A study of the work process of this device showed [6] that the presence of a piston multiplier in the design of the device allows to significantly increase the gas pressure on the surface of the stamped workpiece. When the device is operated without a prechamber, the pressure increases by 2 ... 2.5 times, and when operated with a prechamber, the pressure increases by 4 ... 5 times. This significantly expands the technological capabilities of this device.

A device for sheet stamping was created and mounted according to the scheme shown in Fig.7.

At this installation, experimental studies were conducted. During the experiments, the stamping was done on sheet blanks 1 mm thick of steel 3 using a cylindrical matrix 160 mm in diameter. As a result of the process, sphere-shaped bottoms were obtained. Figure8 shows one of the bottoms stamped at a pressure of the fuel mixture of 0.7 MPa. The obtained bottoms had a smooth surface with not scuffs or corrugations.

Figure 8. Sphericalbottom.

4. Conclusions

1. New types of devices for sheet stamping which provide stamping process with heating of a workpiece have been developed and created.

2. Practical testing of these newly developed devices was carried out, during which rational technological modes of stamping process were developed as well.

3. During testing, parts of good quality were obtained, which allows us to recommend developed devices for use in small-scale manufacture of stamping thin-walled steel and non-ferrous alloys parts.

5. References

[1] Ilin L N, Semenov E I 2009 *Technology of sheet stamping*. (Moscow: Drofa)

[2] Taranenko M E 2014 Possibilities of stamping of automobile sheets from modern high-strength materials at electro-hydraulic presses Forge and Forming Production. Die Forging [in Russian – Kuznechno-ShtampovochnoeProizvodstvo.Shtampovka] **9** 34-40

[3] Bychkov S A 2007 On *the concept of using technological criteria for choosing pulse sheet metal stamping technologies*. Aerospace Engineering and Technology (in Russian – Aviatsionno-kosmicheskayatekhnika i tekhnologiya) **11** 222-31

[4] Pavlenko V N 2002 *Perfection of design parameters and technical features of power units of press guns for hydrodynamic stamping: PhD thesis* (Kharkov)

[5] Musayev A A 2012 Development of initial data for the design of pilot-industrial sample of a two chamber device for gas sheet stamping. *News of HEIs. MechanicalEngineering* **3** 43-8

[6] Sukhov V V 2007 *Creation of gas-explosive systems with multipoint initiation by the influence of methane-oxygen mixture*. Aerospace Engineering and Technology (in Russian – Aviatsyonno-KosmicheskayaTehnika i Tehnologiya) **11** 182-5

[7] Chudin V N, Yakovlev SS, Sobolev Ya A 2004 *Isothermal deformation of high-strength anisotropic metals* (Mashinostroenie-1, Publishing House of TulGU, Moscow) p. 427

[8] Larin S N 2010 Pneumoforming of cellular panels from anisotropic material. *Izvestiya TulGU. Tech Sci* **3**: 51-61

[9] Zausaev A F *2010 Difference methods for solving ordinary differential equations* (In: Proc. Help. Samara State University.Techn.un., Samara) p 100

[10] Erofeev V L, Semenov P D, Pryakhin A S 2008 *Heat engineering: textbook for high schools*. (ICC "Academic Book", Moscow) p 488

[11] Botashev A Yu and Bayramukov R A 2018 Development and study of gas sheet stamping machine with piston hydraulic amplifier Messenger of Samara University. Space Equipment, Technologies and Mechanical Engineering (in Russian – Vestnik Samarskogo Universiteta. Aerocosmicheskaya Tehnika, Tehnologii i Mashinostroenie) **17(2)** 132-44

IOP Conf. Series: Materials Science and Engineering **889** (2020) 012011 doi:10.1088/1757-899X/889/1/012011

Theoretical calculations of mechanotermic destruction process with regard to general rubber goods

I E Ametov[1] and C V Abkhairova[2*]

[1]Associate Professor of the Department of Construction Engineering and Architecture, Academy of Construction and Architecture of V.I. Vernadsky Crimean Federal University, Simferopol, Russia
[2]Associate Professor of the Department of Electrical Engineering and Welding, State Budget Educational Institution of Higher Education of the Republic of Crimea Crimean Engineering and Pedagogical University the name of Fevzi Yakubov, Simferopol, Russia

*sysana70@gmail.com

Abstract. Reference data on the properties of rubbers and some fillers included in their composition are considered. Based on them, the values of electric and thermal power necessary for the processing of waste tires and other rubber products are calculated. The calculations showed the undoubted advantage of the mechanothermal destruction of rubber products in comparison with the classical pyrolysis methods.

1. Introduction

As known, new methods require new theoretical assumptions. The processing of rubber products by pyrolysis (heating to decomposition temperature without air access) has existed since the beginning of the use of rubber on an industrial scale. Relatively recently, about ten years ago, the idea arose to combine the pyrolysis of rubber wastes with their simultaneous grinding [1]. The theoretical prerequisite for our idea was the high value of the viscosity of rubber products, and, therefore, a rather large value of the coefficient of friction caused by the presence of cross-links in the polymers of this class.

Since this method combines mechanical abrasion of the polymer (mechanical destruction [2-5] and pyrolysis (heating the rubber before decomposition without access of air), we propose the following general name for it: mechanothermal destruction of polymers.

All modern technologies and equipment for the processing of worn-out tires and other rubber products have the following unrecoverable flaws:

- the bulkiness and metal consumption of equipment, which increase exponentially with increasing volume of rubber;
- low productivity due to loss of time for loading, unloading and pumping of pyrolysis products from the reactor;
- inefficient heating due to periodic cooling during unloading (heat losses are proportional to the square of the sizes);

- the complexity of the design of the devices of loading, unloading and heating;
- the need for sealing equipment for pyrolysis;
- the need to grind products to the required size;
- a long, expensive process of designing various specialists of the original parts of technological equipment - the reactor, pressurized valves, loading and unloading mechanisms, etc.

The proposed technology of mechanothermal degradation of rubber lacks the drawbacks that are common in the conventional method of rubber pyrolysis. The material is subjected to mechanical stress. In this case, the process of "internal friction" occurs, causing a sharp increase in temperature to values of 450-500 ^{0}C, at which the process of rubber pyrolysis begins [1]. To ensure the irreversibility of the decomposition process, the reaction products are removed from the destruction zone. At the same time, their temperature drops to a value of 120-150 ^{0}C. They are further cooled and condensed in a special device and processed in the usual manner.

In 2009-2010, we created an experimental installation for the processing of automobile rubber, described in [1]. The operation of such installation has not only evidenced the undoubted prospects of the proposed method, but also has revealed a number of significant flaws. One may say that we achieved the results by trial and error. Thus, there is a need to develop theoretical foundations of a new method for rubber goods processing.

1.1. Analysis of the literature.

Since the existing scientific sources do not even mention our proposed method of rubber processing, we came to the conclusion that some reference data should be collected on the composition and construction of rubbers, as well as the effect of fillers, volcanizers and other chemical agents on their properties [2- 6]. Some data are summarized in the table (table 1).

Table 1. Some reference data on rubber processing.

Size, dimension	Rubber	Volcanizate	Volcanizate + C	Carbon
Densityρ, кg/m^3	1050	1200	1227	1400
Coefficient of thermal conductivity λ, W/m·K	0.130	0.167	0.248-0.293	0.85-0.91
Thermal diffusivity, $a \cdot 10^8$, m^2/sek	8.9-9.0	11.5-13.0	17-20	–

The specific heat, density, thermal conductivity and thermal diffusivity are related as follows [4]:

$$C_p = \frac{\lambda}{a \cdot \rho} [\frac{kW \cdot s}{\kappa g \cdot K}] (1)$$

At the same time, the thermal conductivity of rubber varies depending on the content of carbon black according to the following formula:

$$\lambda_{rubber} = \lambda_0 + K_\lambda \cdot \omega \ (2)$$

where ω is the mass fraction of carbon black;K_λ – is a correction coefficient $[\frac{W}{m \cdot K \cdot \omega}]$.

The values of the correction coefficient K_l, which depends on the brand of carbon black, are presented in the table below (table 2).

Table 2. The dependence of the values of the correction factor on the brand of carbon black.

Carbon black brand	K-354	P-234	P-324	P-514	P-803
$K_\lambda, \frac{W}{m \cdot K \cdot \omega}$	0.00151	0.00203	0.00227	0.00235	0.00216

1.2. The purpose of the article.

Based on the analysis of the reference data, we intend to calculate the amount of thermal (electrical) energy required for mechanothermal destruction of rubbers. Based on the heat balance of the process of their processing, we aim to show the prospects of this pyrolysis method.

1.3. Summary.

Using formula (2), we calculated the changes in the coefficient of thermal conductivity of polyisoprene rubbers, depending on the content and type of carbon black used as a rubber filler, to increase the hardness of tire treads, increase their wear resistance and other useful qualities. The results are presented both as a table (table 3) and as a graph (figure 1).

Table 3. Thermal Conductivity Values λ.

Brand of carbon black			K-354		
Mass fraction of carbon black, %	0	10	20	30	40
Coefficient of thermal conductivity λ, W/m·K	0.167	0.182	0.197	0.212	0.227
Brand of carbon black			P-234		
Mass fraction of carbon black, %	0	10	20	30	40
Coefficient of thermal conductivity λ, W/m·K	0.167	0.187	0.207	0.227	0.247
Brand of carbon black			P-324		
Mass fraction of carbon black, %	0	10	20	30	40
Coefficient of thermal conductivity λ, W/m·K	0.167	0.190	0.212	0.235	0.257
Brand of carbon black			P-514		
Mass fraction of carbon black, %	0	10	20	30	40
Coefficient of thermal conductivity λ, W/m·K	0.167	0.191	0.214	0.238	0.261
Brand of carbon black			P-803		
Mass fraction of carbon black, %	0	10	20	30	40
Coefficient of thermal conductivity λ, W/m·K	0.167	0.188	0.208	0.229	0.249

As a result of the analysis of the obtained data, we calculated the difference in the values of the thermal conductivity coefficient. For some types of rubbers, the difference in λ values, depending on the content and brand of carbon black, can reach a maximum of 56% compated to unfilled volcanizate. Accordingly, a significantly greater amount of heat should be spent to heat such a rubber filled with carbon black. According to formula (1), we determined the boundary values of the heat capacity of rubbers of this type:

$$C_P = \frac{\lambda}{a \cdot \rho} = \frac{0.167 \div 0.261}{1.85 \cdot 10^{-7} \cdot 1227} = (736 \div 1150)[\frac{kW \cdot s}{\kappa g \cdot K}] \quad (3)$$

With an increase in temperature for every 100 °C, the heat capacity increases by approximately 15% [6]. Thus, since the temperature of the rubber increases by approximately 400 K, it is possible to calculate the change in heat capacity during the heating process according to the following formula:

$$\overline{C_P} = \frac{C_P^0 + 1.15C_P^0 + 1.30C_P^0 + 1.45C_P^0 + 1.60C_P^0}{5},$$

$$C_P^0 = 736 \div 1150[\frac{kW \cdot sek}{\kappa g \cdot K}]. \quad (4)$$

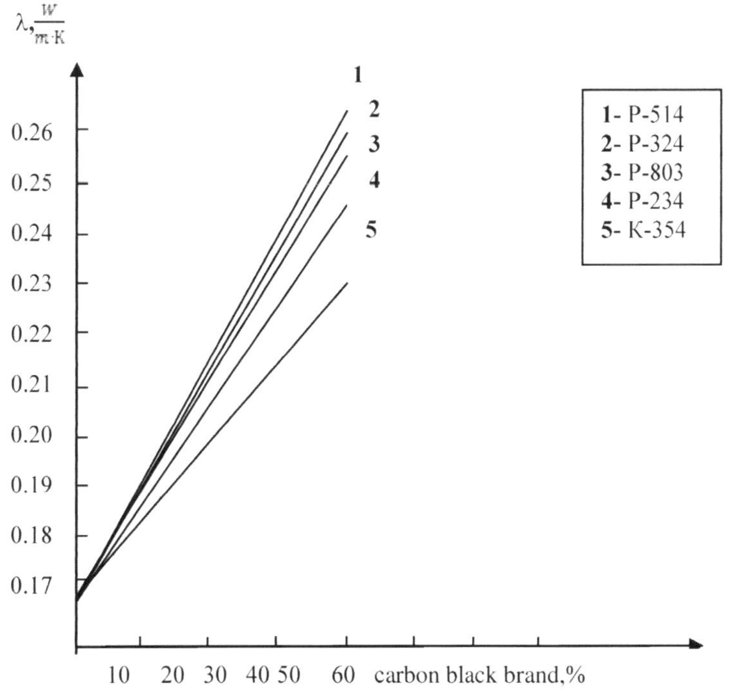

Figure 1. Graphic dependence of the coefficient of thermal conductivity on the content and brand of carbon black.

Substituting in formula 3 the boundary values of heat capacity of rubbers obtained from the calculation according to formula 1, we obtain the following range of heat capacity for this type of rubber:

$$\overline{C_p} = \frac{[736 \div 1150] + [846 \div 1323] + [957 \div 1495] + [1067 \div 1667] + [1177 \div 1840]}{5} = 957 \div 1495 \, [\frac{kW \cdot s}{\kappa g \cdot K}] \quad (5)$$

The obtained range of heat capacity of filled rubbers is quite wide, but since values of the same order are obtained, it is possible to calculate the energy consumption necessary for processing a unit mass of rubber goods for depolymerization products. To do this, we convert W·s to kW·h and get the following range of values:

$$W_1 = \frac{(957 \div 1495)}{1000 \cdot 3600} = (2.66 \cdot 10^{-4} \div 4.15 \cdot 10^{-4})[\frac{kW \cdot h}{\kappa g \cdot K}] \quad (6)$$

Therefore, it takes 0.266 to 0.415 watts of electrical energy to heat 1 kg of rubber per 1 K. Thus, to heat 1 kg of rubber to 500 ^{0}C (note that depolymerization process does not require more than this), a minimum of 0.133 to 0.208 kW·h of electrical energy would have to be spent.

The process of mechanothermal destruction has another significant difference from conventional rubber pyrolysis. With a gradual increase in temperature due to mechanical stress, those bonds in the polymer that have the least strength are the first to break. In this class of polymer compounds, a disulfide group, which crosslinks the polymer molecules into a three-dimensional network, has the lowest value of dissociation energy of in the. Some values of dissociation energy are given in the table (table 4).

Table 4. Dissosiation energies [7].

Type of bond	Dissociation energy, kJ / mol
Hydrogen-hydrogen	432
Carbon-carbon (single)	345.6
Carbon-carbon (dual)	602 ± 21
Carbon-sulfur (single)	272
Methyl-carbon (–CH_3→ –CH_2– + H)	142 ± 4
Methylene-carbon (–CH_2– → –CH= + H)	≈ 382

As can be seen from the data in the table, the carbon-hydrogen bonds of the terminal methyl group and the single carbon-sulfur bond have the smallest breaking energy. The percentage of bonds of the first type is extremely small, since such groups are found in polymer molecules, at best, one in 10,000 groups or less. The number of bonds of the second type is directly proportional to the amount of sulfur in the polymer.

For example, with a sulfur content of 1% in rubber, each kilogram of polymer contains 10 g of sulfur. Considering that each sulfur atom is associated with one carbon atom, we find that the amount of substance of the carbon-sulfur atomic groups is 0.3125 mol for each percentage of sulfur. Since it takes 272 KJ of energy to break each mole of carbon-sulfur bonds, only 85 KJ is consumed per percent of sulfur. This amounts to 0.0236 kW·h of electrical energy, respectively. The figure (drawing2) shows the boundary values of the increase in the values of electric energy for the pyrolysis of rubber, depending on the sulfur content. From the figure it follows that at high sulfur contents (from 5-6% and higher), the energy consumption can double, and even triple at a sulfur content of more than 10%. Since automotive rubber can contain up to 18-20% sulfur (in truck tires), and for heavy vehicles the mass fraction of sulfur can reach 28-30%, a sharp increase in power consumption should be taken into account when pyrolyzing such rubber.

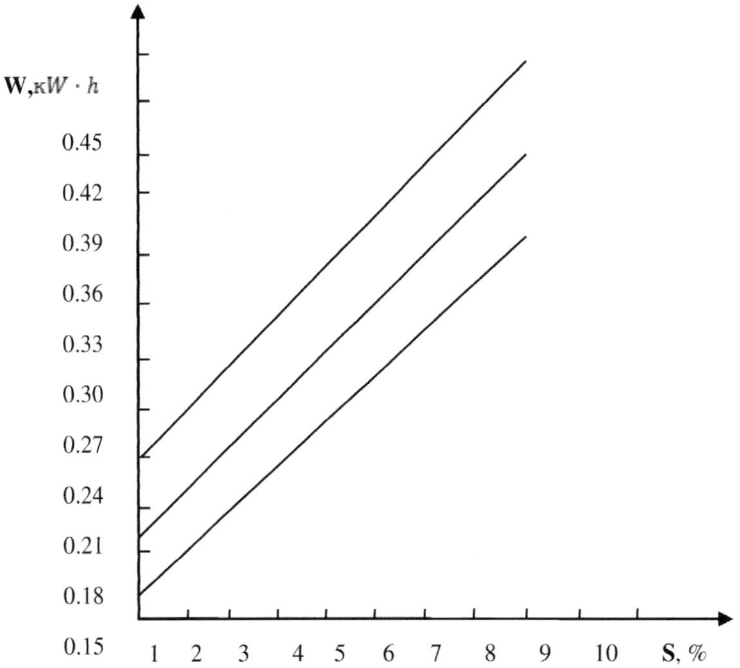

Figure 2. Dependence of the average power of rubber pyrolysis on the sulfur content.

IOP Conf. Series: Materials Science and Engineering **889** (2020) 012011 doi:10.1088/1757-899X/889/1/012011

Since the boiling point of sulfur is 440 ^{0}C, all atomic sulfur obtained in the pyrolysis process is released in the form of vapors into the expansion chamber of the unit, i.e., the energy spent on dissociating carbon-sulfur bonds is irrevocably consumed.

As for the dissociaton energies of single carbon-carbon bonds (table 4, No. 2) and the removal of hydrogen from the methylene units of the polymer (table 4, No. 6), this energy is compensated by the formation of hydrogen molecules (table 4, No. 1) and the formation of large quantities of carbon black, in which only carbon-carbon bonds are present. The energy of such bonds is lower than in the polymer, but a large number of them are formed.

Another important factor affecting the energy consumption during mechanothermal decomposition of rubber is the heat loss by pyrolysis products of the heat generated as a result of the polymers destructionSince the temperature of the pyrolysis products that are removed from the expansion chamber is 100-120 ^{0}C, then, therefore, about 25% of the power spent on processing rubber goods is lost in the form of thermal energy. Thus, based on the target of research (worn-out automobile tires) that contain, in addition to rubber, from 3 to 5% sulfur and about 15-20% of carbon black, and also, taking into account heat transfer by pyrolysis products, we can calculate average power consumption of electric energy required for mechanothermal processing of this typr of rubber goods.

The average value of the heat capacity of rubbers of the indicated content with a volcanizing agent (sulfur) and carbon black is:

$$\overline{C_p} = 1226 \left[\frac{W\cdot s}{Kg\cdot K}\right] \quad (7)$$

which corresponds to 0.31 ± 0.04 kW·h of electrical energy consumption per 1 kg of rubber if heated to 450 K.

As a comparison, with the usual pyrolysis method, the energy consumption is about 1.75 kcal for heating 1 kg of rubber per 1 K [8]. If converted to electric power units (when heated to 450 K), it is about 0.9 kWh per 1 kg of rubber.

2. Conclusion

The calculations evidenced the undoubted promise of the new method of processing worn rubber products. The obtained values of power consumption are 3 times less than in the conventional rubber pyrolysis, which confirms the effectiveness of this new method of solving the global environmental problem of our time.

In the future, we contemplate to create a full cycle installation for mechanothermal destruction ofused car tires and other rubber products obtaining clean end products.

3. References

[1] Ametov I E and Obolonsky V V 2011 Pyrolysis of rubber by the method of "internal heating" *Scientific notes of the Crimean Engineering and Pedagogical University (*Engineering: Simferopol) **29** 8-11
[2] Tager A A 1968 *Physicochemistry of polymers* (Moscow: Chemistry) p 536
[3] Shur A M 1981 *High molecular compounds* (Moscow: Higher School) p 656
[4] Tugov I I andG I Kostrykina 1989 *Chemistry and physics of polymers* (Moscow: Chemistry) p 432
[5] Baramboim N K 1978 *Mechanochemistry of macromolecular compounds* (Moscow: Chemistry) p 362
[6] Makhlis F A 1989 *Terminological reference for rubber* (Moscow: Chemistry) p 400
[7] Efimov A I Belorukova et al 1983 *Properties of inorganic compounds* (Moscow: Chemistry) p 392

IOP Conf. Series: Materials Science and Engineering **889** (2020) 012012 doi:10.1088/1757-899X/889/1/012012

Biopolymer bindings for casting manufacturing processes on the technological lignin basis

Yu A Svinoroev[1], V V Dyadichev[2*], O A Ternovskoy[1], I V Dyadicheva[2], A Yu Melnik[2] and S Ye Chornobay[2]

[1]Lugansk national University named after V. I. Dal, Lugansk, Ukraine
[2]V.I. Vernadsky Crimea Federal University, Simferopol, Russia

*mr.dyadichev@mail.ru

Abstract. The paper analyzes the problem of using technological lignins to make new bindings for casting manufacturing processes. The reasons, restraining their application in casting manufacture, are considered. The objective (based on the physical nature of the material) and subjective (based on commercial benefit) reasons are presented. The paper considers the perspectives of application of technological lignins by the example of technological lignosulfonates use which is the main representative of the technological lignins family on the Russian market of bindings.

1. Introduction. State analysis

Technological lignin is the product of delignification of vegetable raw materials, which can be any vegetable tissues, from wood to seaweeds. This material is present in a vegetable cell invariably in different states, performing the function of binding which guarantees a general durability of the construction. That is why during processing it is generated as a by-product, as a rule, as a waste. The scale of the problem is so huge that in 1992 in Lausanne (Switzerland) the International Lignin Institute was established.

According to the Institute data [1], annually from 40 to 50 mln tons of technological lignin are generated around the world; the utilization of lignin is a matter of considerable difficulty and does not have a rational solution at present.

Russia, having the largest timber resources, is one of the main countries, generating technological lignin. Pulp-and-paper industry is the leader in this process. According to the estimations, presented in works [2, 3], the controlled volumes of annual generation of this product in its different forms (sulfite alkaline solution, technological lignosulfates, hydrolytic lignin, etc.) fluctuate in the limit of 4-5 mln tons. According to the data [4], in Russia about 95 mln tons of technological lignin have been generated for the recent period of the economical activity. According to the experts' estimation, the volume of the vegetable raw materials processing will be increasing, it is expected that this index will have increased by 50-60% by 2050 [1, 4], in connection with this the problem of rational technological lignin utilization will be very acute both in the nearest perspective and in the future.

One of the most perspective areas of rational lignin use is application of its modifications as binding materials in casting manufacture.

The objective of the work was to estimate the possibilities of making binding materials meeting the modern requirements of casting manufacture on the basis of technological lignosulfates.

2. The course of doing the research

Technological lignosulfates (LST) are the modification of the technological lignin widely and traditionally presented on the Russian market of bindings. They are large-tonnage waste when processing timber into sulfite cellulose. This material was traditionally used in different productions, including foundry goods manufacture, as a binding material.

LST as a binding was widely-used in the USSR [5], for example in the work [6] it is pointed out that this material is the third in volume by being used in casting manufacture after clay and liquid glass. On the basis of the conducted researches, dedicated to the mixtures on the lignosulfonate basis, theses, including doctoral ones, were defended [7]. However, since the end of 90-ies the use of LST in casting manufacture has dramatically decreased, being substituted by various resin bindings. Nowadays this material is almost not used in modern casting technologies. The sphere of its use is limited to the production of simple castings, such as cast-iron moulds. The reasons for this are both of objective and subjective character.

Objectively, the binding on the LST basis do not meet the following modern requirements:

1) they do not have high strength properties, which are inherent in resin bindings;

2) they possess non-stable properties, this is manifested by abrupt (up to 30% of nominal values) fluctuations of material properties of different batches of the same manufacturer; and non-correspondence of the properties in case of buying material from another manufacturer; as well as the dependence of properties on seasonal temperature fluctuations;

3) they do not have enough thermal stability, being limited by temperatures 300-350^{0}C;

4) casting rods and forms on their basis have high hygroscopicity;

5) the technologies of their use require thermal hardening, and this increases power inputs and prolongs the production cycle.

The subjective character of reasons for abrupt fall of lignin bindings' use is that since the mid of 90-ies of the last century the casting sector of Russia has changed greatly, especially in the aspect of technological modernization. The equipment and technologies of leading western companies, prevailing now on the market, replaced domestic out-of-date technologies. It led to the replacement of domestic resource base of binding materials by imported bindings, as a rule, resin ones. On the one hand, the domestic materials were not adapted to work on the imported equipment (there was no interest in it); on the other hand, the market relations – commercial benefit, dictated the necessity of "tying" newly gained markets by foreign companies to their supplies. This was the reason, merely commercial one, for excluding domestic materials from the casting bindings market.

That time the chemical plant "Uralchimplast" was modernized, in practice it was rebuilt from the start, with the help of German investments of "Hüttenes-Albertus" [8], which is the leading manufacturer of chemical production for casting industry. Nowadays the enterprise is the leader in the production of casting binding materials in Russia; it is able to satisfy any requirements of the casting sector with modern resin bindings produced on the basis of the raw materials imported from Germany (!) [9]. A reasonable question arises, "What will happen if the supplies of raw materials from Germany stop!?".

The disadvantages of the lignin containing bindings have already been mentioned, but they possess a complex of positive properties which are withheld. They substantially excel the existing imported resin binding in these characteristics:

1) the source of the raw material for their production are domestic enterprises [3,7];

2) the cost of lignosulfonate bindings is incommensurably less; 4.5 – 6.0 rub/kg of liquid LST as against 3.5 – 5.0 euro/kg (265 – 378 rub/kg according to the present currency rate of ruble) for resin materials [2];

3) the technologies based on lignins are environmentally friendly: during transportation, during usage and during utilization [7,10,11];

4) the simplicity of storage: they do not require special regimes, have a long shelf life [7,10];

5) domestic casting enterprises have many years' positive experience of their usage [7].

To understand the significance and perspectives of using lignin products for the time being it is reasonable to analyze the main tendencies in the development of casting bindings.

The bindings on the basis of technological lignin can be related to the biopolymer group of binding materials which are traditionally and especially actively developed in the countries of the European Union [12, 13]. The reason for the activity is connected with toughening of the requirements of sanitary and hygienic working conditions in the foundries and of the general ecological working safety of casting enterprises. This is presented in the materials regulating this area of their functioning [14] and reinforced in the Directive of 2010 year [15], which strictly regulates the application of resin materials on the phenol basis, which constitute 90-95% of modern casting bindings. This is connected with numerous researches having found out a negative influence of the mentioned above materials on the health of the factory personnel [16] and the population, living in the areas where the enterprises, using these materials, are located [17, 18].

The mentioned Directive [15] assumes a complete removal of similar resin binding materials from the production technological cycle. This conditioned the researches aimed at searching for the alternatives. In particular, compositions on the basis of different materials are offered: on the basis of protein materials [19], on the basis of starch [20], on the basis of different production wastes [21, 22]. Their common disadvantage is limitedness of the raw materials source in combination with high initial cost of separate initial components (starch); in connection with this the binding on the basis of technological lignin, in particular LST, could compensate these disadvantages.

3. Perspectives and possibilities

By its genesis lignin performs the function of binding separate elements of the vegetable tissue and provides leaktightness of the cytoderms, and due to its pigments it conditions the hue of the lignificated tissue. It is located in the nodes of intercellular space and in the cytoderms, providing a strong coupling of cellulose fibers. Such positioning leads to guaranteeing the strength – "lignification" of the vegetable tissue.

Lignin polymer is formed at the growing phase – plant's lignification after the formation of the vegetable cell in the intercellular space and directly in the cytoderms of the cell itself [23].

The genesis of lignin explains the specificity of its structure and multi-factor of the influence on structure and chemical composition formation. This involves difficulties in the process of vegetable raw material technological processing, explains the impossibility of extracting natural lignin from the vegetable tissue in its natural state. That is why they distinguish between "protolignin" – the substance being directly in the plant and the material – the product generated as a result of the totality of technological influences for the delignification of the vegetable raw material.

It is necessary to point out that lignin is not purposefully produced according to rigid standards; lignin containing substances are obtained, as a rule, as large-tonnage wastes from different productions. In its turn this predetermines the indefiniteness of the chemical structure, multilayer structure and molecular-mass distribution of the generating lignin products. In the process of physico-chemical influences on the vegetable tissue, regulated by the specifics of the processing technology and the type of the initial vegetable raw material, the molecular mass of lignin is transformed; due to several times decrease its chemical activity undergo changes, the consistency and component chemical composition also change.

In chemical respect, lignin is not a single chemical substance; it is a composition of aromatic polymers of related structure. The true, empiric, or gross-formula of lignin can be presented as $C_{288}H_{318}O_{102}$ [24].A general view of the lignin material can be presented as a scheme – a structural formula (see figure 1). Its structural elements are phenyl-propane links, i.e. structural units which are derivatives of phenyl propane.

They distinguish technological forms of lignin materials, which can be found on the market, and thus being potential objects for further development: sulfate lignin, sulfite lignin and hydrolytic lignin. Exactly these materials can be the initial raw material for developing new competitive products for different needs. In this respect lignin containing materials are a typical representative of the functional materials, a promising raw material for developing new binding materials for casting manufacture [25].

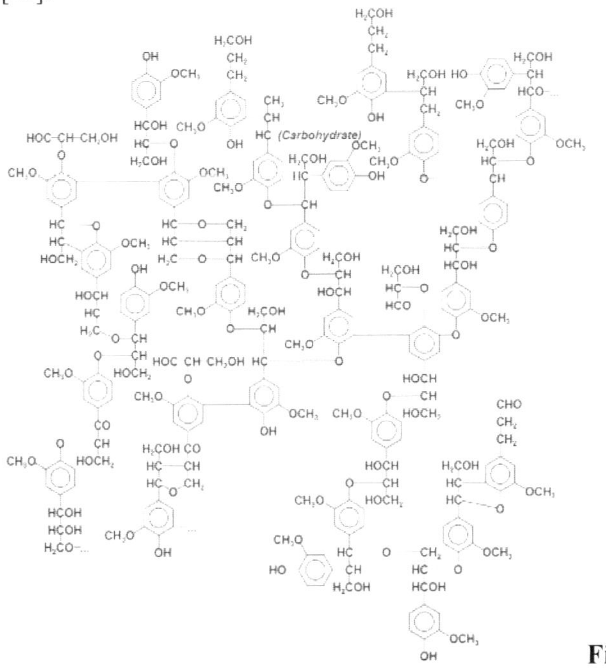

Figure 1. The structural formula of lignin [24].

The researches and results, obtained during preliminary approbation, allow claiming that the main difficulties can be overcome.

The problems of stability and strength properties' improvement are solved by selecting modifiers [10,17,25], see figure 2 and figure 3.

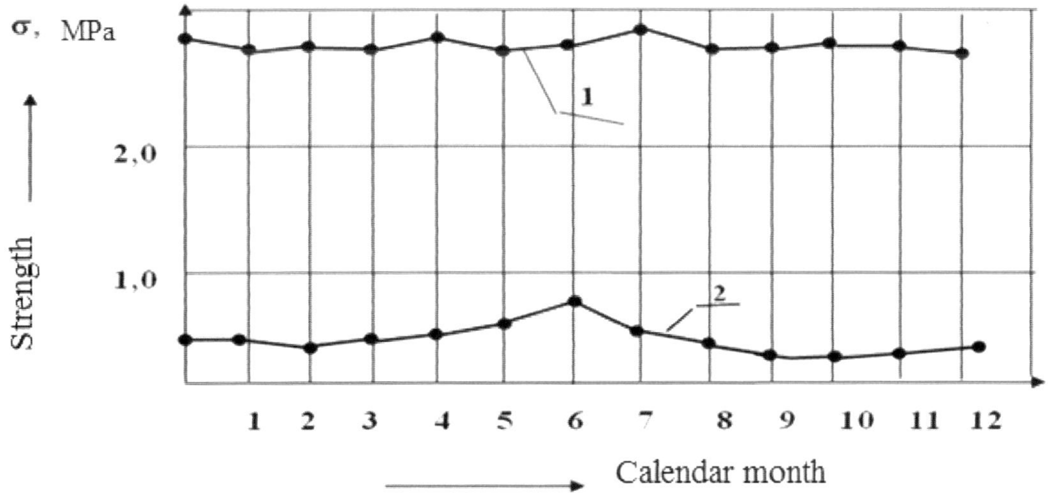

Figure 2. The stability of binding property of LST compositions when testing during a calendar year: 1 – composition LST – NSAS, 2 – LST (without any processing).

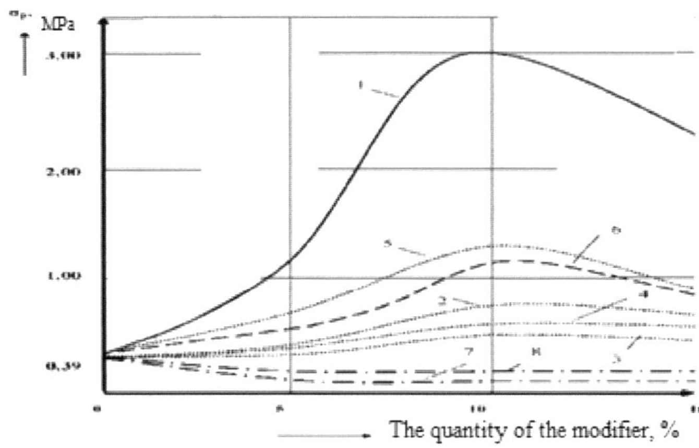

Figure 3. The indicators of the effect of different classes' substances (acids, salts, bases) on the binding LST ability, curves: 1. – nonionic surface active substances (NSAS); 2. – hydrochloric acid; 3.- sulfuric acid; 4.- nitric acid; 5.- sulfonic acid; 6- ammonium sulfate (12% water solution); 7- solution of caustic potassium; 8 – solution of caustic sodium.

The duration of hardening cycle can be considerably reduced due to the usage of blowing technologies, in particular, hardening of the rod, formed in a core box, with further blowing by hot air (see figure 4) and extraction of the finished rod. During 50 sec the necessary strength is obtained.

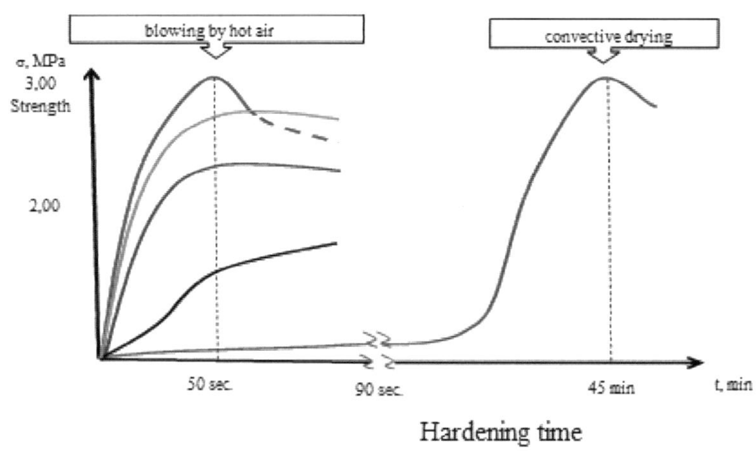

Figure 4. The possibilities of guaranteeing strength of binding compositions on the LST basis by different hardening methods.

4. Conclusions

The presented results of casting binding material development on the basis of technological lignin (LST) allows claiming the technological facilities of making binding materials on the basis of lignin materials which are capable to compete successfully with widely-used now synthetic resins on the phenol basis. Lignosulfonate materials can be considered as a typical example of resource-saving technology and rational use of the secondary raw material, generated from the renewable natural resource.

References

[1] *Information database of International Lignin Institute*(The electronic resource):http://www.ili-lignin.com/aboutlignin.php

[2] *The survey of lignosulfates market in Russia* 2015 (The information resource): http://www.infomine.ru/research/36/500

[3] Varfolomeev A A 2009 The development of environmentally friendly phenol-formaldehyde resins, modified by technological lignins *The thesis for obtaining the PhD degree in Chemistry: qualification 05.21.03 "Technology and equipment of chemical processing of tree biomass; chemistry of timber"*(Bratsk, Russia) p 160

[4] Thegovernmentreportofthestateandabout the protection of the environment of Irkutsk region in 2003 (Irkutsk, Russia: Oblmachinform) p 296

[5] IllarionovIE and Vasin Yu P 1992*Moulding materials and mixtures: monograhy*(Cheboksary, Russia: The publishing house of Chuvash University) part 1, p 223

[6] Doroshenko S P, Avdokushin V P and Matsishek I 1990 *Moulding materials and mixtures*(Kiev, Ukraine: High School/ Prague: SNTL) p 415

[7] SemikAP 1987Moulding and rod mixtures with binding materials on the basis of technological lignosulfonate *The theses for obtaining the degree of Doctor of technical sciences: qualification 05.16.04 "Casting manufacture"* (Kiev) p 225

[8] *The information resource, the leading manufacturer of casting binding in Russia* («Hüttenes-Albertus», Germany) access mode: http://www.huettenes-albertus.ru/

[9] DibrovIA 2017About conducting XIII International Congress of foundry workers and the international exhibition "Casting – 2017" 19-21 September 2017*The foundry worker of Russia***10** 7-14 (Chelyabinsk)

[10] RodionovO V andYu O Svinoroev 2011 *The formation of the environmental safety of region's enterprises* (Monography,Lugansk, Ukraine: "Knowledge") p 232

[11] SuvorovB L, Ogloblina R I,KoravoevES and et al 1987 *Castingbindingsinthemassproduction: catalogue* (Sverdlovsk) p 36

[12] Tackes G 2001 Core Binders: A Look into the Future*Modern Casting***10**24 –7

[13] Serghini A 2011 The future of organic foundry binders in Europe *Przegląd Odlewnictwa***11-12**, 548-53 (in Polish)

[14] European commission 2005*Integrated Pollution Prevention and Control Reference Document on Best Available Techniques in the Smitheries and Foundries Industry* p 363

[15] *The Directive of the European Parliament and Council of the 24-th November 2010 concerning industrial emissions*2010

[16] Löfstedt H, Westberg H, Seldén A I, Lundholm C and Svartengren M2009 Respiratory symptoms and lung function in foundry workers exposed to low molecular weight isocyanates*American Journal of Industrial Medicine***52** 455-63

[17] Löfstedt H, Westberg H, Seldén A I, Rudblad S, Bryngelsson I L, Ngo Y and Svartengren M 2011 Nasal and ocular effects in foundry workers using the Hot Box method*Journal of Occupational and Environmental Medicine***53** 43-8

[18] Löfstedt H, Westberg H, Seldén AI, Bryngelsson I L and Svartengren M2011 Respiratory symptoms and lung function in foundry workers using the Hot Box method – a 4-year follow-up*Journal of Occupational and Environmental Medicine***53** 1425-9

[19] Eastman J 2000 Protein – based binder updale: Performance put to the Test *Modern Casting***10** 32-4

[20] Zhou X, Yang J and Guohiu Q 2007 Study on synthesis and properties of modified starch binder for foundry*Journal of Materials Processing Technology***18** 407-11,doi: 0.1016/j.jmatprotec.2006.11.001

[21] Grabowska B2009Cross-linking of polyacrylan compositions with biopolymers with use of selected chemical and physical agents *Polimery***54**, 7-8, 19-20 (in Polish)

[22] Patterson M and Thiel J2010 Developing Bio-Urethanes for No-Bake. Foundry Management

and Technology
[23] Antonio Pizzi, Milena Properzi, Jean-Michel Leban, Michela Zanetti and Frederic Pichelin 2003 Mechanically-induced wood welding *Maderas. Cienciaytecnologia***5(2)** 101-6
[24] Kuleznev VN and Shershnev V A 2014*Chemistry and physics of polymers. A tutorial for chemical-technological higher educational establishments*(Moscow, Russia: "Lan") p 312
[25] Svinoroev Yu, Kostrub V and Klimova O 2010 New ecological binder materials based on vegetative rawmaterials processintg products *Teka commission of motorization and power industry in agriculture Lublin university of technology Volodymyr Dal East-Ukrainian national university of Lugansk* in 2010 **XB** 227 – 31

Acknowledgments
The study was carried out with the financial support of the Ministry of Education and Science of the Russian Federation within the framework of the scientific project FZEG-2020-0030.

IOP Conf. Series: Materials Science and Engineering **889** (2020) 012013 doi:10.1088/1757-899X/889/1/012013

Interaction of components features of the reaction mixture $ZrO_{2(m)}$–Al–C systems when heated

S I Dzhelialov[1]

[1]Crimean Engineering and Pedagogical University the name of Fevzi Yakubov, Department of Electromechanics and Welding, 295015, Simferopol, Uchebny side street, 8, Russia.

E-mail: dzhelyalov@mail.ru

Abstract. By the complex analysis basis of the interaction theory of the reaction mixture components $ZrO_{2(m)}$–Al–C when heated to temperatures of 1650–1775°C and experimental data (DTA, x-ray diffraction analysis), the nature of the formation of final synthesis products is studied. The probability of multistage interaction of components in the heating process with the formation of aluminum oxide Al_2O_3 and zirconium carbide ZrC as final components, which is accompanied by an intermediate stage of formation of intermetallide based on Al and Zr, is noted. The results obtained indicate that it is advisable to further study the laws of phase and structure formation of materials in the considered system in order to obtain high-density functional ceramics of the oxide-carbide type.

1. Introduction

The reaction synthesis of many types of ceramic materials can significantly improve the manufacturability of products of various uses. The types of tool ceramics that have a set of properties that allow it to be used as the basis of a metal-cutting tool that can work at high cutting speeds were made and invented by the authors of the work [1–2]. The characteristics of such materials directly depend on the properties of their components formed during the chemical interaction of the components of the initial charge. The structure of tool ceramic usually consists of oxides, carbides, nitrides, and other high-hardness and high strength compounds. Known types of oxide-carbide ceramics for tool use mainly contain aluminum oxide, reinforced with dispersed particles of carbides, as well as zirconium oxide, which exhibits transformational hardening [3]. It is advisable to start creating ceramics of new compositions in order to increase productivity and reduce the energy intensity of synthesis processes by predicting the possibility of interaction of the components of the initial reaction mixtures. In this regard, it is of scientific interest to study the kinetics of the interaction of components of a reaction mixture of the composition $ZrO_{2(M)}$– Al-C, which, by analogy with the reaction of titanium oxide with aluminum and carbon [4–7], can be generally represented as:

$$3ZrO_{2\,(m)} + 4Al + 3C \rightarrow 3ZrC + 2Al_2O_3. \quad (1)$$

Such studies become even more relevant when the synthesis of the material is carried out using energy-intensive technologies, in particular — hot pressing, which is the most common method for producing ceramics, including instrumental purposes.

IOP Conf. Series: Materials Science and Engineering **889** (2020) 012013 doi:10.1088/1757-899X/889/1/012013

The purpose of this work is to establish the kinetics of the interaction of components of the reaction mixture $ZrO_{2(m)}$–Al–C during heating and to identify the possibility of using this system to produce ceramics of the oxide-carbide type for instrumental purposes, based on the analysis of the results of thermodynamic calculations, differential thermal and x-ray structural analyses.

2. Methods of study

The material test samples were prepared using a powder mixture which consists of zirconium dioxide of monoclinic production from SRPE «Tsirkonii» (Dneprodzerzhinsk, Ukraine), aluminum powder, and lampblack (table 1).

Table 1. Powders used to prepare experimental samples.

Material	Technical conditions and standards
Zirconium dioxide powder (monoclinic)	TU–6–09–24–86 TUU 14–10–098–99
Aluminum powder PAP-4	GOST 5494–71
Lampblack powder S-1	GOST 18307–78

The results of differential thermal analysis (DTA) were obtained on a high-temperature differential thermal analyzer HDTA–7, designed for the study of refractory materials in the temperature range of 20-2500 °C. The temperature rise is performed automatically according to a specified program with simultaneous registration of measurement results in coordinates: sample temperature – the temperature difference between the sample and the standard.

X-ray diffraction studies were performed using the DRON-2 diffractometer. Scanning of samples subjected to differential thermal analysis was performed in $Cu_{K\alpha}$-radiation with a Ni filter.

The opportunity of reactions in the temperature range 100–1775 °C, taking place in a specified system, was established based on the calculation results of Gibbs free energy (ΔG). Thermodynamic quantities used in the calculations are taken from sources [8–11].

3. Discussion of results

According to the calculations, the values of the Gibbs energy change for the reaction (1) in the temperature range of 100–1700 °C are in the range of 13.8–225.6 kJ/mol. This indicates that the reaction being tested should not take place. However, we should consider the possibility of passing other chemical reactions between the components of the mixture.

The formation mechanism of the final reaction components (1), by analogy with [4] (where the interaction in the TiO_2–Al–C system is considered) can be determined by the sequence of reactions occurring in the systems:

$$Al - C \tag{2}$$

$$ZrO_{2(m)} - C, \tag{3}$$

$$ZrO_{2(m)} - Al. \tag{4}$$

According to the aluminum-carbon state diagram [12], the only stable compound formed in system (2) is Al_4C_3 carbide, which is released as a result of a peritectic reaction at 2300°C. At lower temperatures, the solubility of carbon in aluminum is extremely small: 0.14% (wt %) at 1000°C and 0.32 (wt %) at 1200°C.

The results of thermodynamic calculations obtained for the equilibrium equation:

$$ZrO_2 + 3C \rightarrow ZrC + 2CO \tag{5}$$

characterize the interaction in the system (3). The change in the Gibbs energy in the temperature range of 100–1700°C, according to the calculations, is in the range of – 341.9–(–64.7) kJ/mol and indicates the possibility of passing this chemical reaction. The obtained data correlate well with the data [13], where it is also noted, that the temperature of the beginning of the interaction between ZrO_2 and C by the reaction (5) is 1400 °C. It is also known, that one of the methods of industrial synthesis of ZrC in large quantities is the carburization of a metal zirconium powder [14–16]. This may be possible as a result of the release of Zr in the reaction:

$$3ZrO_2 + 4Al \rightarrow 3Zr + 2Al_2O_3. \quad (6)$$

The lower position of zirconium dioxide compared to aluminum oxide in the diagram of oxygen affinity for metals presented in [17] suggests, that such an interaction may take place. The presence of such interaction is also stated in the work [18]. These data are confirmed by DTA results (fig. 1) and x-ray diffraction analysis (fig. 2), taken from a sample obtained from a mixture of ZrO_2–22Al (wt %) (stoichiometric composition calculated from the equilibrium equation (7)). The first endothermic effect (652–685°C) on the DTA-thermograms (see fig. 1) corresponds to the melting of aluminum. Next, the Al melt begins to interact with zirconium oxide by reaction (7), which indicates itself as an exothermic effect at 770°C. The formed aluminum oxide (see the reaction (7)), in accordance with the results described in [19], has the structure θ-Al_2O_3 at this temperature. Subsequently, at 1275 °C (see fig. 1), aluminum oxide will undergo polymorphic transformation into the α-phase, which is stable for normal conditions.

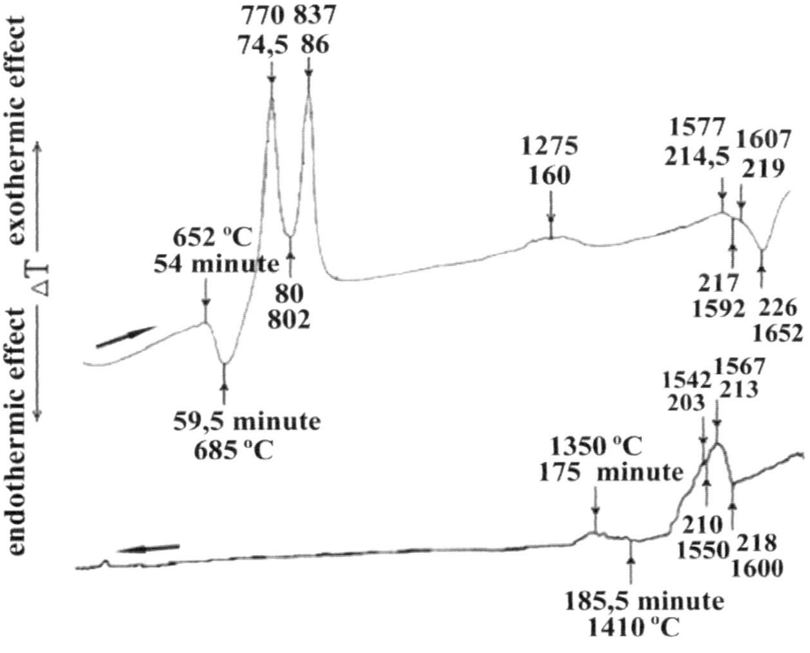

Figure 1. DTA-thermogram, obtained by heating and cooling a sample from a mixture of $ZrO_{2\,(m)}$ – 22Al (wt %).

Since we are dealing with a powder system, there is a possibility, that not all aluminum is involved in the zirconium release reaction. The part of aluminum that is on the border with zirconium released during the reaction (6) interacts with it to form intermetallides. This interaction corresponds to the peak of the DTA-diagram at 837°C.

The presence of zirconium dioxide peaks on the roentgenogram taken from the test sample (see fig. 2) indicates the presence in the composition of the material of the part, which did not react with Al, since before this aluminum was partially spent on the formation of intermetallides.

In [20] an assumption is made about the possibility of formation of an Al_3Zr_5 compound in this system with subsequent decay reaction:

$$Al_3Zr_5 \rightarrow Zr_2Al + Zr_3Al_2 \quad (7)$$

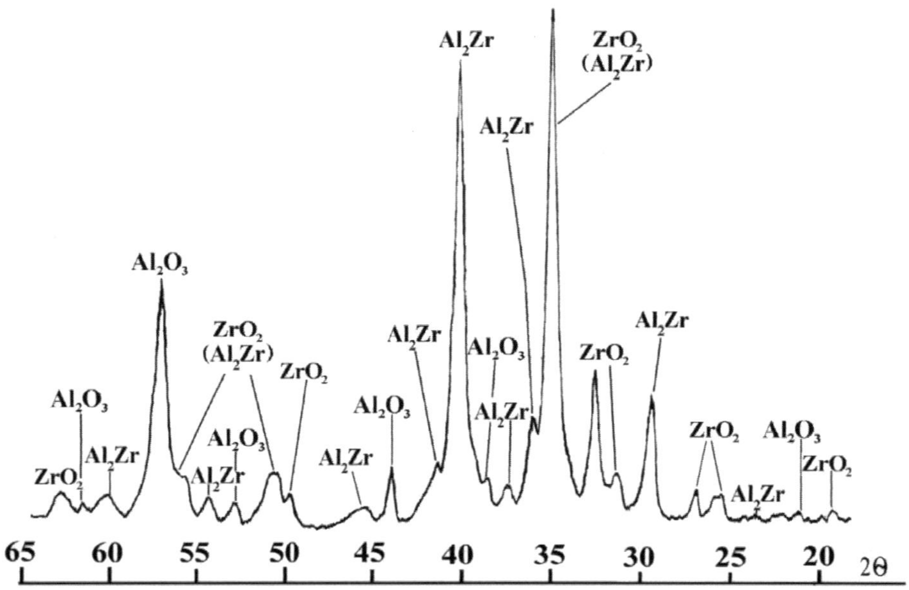

Figure 2. X-ray diffraction pattern taken from a sample of $ZrO_{2(m)} - 22Al$ (wt %) composition after heating in a DTA-analyzer.

At the same time, the authors of [21], who studied the interaction of zirconium with aluminum, found the presence of two intermetallides Al_3Zr and Al_2Zr in the final composition of the material along with other components. The possibility of formation of any of the above compounds should be related with the concentration of aluminum in the initial composition of the powder mixture in combination with temperature conditions. Many exo-and endoeffects that appear when heated above 1500 °C are probably the consequences of recrystallization of previously formed intermetallides, which eventually led to the formation of a stable phase of Al_2Zr, found in the composition of the final material as a result of x-ray diffraction studies (see fig. 2). Some of the temperature effects occurring in the range of 1500→1652→1350 °C may also indicate possible polymorphic transformations characteristic of previously formed oxides.

With the appearance of carbon in the reaction mixture ($ZrO_{2(m)}$–21Al–7C (wt %) – stoichiometric composition calculated by equation (1)), the exoeffect on the DTA-diagram related with the formation of intermetallides shifts to the right (peak at 915 °C in fig. 3).

At the same time, the temperature of the assumed $\theta \rightarrow \alpha$ transformation of aluminum oxide remains in the range close to the previous system – 1170 °C, slightly shifted to the left. According to [13], the presence of an exothermic peak at 1417 °C on the DTA-diagram characterizes the interaction of zirconium dioxide and carbon by reaction (5) (see fig. 3). Note, that some of the zirconium oxide has already interacted with aluminum.

Given that the amount of C in the initial mixture was calculated to interact with the entire volume of ZrO_2, a certain amount of carbon still remains in the mixture. The system also released the CO gas phase from the reaction (5). With the temperature increase C and CO interact with previously formed intermetallide and the free zirconium, is also forming a carbide. In this case, the oxygen separated from the CO-gas interacts with the aluminum released from the intermetallide, forming Al_2O_3.

It can be assumed that the presence of some exo- and endoeffects at temperatures of 1510 °C and higher (see fig. 3) is associated with these chemical reactions. Compliance with their above assumptions in the framework of these studies has not been established.

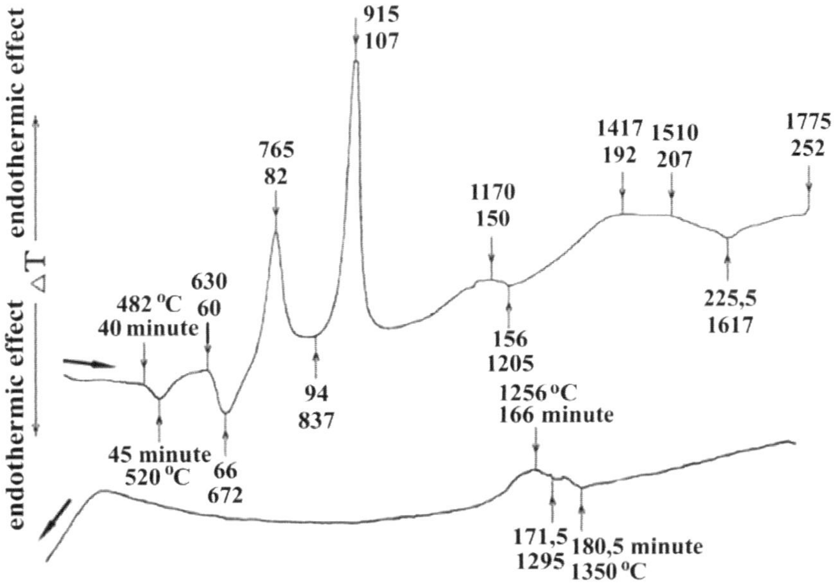

Figure 3. DTA-thermogram, obtained by heating and cooling a sample from a mixture of $ZrO_{2(m)}$–21Al–7C % (wt %).

The roentgenogram taken from the sample of $ZrO_{2(m)}$–21Al–7C% (wt %) heated in a DTA analyzer contains peaks of two types: alumina and zirconium carbide (fig. 4). This indicates that the overwhelming amount of the intermediate phases formed in this system, including free zirconium, intermetallides, as well as the CO gas phase actively interact with the formation of stable compounds included in the final material.

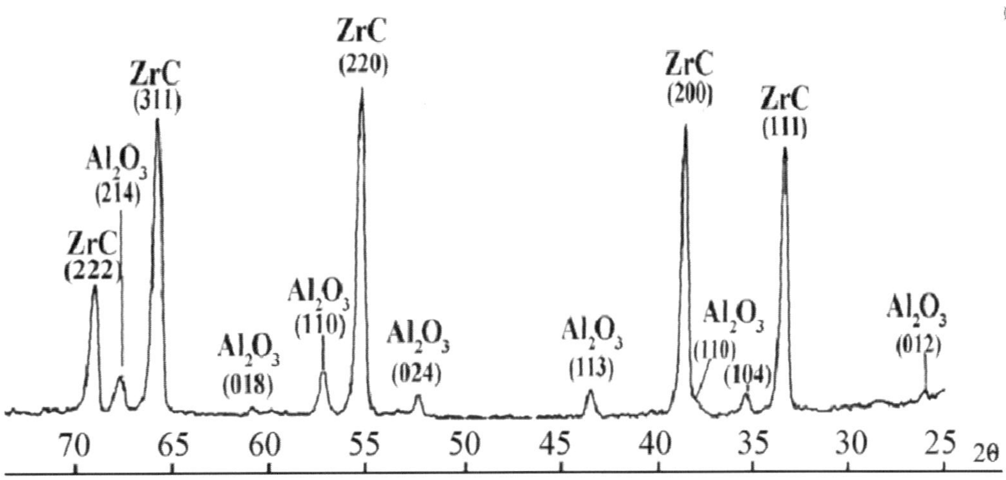

Figure 4. X-ray diffraction pattern taken from a sample of $ZrO_{2(M)}$–21Al–7C (wt %) composition after heating in a DTA-analyzer

IOP Conf. Series: Materials Science and Engineering **889** (2020) 012013 doi:10.1088/1757-899X/889/1/012013

4. Conclusions

Studies based on the analysis of the interaction theory of reaction mixture components $ZrO_{2(m)}$–Al–C when heated to temperatures of 1650–1775°C, as well as experimental data of differential thermal and x-ray structural analyses of samples of stoichiometric compositions $ZrO_{2(m)}$–21Al–7C (wt %) and $ZrO_{2(m)}$–22Al (wt %), calculated from thermodynamically possible reactions of their interaction, allowed us to establish that:

1. The formation of Al_2O_3 and ZrC by heating the mixture $ZrO_{2(M)}$–Al–C system is a prerequisite for further study of the nature of structure formation in it, in order to synthesize dense functional oxide-carbide ceramic materials types, including instrumental purposes;

2. Formation Al_2O_3 in the system $ZrO_{2(m)}$–Al–C possibly as a result of direct interaction of Al with $ZrO_{2(m)}$, and by the decay of the intermetallide Al_2Zr, which is formed as an intermediate phase, subsequently interacting with the oxygen released from the previously formed gas phase CO; while ZrC is also synthesized, which formation is possible as the result of a interaction of zirconium oxide with carbon;

3. The formation of the final composition of components when heating a mixture of $ZrO_{2(m)}$–21Al–7C (wt %), is accompanied by a variety of exo- and endoeffects, which, in addition to the above interactions, are presumably associated with polymorphic transformations of aluminum oxide, as well as a some of chemical reactions caused by the formation of various compounds based on aluminum and zirconium.

5. References

[1] Merzhanov A G 1998 *Combustion processes and material synthesis* (Processy goreniya i sintez materialov) (Chernogolovka. ISMAN) p 512

[2] Gevorkyan E S 1989 *Development of a composite material based on chromium oxide for a* blade tool: avtoref. dis. Ph.D. (Kiev: Technical Sciences) p 26

[3] Prokopiv M N and Dzhelyalov S I 2005 Cutting ceramics. Status and development prospects *Vestnik of SumGTU* Sumy **11 (83)** 129–34

[4] Chia T D, Vang T M, Zhao V and Liu T E 1999 Microstructure of Al_2O_3–TiC ceramics sintered by SHS-synthesis. *J. Perspektivnye Materialy.* **6** 57–61

[5] Choi Yoon and Rhee Shi-Woo 1995 Reaction of TiO_2–Al–C in the combustion synthesis of TiC–Al_2O_3 composite. *J. Amer. Ceram. Soc.* 78 **4** 986–92

[6] Xia Tiandong Liu Tianzuo and Zhao Wenjun, Wang Tianmin 1997 The form of phases in the TiC – Al_2O_3 composites combustion synthesized from TiO_2–Al–C *J. Gansu Univ. Tehnol.* **1** 16–20.

[7] Boudebane S, Boudebane A, Lemboub S, Mechachti S, Benchiheb O and Boutefnouchet H 2002 The influence of additions on the self propagating high-temperature synthetis of TiC–Al_2O_3 ceramics. *Annales de chimie.* **27 (2)** 1–15

[8] *Thermodynamic Properties of Individual Substances.* Fourth Edition 1990 Vol. 2 (Editors: Gurvich L V, Veyts I V, Alcock C B. Tab. Ed: Iorish V S). Elements C, Si, Ge, Sn, Pb, and Their Compounds. Hemisphere Publishing Corporation. Part One. Methods and Computation. XII+ 569 pp. Part Two. Tables. XII+372 pp

[9] *Thermodynamic Properties of Individual Substances*. Fourth Edition 1994 (Ed.: Gurvich L V, Veyts I V, Alcock C B. Tab. Ed.: Iorish V S, vol 3) Elements B, Al, Ga, In, Tl, Be, Mg, Ca, Sr, Ba, and Their Compounds. CRC Press and Begell House. Part One. Methods and Computation. VIII+707 pp. Part Two. Tables. XV+448 pp

[10] Chase M W, Davies C A, Downey J R, Frurip D J, McDonald R A and Syverud A N 1985 JANAF Thermochemical Tables. Third Edition. *J. Phys. Chem.* vol 14 **1** 1856

[11] Barin I, Knacke O and Kubaschewski O 1977 *Thermochemical Properties of Inorganic Substances*. (Berlin: Springer-Verlag) p 861

[12] Lyakishev N P, Bannih O A and Rohlin L L oth. 1996 Binary state diagrams (Diagrammy sostoyaniya dvojnyh sistem) Directory In 3 v vol 1 (Moscow, Mashinostroenie) p 992
[13] Rogajlin, M I and Chalyh E F 1974 *Carbon Graphite Handbook* (*Spravochnik po uglegrafitovym materialam*) (Himiya) p 208
[14] Samsonov G V and Latysheva V P 1956 The study of the diffusion of boron and carbon in some metals of transition groups *Physics of Metals and Metallography* **II (2)** 309–11
[15] Samsonov G V and Epik A P 1962 To the question of the parameters of the reaction diffusion of boron and carbon into refractory transition metals *Physics of Metals and Metallography* vol XIV, **3** 479–80
[16] Pavlinov L V and Bykov V N 1965 *Carbon Diffusion in β-Zirconium* (Physics of Metals and Metallography vol XIX) pp 397–400
[17] Kosolapova T Ya 1968 *Carbides* (Moscow, Metallurgiya) p 300
[18] Matrenin S V, Il'in A P, Tolbanova L O and Zolotareva E V 2010 Activation of sintering of oxide ceramics by additives of nanodispersed powders *Bulletin of the Tomsk Polytechnic University* **317 (3)** 24–8
[19] Zharnylskaya A L and Volhin V V 2009 Synthesis of a ceramic precursor of the Al_2O_3 – ZrO_2 system by the sol – gel method *J. Vestnik Bashkirskogo universiteta (Bulletin of Bashkir University)* **14 (3)** 753–6
[20] Shishkovskij I V, Shcherbakov V I and Petrov A L 2001 Laser synthesis of refractory ceramics from Al and ZrO_2 powders *S.-T. J. Physics and Chemistry of materials treatment* **3** 45–8
[21] Natanzon Ya V and Petrishchev V Ya 1992 The interaction of zirconium with liquid aluminum saturated with zirconium. *J. Poroshkovaya metallurgiya (Powder Metallurgy)* **8** 80–5

Optimization method of the extrusion mixture components formulation and consumption in the processing of the secondary heterogeneous polymeric materials

V V Dyadichev[1], A V Kolesnikov[2], A V Dyadichev[1], S G Menyuk[1], O F Babushkina[1] and S Ye Chornobay[1]

[1]V.I. Vernadsky Crimean Federal University, 295000, Pavlenko Str. 3, Simferopol, Crimea, Russia
[2]Lugansk National University named after Vladimir Dal, Department of Automation and Computer-Integrated Technologies, 20-A Molodezhnyi Sq., Lugansk, 91034, Ukraine

*mr.dyadichev@mail.ru

Abstract. The paper scrutinizes the main methods, applied practically, for correcting the consumption of the components considering their fullness in the technological extrusion process. The method, chosen as the basic one, includes the following: the common fullness of the components, getting into the mixer, is subtracted from the common consumption of the filler. A mathematical model for determining the values of the components for setting the dosing devices is designed. For the operational control of the filler's components content we determined the dependencies of these values correction on the period of the filler content evaluation and transport delay of the components supply. The algorithm of the method for controlling the extrusion mixture components' consumption, considering the content of the filler in it, is made.

1. Introduction

In the beginning of XX century the world polymer production was only several thousand tons of different polymeric materials [1]. If compare to the classical construction materials – metals, cement, and glass – it was a small percentage. Nowadays modern developed countries produce polymers in the amount of more than melt steel, not mentioning aluminum, copper and other non-ferrous metals taken together. The general tendency of plastic production development is that every year the increase of polymeric materials production goes on with a stable speed, while on the contrary, the metal production is constant. In 2016 122, 800 thousand tons of different polymeric materials were produced [2]. The growth of polymeric materials continues and the amount of the generated waste, unfortunately, passes ahead their production increase. Nowadays the polymeric materials waste in the world is hundreds of millions tons. The reuse of polymeric waste as the secondary raw material allows efficiently using the materials during the production, as well as solving the urgent problem of waste recycling, the most problematic of which is mixed heterogeneous polymeric waste [3-5]. One of the most promising technology for this purpose is the extrusion due to combination of the modification phase and the main processing, due to a wide range of the obtained products with the prospect of their expansion for account of the secondary deformation additional phases; however, it is necessary to research in details the process for optimization of the extrusion mixture components' consumption, considering the content of the filler in it [6-7].

Polymeric materials as the initial materials of the extrusion mixture are used; these materials are mainly heterogeneous in compatibility and have been used in the production and the processing spheres, plastic masses from the sphere of individual and common use (the filled waste, secondary polymers). That is why the content of these components' filler varies from 7 to 15 %, and in some cases reaches up to 20 – 30 % [8-9]. This cannot help affecting the strength and rheological properties of the mixture. And if the decrease of the shift limit stress τ_0, caused by the presence of additional filler in the mixture, leads only to the lessening of the extruder and the extrusion head obstruction probability, the decrease of the product strength requires taking particular measures to correct the components consumption [10-11].

2. Method of controlling the consumption of the extrusion mixture components considering the presence of a filler in it

Let us consider the main methods applied practically to control the consumption of the components considering their fullness [12-15].

One of the methods is as follows: the common fullness of the components, supplied into the mixer, is subtracted from the common filler consumption. This method is the most used in practice [16-19]. The total calculating amount of the filler in the mixer is preserved, as the fullness of the components is considered. However, in this case the consumption of the components, occupied by the filler, is not corrected. As the content of the filler in the initial polymer is insignificant, then the initial dry polymer consumption can be accepted as constant. Consequently, the volume of the obtained mixture turns out to be less than of the set volume, and the consumption of the initial polymer for 1 m^3 of the mixture is overrated. The application of this method meets the requirement of the durability, but does not comply with the aim of the set function – minimum consumption of the initial polymer. The actual efficiency of the equipment V_f will be less than the set one.

We offer the method for correcting the filler in the inert materials according to the content of the filler in them while making the mixture; this method consists of measuring the filler in the materials ω_i and of identifying the amount of the filler in the material $Q_{нi}$:

$$Q_{нi} = \omega_i \frac{Q_з}{100}, \tag{1}$$

where ω_i is the content of the filler in the component;
$Q_з$ is the set consumption of the component.

Then the amount, occupied by the filler $Q_н$ is subtracted from the set component consumption, and the component consumption increases for this value. However, in this case it is not taken into account that there is a filler in the added portion of the material while dosing.

For instance, if the components consumption (durability 55.0 MPA) for 1 m^3 of the prepared mixture is 100 kg of the filled wastes, 730 kg of the secondary waste, 540 kg of the initial polymer and 650 kg of the filler, then the content of 10 % filler in the secondary polymer, which is not considered in the addition of the secondary polymer, and 15 % in the filled wastes, gives additionally $\Delta m_н$ 88 kg of the filler, and its total consumption for 1 m^3 of the mixture $m_н$ will be 738 kg [20-21]. The increase of the filler consumption $\Delta m_н$ by 88 kg decreases the durability of the product by 0.27 MPa. For maintaining the permissible durability the consumption of the initial polymer has to be decreased, approximately, by $\Delta m_{пп} = k\Delta m_н = 0{,}3332 \cdot 88 = 29{,}32\ кг$.

The considered above allows us concluding that for the accuracy increase of correcting the components' consumption depending on the content of the filler in them, it is necessary to consider the filler in the correcting additions while weight dosing. Further the method of influencing the materials consumption under the change of the filler content in them is presented; this method considers the complete content of the filler in the main mixture and in the additions.

If the consumption of the non-filled components is set by the basic values m_{0ip}, m_{0fw}, m_{0sm}, and the content of the filler in the components is determined by the values of the fullness coefficients ω_{ip}, ω_{fw}, ω_{spw}, then to provide the required value of the consumption of the non-filled components (basic), the dosing devices have to increase the setting for the filler value, contained in the actual material:

$$\Delta m_{нi} = m_{0i} \frac{\omega_i}{100}, \quad (2)$$

where i are 'но' (filled waste), 'вп' (secondary polymeric waste), 'пп' (initial polymer).

The formula (2) is got form the analysis of the known formula of the filler content:

$$\omega = \frac{m_н}{m_{см}} \cdot 100\%, \quad (3)$$

where $m_н$, $m_{см}$ are the mass of the filler and the mass of the constituent material without the filler respectively.

Then the actual dosing value of the mixture components, considering the content of the filler in the materials, has to be at the level:

$$m_{Фпп} = m_{0пп} + \Delta m_{нпп} = m_{0пп} + m_{0пп} \frac{\omega_{пп}}{100} = m_{0пп} \frac{100 + \omega_{пп}}{100}, \quad (4)$$

$$m_{Фно} = m_{0но} + \Delta m_{нно} = m_{0но} \frac{100 + \omega_{но}}{100}, \quad (5)$$

$$m_{Фвп} = m_{0вп} + \Delta m_{нвп} = m_{0вп} \frac{100 + \omega_{вп}}{100}. \quad (6)$$

At the same time the actual consumption of the filler will be determined:

$$m_{Фн} = m_{0н} - \sum_i \Delta m_{нi}, \quad (7)$$

where i is '*но*', '*вс*', '*пп*'

Thereby, if the content of the filler ω_i component is known, then the task for giving i-component has to be corrected according to the expressions (4) – (6), and for the filler – (7).

Along with this, the addition $m_{дi}$ of the dry component, being calculated by the expression (2) in its turn contains the filler, which has to be calculated:

$$\Delta m_{нi}^{к+1} = \Delta m_{нi}^{к} \frac{\omega_i}{100}; \quad (8)$$

and considered in the addition:

$$m_{дi} \Leftarrow m_{дi} + \sum_{к=1}^{l} \Delta m_{нi}^{к+1}, \quad (9)$$

where 'k' is a step of the correction of the filler's content in the additions to the component consumption, k is 1,2,..,l. When k is 1, $\Delta m_{нi}^{1} = \Delta m_{нi}$, calculated by the expression (2). Thus, the expressions (4 – 7) are specified due to their iteration considering index k.

The expressions (8 – 9) describe the iterative process of identifying the value of the addition $m_{дi}$ to the consumption of i-component, compensating the content of the filler in dry material. The given iterative process is fast converging with different amount of the steps for each component and continuing till the condition is satisfied:

$$\left| \Delta m_{нi}^{к+1} - \Delta m_{нi}^{к} \right| > \varepsilon_i, \quad (10)$$

where ε_i is the value set beforehand for i-component, conditioned by the facilities and characteristics of the equipment used for dosing.

The obtained after the iterative processes values $m_{дi}$ are considered when setting the tasks for the consumption of dry components and the filler for 1 m^3 of the mixture:

$$m_{зi} = m_{0i} + m_{дi} = m_{0i} \frac{100 + \omega_i}{100} + \sum_{к=1} \Delta m_{нi}^{к+1}; \quad (11)$$

$$m_{зн} = m_{0н} - \sum_i m_{дi} = m_{0н} - \sum \left(m_{0i} \frac{\omega_i}{100} + \sum_{к=1} \Delta m_{нi}^{к+1} \right), \quad (12)$$

where i is '*но*', '*вс*', '*пп*'.

As the mixture volume has to be invariable while doing corrections, the sum of the components mass portions has to be equal to th Start t volume:

$$V_{см} = 10^{-3}\left(\vartheta_{пп} m_{зпп} + \vartheta_{вп} m_{звс} + \vartheta_{но} m_{зно} + \vartheta_{н} m_{зн}\right) = 1м^3, \qquad (13)$$

where ϑ_{ip}, ϑ_{sm}, ϑ_{fw}, ϑ_f are specific volumes of the initial polymer, the secondary mixture, the filled waste and the filler, respectively, m^3 /kg.

The expression (13) may be used as the referent one while correcting the consumption of components considering the content of the filler in them.

For the values of the components (11) – (12) the dosing devices are adjusted. If the control of the content of the filler components is done promptly, the correction of these values is done according to the formulae (11) – (12) with the period of filler content evaluation and transport delay of the component supply (it depends on the location of the device for determining the filler content in the transport feed gear).

The algorithm of the method for controlling the extrusion mixture components' consumption considering the presence of the filler in it is presented in Figure 1.

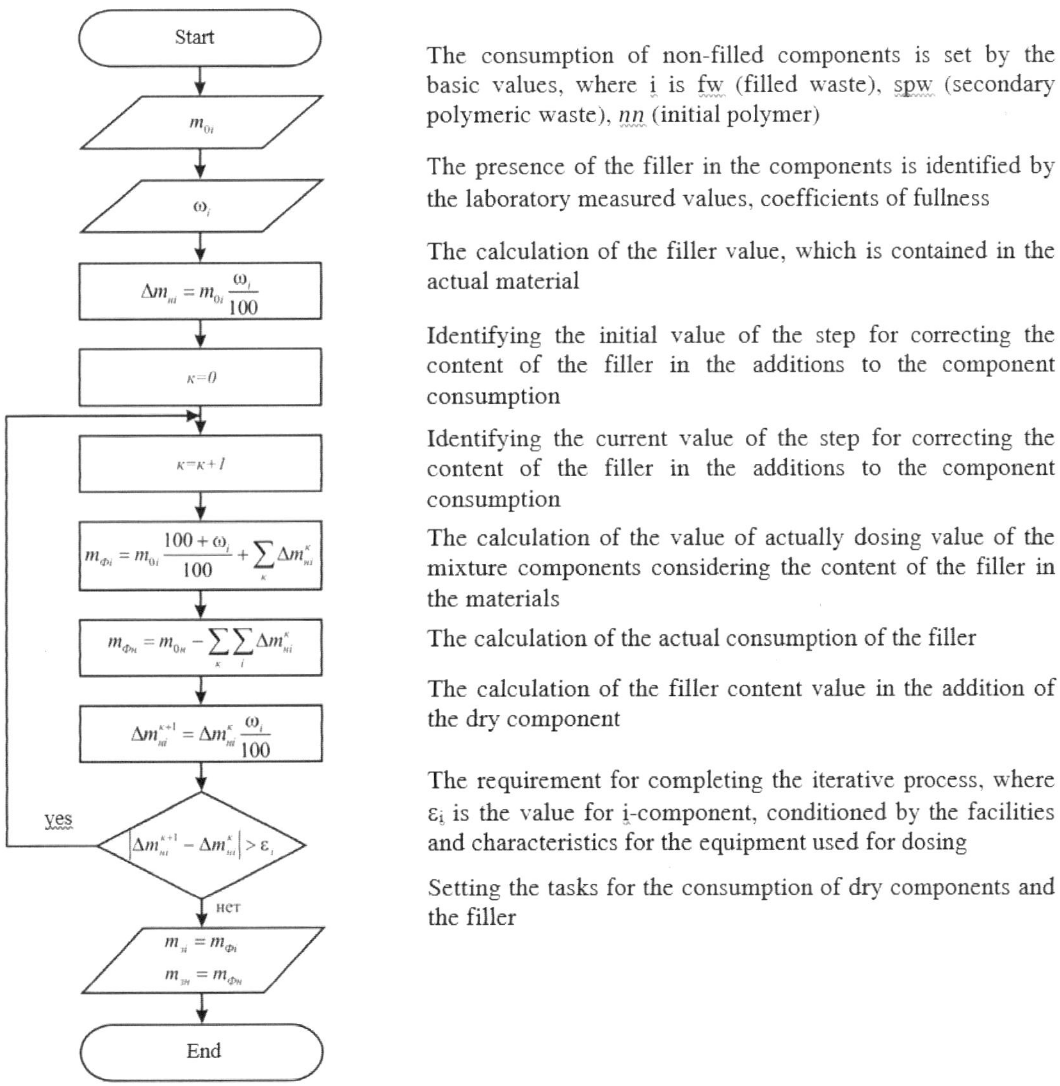

Figure 1. The algorithm of the method for controlling the extrusion mixture components' consumption considering the content of the filler in it.

3. Conclusion

A method is developed and analytical dependencies are obtained for correction of the tasks on the consumption of mixture components, considering the content of the filler in the main material and additional portions, and allowing eliminating the influence of the mixture materials fullness on the durability of the extrusion product.

We also developed the algorithm for accounting the content of the filler in the consumption of the extrusion components of a multicomponent mixture, which provides the implementation of the suggested mathematical model of the process and method for controlling the extrusion equipment.

References

[1] Dyadichev V V, Kolesnikov A V, Menyuk S G and Dyadichev A V 2019 Improvement of extrusion equipment and technologies for processing secondary combined polymer materials and mixtures *IOP Conf. Series: Journal of Physics: Conf. Series* **1210** Art. No 012035 doi: 10.1088/1742-6596/1210/1/012035

[2] Kim V S 1988 *Dispersion and mixing in the production of plastic* (Moscow: Chemistry) p 293

[3] Davis B. and et al. 1998 Grooved Feed Single Screw Extmders – Improving Productivity and Reducing Viscous Heating Effects. *Polym. Eng. Sci.* **38** (7) 1199

[4] Han C D 1981 *Multiphase flow in polymer processing* (New York: Academic Press)

[5] Steller R T 1990 Theoretical Model for Flow of Polymer Melts in the Screw Channel, *Polym. Eng. Sci.* **30** (7) 400-7

[6] Rauwendaal C 2001 *Polymer extrusion* (Munich, Hauser Garduer) p 777

[7] Pape J, Potente H and Obermann C 1999 Influence of Model Simplifications on the Accuracy of Simulation Results in Single Screw Extruders *15th Annual Meeting of the Polymer Processing Society* (Den Bosch, the Netherlands)

[8] Elemans P 2000 Enhancing Dry-Color Efficiency in Starve-Fed Injection Molding *58th SPE ANTEC* 2582-6

[9] Elemans P and van Wunnik J M 2000 The Effect of Feeding Mode on the Dispersive Mixing Efficiency in Single-Screw Extrusion *58th SPE ANTEC* 265-7

[10] Shvetsov P A 1988 *The technology of processing plastic masses* (Moscow: Chemistry) p 462

[11] Dyadichev V, Kolesnikov A, Dyadichev A, Menyuk S and Dyadicheva E 2018 Model of creation a multilayer structure for production of building pipes *MATEC Web Conf., VI International Scientific Conference "Integration, Partnership and Innovation in Construction Science and Education" (IPICSE-2018)* **251**, Art. No 251 01006 doi: 10.1051/matecconf/201825101006

[12] Dyadichev A V, Dyadicheva E A, Kolesnikov A V, Dyadichev V V, Menyuk S G and Chornobay S Ye 2019 Mathematical model of adhesion junction of layers during coextrusion *Journal of Physics: Conf. Ser. (JPCS): Int. Scientific Conference APITECH 2019* **1399** (4) Art. No 044112 doi: 10.1088/1742-6596/1399/4/044112

[13] Lyubomirskiy N V, Fedorkin S I, Bakhtin A S and Bakhtina T A 2017 Structuring of composite systems based on lime harden through carbonation and secondary limestone raw materials *Malaysian Construction Research Journal (MCRJ)* **23** (3) 15-26

[14] Fic S, Lyubomirskiy N and Barnat-Hunek D 2018 The Influence of the Natural Aggregate Roughness on the ITZ Adhesion in Concrete *Mater. Sci. Forum: Materials and Technologies in Construction and Architecture* **931** 564-7

[15] Gusev A, Shul'gin V, Braga E, Zamnius E, Starova G, Lyssenko K, Eremenko I and Linert W 2018 Luminescent properties of zinc complexes of 4-formylpyrazolone based azomethineligands:Excitation-dependent emission in solution *J of Luminescence* **202** 370-6

[16] Gusev A N, Shul'gin V F, Braga E V, Nemec I, Minaev B F, Baryshnikov G V, Trávníček Z, Ågren H, Eremenko I L, Lyssenko K and Linert W 2018 Synthesis and photophysical properties of Zn(II) Schiff base complexes possessing strong solvent-dependent solid-state fluorescence/ A. N. Gusev *Polyhedron*

[17] Shul'gin V, Pevzner N, Gusev A, Sokolov M, Panyushkin V, Devterova J, Kirillov K, Martynenko I and Linert W 2018 Tb(III) complexes with 1-phenyl-3-methyl-4stearoyl-

pyrazol-5-one as a material for luminescence Langmuir–Blodgett films *J. of Coordination Chem.*

[18] Yankovskaya V S, Dovhyi I I, Bezhin N A, MilyutinV V, Nekrasova N A, Kapranov S V and Shulgin V F Sorption of cobalt by extraction chromatographic resin on the base of di-(tert-butylbenzo)-18-crown-6 *Journal of Radioanalytical and Nuclear Chemistry* **318** (2) 1085–97

[19] Dyadichev V V, Kolesnikov A V and Dyadichev A V 2020 Determination of rational technological parameters of co-extrusion processing of secondary polymeric materials by the method of physical modeling *Solid State Phenomena: 5th International Conference on Industrial Engineering, ICIE 2019* **299** 43-8 https://doi.org/10.4028/www.scientific.net/SSP.299.43

[20] Malitskova E A and Potapov I I 1997 *Processing of plastic waste* (Moscow: Avis Original) p 159

[21] Fogarty J, Rauwendaal C, Fogarty D and Rios A 2001 Turbo-Screw, New Screw Design for Foam Extrusion *SPE ANTEC Techn. Papers*

Acknowledgments
The study was carried out with the financial support of the Ministry of Education and Science of the Russian Federation within the framework of the scientific project FZEG-2020-0030.

IOP Conf. Series: Materials Science and Engineering **889** (2020) 012015 doi:10.1088/1757-899X/889/1/012015

The structure and properties of laminated aluminum-glass reinforced plastics

M I Gordeeva[1] and Yu A Knyazeva[1]

[1]MAI-Moscow Aviation Institute (National Research University), Moscow, Orshanskaya St. 3, 121552, Russia

E-mail: nauka@mati.ru

Abstract. The current state of the problem of the use in aviation of new laminated aluminum glass-reinforced plastics (Glare) is given in this review. It is shown that Glares are characterized by low density (2.4 g / cm^3), high strength (> 600 MPa), but at the same time they are inferior to aluminum alloys by the elastic modulus (<60 GPa).Methods for evaluating Glares service characteristics are considered, it is indicated that replacing traditional aluminum alloys in Glares with lithium alloys that have unique elastic properties can significantly improve their service characteristics.

1. Introduction

Significant progress in the use of new materials in aviation is associated with composite materials, such as layered aluminum-glass-plastics-Glares, which are distinguished from monolithic sheets from aluminum alloys by a density reduced by 10-15% (2.4 g/cm^3), high strength (σ_B> 600 MPa), fire resistance [1-3]. The use of Glares in the fuselage skin of the A-380 airbus led to a weight saving of more than 500 kg. At the same time, Glares are inferior to aluminum alloys in such an important modulus as elastic modulus [4]. VIAM [5, 6] proposed a new class of Glares based on aluminum-lithium alloys, which made it possible to increase the elastic modulus by 8–10% and decrease the density by 5–7%. In this paper, the structure and properties of Glares are considered.

2. Results and discussion

Figure 1 shows the appearance and "construction" of Glare, which usually consists of three layers of aluminum alloy, between which are packages of epoxy prepreg layers reinforced with fiberglass, which can be located in each of the two layers of both packages in one, shared direction ($0^0/0^0$) or in the fractional and transverse ($0^0/90^0$) directions (table 1).Characteristics of Glares and the corresponding preferential properties are also given in table. 1. Sophisticated complex manufacturing technology of Glares requires various testing and control methods for product quality control. A significant number of studies of composite materials, including Glares, are devoted to the determination of their mechanical and elastic characteristics, including in conditions of multi axial stress states.This is due to the strong anisotropy of Glares, the high sensitivity of their properties to technology parameters and the presence of complex correlations of their mechanical properties with structural and geometric parameters. One of the important areas of research is also the development of non-destructive methods of quality control of Glares. Without the creation of such methods, their use in critical aircraft components would be impossible, since in most cases the material itself and the part are made of it at the same time and a separate material test cannot be carried out.

One of the most important advantages of Glares in comparison with metallic materials is its high resistance to fatigue crack growth. For this reason, much attention is paid to the modeling of this process [7-9]. In [7], analytical approximations are proposed for predicting the fatigue crack growth rate based on empirical relations of the Paris type, in which the effective stress intensity factor (SIF) is adjusted based on the bridging effect.It is implied that the decrease in fatigue crack growth rate in Glares occurs due to a zigzag change in the trajectory of thc crack, leading to its branching and a decrease in the SIF by K Quantitatively, this effect is determined by the dimensionless coefficient β equal to the ratio of the K_{br} value to the SIN value that a crack would have in the absence of overlap (K_0). Figure 2 shows the stress values in each layer of the Glare 4A-3/2.

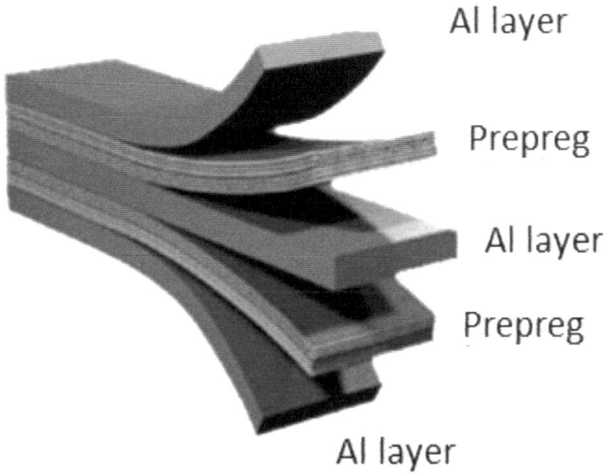

Figure 1. Type of Glares

It is seen that tensile stresses act in the metal layer, and curing stresses in the prepreg are absent in the fractional direction and compressive in the transverse direction. As a result, the crack under cyclic loading is revealed by the applied load in the metal layer and is closed by the overlapping stresses in the prepreg.In [9], a critical analysis of all existing forecasting models of fatigue crack growth rate Glares was carried out, the main drawback of which the author considers the underestimation of such an important factor as the stratification effect. The author believes that the consistency of predictive models with experiment is largely due to an insufficient amount of experimental data. When developing his own model, the author was based on two fundamental positions. Firstly, on a detailed study and description of real fracture mechanisms in Glares, and secondly, on the development of such a model of crack growth and stratification in Glares under fatigue loading, which would exactly correspond to the fracture mechanisms mentioned above. When developing his model, the author adhered to the following important criteria:

- analytical relationships should be based on physical mechanisms without the use of adjustable parameters;
- the model must accurately predict the growth rate of the fatigue crack compared with experimental results in a fairly wide range of experimental parameters;
- the results should not depend on the initial parameters, such as the length of the crack and the shape of the bundle.

As a result, the author proposed a model that allows one to take into account the stratification effect when calculating the energy balance, which is in good agreement with the experimental results. For technologists in his approach, it is important that the author introduces a parameter into his model, which in practice has shown its importance for the performance of Glares, namely the parameters of the interface between the metal component and the prepreg. From this point of view, the model proposed in [7] makes it possible in the future to search for quantitative correlations between technological parameters characterizing the adhesion characteristics of the interface with service properties, of which the fatigue crack growth rate parameter is one of the most important for ensuring the reliability of Glares in aircraft.

Table 1. Standard assortment of Glares and advantageous properties.

Glare grade	Sub	Metal sheet thickness [mm] and alloy	Prepregorientation[a)]in each fibrelayer[b)]	Main beneficial characteristics
Glare 1	-	0.3-0.4 7475-T761	0/0	Fatigue, strength, yieldsstress
Glare 2	Glare 2A	0.2-0.5 2024-T3	0/0	Fatigue, strength
	Glare 2B	0.2-0.5 2024-T3	90/90	Fatigue, strength
Glare 3	-	0.2-0.5 2024-T3	0/90	Fatigue, impact
Glare 4	Glare 4A	0.2-0.5 2024-T3	0/90/0	Fatigue, strengthin 0^0direction
	Glare 4B	0.2-0.5 2024-T3	90/0/90	Fatigue, strengthin90^0direction
Glare 5	-	0.2-0.5 2024-T3	0/90/90/0	Impact
Glare 6	Glare 6A	0.2-0.5 2024-T3	+45/-45	Shear, off-axisproperties
	Glare 6B	0.2-0.5 2024-T3	-45/+45	Shear, off-axisproperties

a)All aluminium rolling directions in standard laminates are in the same orientation; the rolling direction is defined as 0^0, the transverse rolling direction is defined as 90^0.
b) The number of orientations in this column is equal to the number of prepreg layers (each nominally 0.133 mm thick) in each fibre layer.

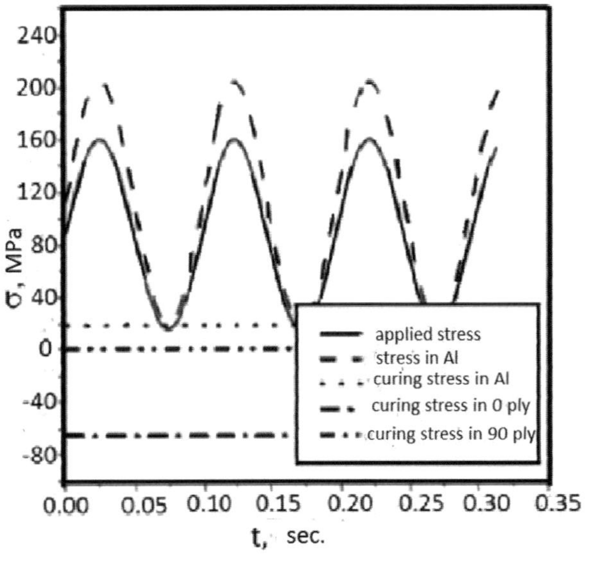

Figure 2. The stress levels in Al layer and curing stress in each layer for Glare4A-3/2laminate under cyclic loading at RT with a frequency of 10 Hz. The maximumapplied stress was 160 MPa with a stress ratio of 0.1.

Interesting results were obtained in [10] when studying the elastic properties of composites. The experimental values of the elastic characteristics for GF/E and Glares turned out to be close to the calculated values calculated for the GF/E composite based on micromechanical modeling ($E_{GF/E}$), and for Glares based on the additivity rule (table 3):

$$E_{Al/GF/E}=E_{Al}\cdot V_{Al} + E_{GF/E}\cdot(1-V_{Al}) \quad (1)$$

Table 2 shows the characteristics of the components of SIALs.An important circumstance associated with the results of modeling the elastic characteristics of Glares should be noted. If the additivity rule is used for theGF/E composite, then based on the data in Table 3 (the volume fraction of the epoxy filler is 40% and the module is 5 GPa, and the glass fiber module is 72 GPa), we obtain that the Young's modulus of the GF/E composite will be 45.2 GPa. Nevertheless, the micromechanical calculation gives 30.6 GPa, and the experiment 27.4 GPa. This is a consequence of the fact that, unlike epoxy resin and aluminum, fiber glass has strong anisotropy and the Young's modulus in the normal axis of the fiber is much lower [10] (table 2). An important conclusion follows from this that an increase in the Young's modulus of a fiber is less effective than an increase in the modulus of a metal component, in this case aluminum. Even in the case of using the most effective Glare structure for shells (two layers with parallel and one with perpendicular fibers), an increase in the axial modulus of 10% will correspond to an increase in the modulus of the metal by only 3%.

Therefore, the most effective way to increase the elastic modulus of Glares is to increase the elastic modulus of the metal component. An effective way to increase the elastic moduli is to form intermetallic phases in the alloy particles, which usually have higher elastic properties compared to the matrix.For an Al_3Li intermetallic with $L1_2$ lattice, the anisotropy of the elastic moduli can be quite high, and since it is known that it has a pronounced texture (coinciding with the alpha-solid solution), Mg-Li system alloys can have a noticeable anisotropy of elastic properties in the sheet plane, which must be taken into account in the calculations and cutting the sheet, and the magnitude of the anisotropy is due to the type and intensity of the texture and the amount of intermetallic.

Table 2. Properties of Glare components.

Properties	Al (2024-T3)	Prepreg(Glass fiber/Epoxy)
E_x (ГПа)	72.4	55.5
E_y(ГПа)	72.4	9.5
E_{xy}(ГПа)	27.6	5.55
ν_{xy}	0.33	0.3
ν_{yx}	0.33	0.0575

Table 3. Elastic properties of Glare and epoxy fiber composite.

Material	Fraction content	Parameters used in the FGM[a)] program and the mixtures rules				Calculation (experiment)		
		E_x	E_y	G_{12}	ν_{12}	E_x	E_y	ν_{12}
	(%)	GPa				GPa		
Epoxy	40	5	5	1.85	0.3	-	-	-
Glassfiber	60	72.0	72.0	28.8	0.14	-	-	-
Aluminum 2024-T3	57	72.4	72	28.0	0.33	-	-	-
Gf/E	43	-	-	-	-	30.6(27.4)	30.6(27.4)	0.38
Glare	100	-	-	-	-	54.8(57.3)	54.8(57.3)	0.25

[a)] FabricGeometry Model software

3. Conclusions

1. The article presents an overview of the state of the problem of using new laminated aluminum glass-reinforced plastics - Glare in aviation.

2. It has been shown that, along with the significant advantages of composites (with a density of 2.4 g/cm^3, σ_B> 600 MPa, significantly higher crack resistance), they are inferior to monolithic aluminum alloys by the value of the elastic modulus.

3. It is shown that the elastic properties can be improved only due to the metal component of the composite, which can be achieved by replacing the traditional aluminum alloys in Glares with aluminium-lithium alloys, which have unique elastic properties.

4. References

[1] Alderliesten R, Campoli G and Benedictus R 2007 *Comp. Sci. and Technol.* **67** 2545–55
[2] Khan S, Alderliesten R and Benedictus R 2009 Composites. *Sci. and Technol.* **69** 2604–15
[3] Rodi R, Alderliesten R and Benedictus R 2010 *Fracture Mechanics* **77** 1012–24
[4] Chang P and Yang J 2008 *Int. J of Fatig.* **30** 2165–74
[5] Shestov V, Antipov V, Senatorova O and Sidelnikov V 2013 *Metalloved Term. Obrab. Met.* **9** 28–31
[6] Antipov V, Kolobnev N and Khokhlatova L 2013 *Metalloved. Term. Obrab. Met.* **9** 5–11.
[7] Newman 1984 J *Int J Fract* **24** 131–35
[8] Guo Y and Wu X 1998 *Fatig.Fract.Eng.Mater.Struct.* **21** 1133–45
[9] Guo Y and Wu X 1999 *Eng Fract Mech* **63** 147–63
[10] Parka S, Choi W, Choi H and Kwon H 2010 *J Mater Proc.Techn.* **210** 1008–16.

Optimisation of electro-spark coating method

N Yu Loginov[1,3], V A Gulyaev[1] and O Yu Khudyakova[2]

[1]Department of equipment and technologies of mechanical production, Togliatti State University, 14 Belorusskaya Street, Togliatti, Russia
[2]Department of World Economy, Diplomatic Academy of the Ministry of Foreign Affairs of the Russian Federation, 53/2 Ostozhenka Street, Moscow, Russia

[3]E-mail: loginovnik@mail.ru

Abstract. Application of wear resistant coatings on cutting tools in a technological process requires taking into account multiple factors. Optimisation of the hardening process presents one of the main steps that needs to be addressed. This paper describes a method of optimisation of the electro-spark coating (ESC) process. The performance of the process was chosen as the target function. A parametric function was defined based on the previously developed mathematical model of the process whereby the condition of the strong connection of coating material with the foundation was used. Another parametric function is a prerequisite for a suitable coating thickness. Additional limitations are the characteristics of the equipment for the reinforcement

1. Introduction

Given the economically unstable situation in the world, the efficient use of available resources becomes a first priority. Minimising the cost of manufacturing new products leads to a significant increase in profits. [1] One of the tools to achieve profit growth is the introduction of lean manufacturing methods (Lean Production). The basic idea of lean production is to increase the efficiency of the producers within existing resources through the modernisation of all processes. [2] The first lean production practices have been applied by Toyota. The basic idea of this principle was the elimination of some losses in the existing technological process: loss of time and budget due to standby, overproduction, extra processing steps, unnecessary transportation, redundant inventory, and defective goods. [3]

Implementation of lean ideas in large-scale and mass production instantly brings significant profit to companies; however, it is a complicated task since one has to operate in the already running process.

One of means of increasing the wear resistance of the cutting tool is a method of electro-spark coating (ESC). The organisation of this method in the production environment for hardening of tools not only requires special equipment, but also optimisation of the process. This will reduce the operational time spent on resurfacing and the costs associated with it.

The process of ESC in the real environment is defined by the following parameters [4, 5, 6]:

1) the power of a pulse generator, W;
2) the speed of an electrode, υ;
3) the thickness of a coating, h.

2. Mathematical model of the process

To obtain a coating with a durable connection of the anode material (tool) with the cathode material (product), it is necessary to comply with a condition stating that the surface temperature of the cathode equals the melting temperature [7, 8]. Thus, this condition can be written as

$$T_M = \frac{q_0}{2\pi\lambda} \cdot F(k) \tag{1}$$

In this equation, the unknown variables are the intensity of the heat source, q_0, and the velocity of the source, which is the argument of function $F(k)$.

In [7], it was determined that the heating of the cathode is supplied from two sources, the total power of which is determined by the power pulse generator

$$q_2 + q_3 = q_0 = \eta \cdot \frac{W}{S_E} \tag{2}$$

where η is the efficiency of the pulse generator; $\eta \approx 0{,}8$; W is the discharge power determined by the type of a generator used in a particular case.

Equation (1) is the lower technical limit that is a function of equal temperatures which should provide a set of parameters: generator power W and the velocity of the source υ, which is taken into account by function $F(k)$.

The speed of the source determines the performance of the coating process and in the calculation of the optimal mode should be selected as large as practical. This expression($\upsilon \rightarrow$ max) will be the objective function for the system optimisation.

Equation (2) substituted into (2) yields

$$W = \frac{2\pi \cdot \lambda \cdot S_E \cdot T_M}{\eta \cdot F(k)} \tag{3}$$

To determine the third process parameter, the coating thickness, we assume that in a steady process of electro-spark alloying all the energy supplied to the anode (electrode) is spent on its partial melting and is transferred to the cathode. This assumption is fair since in a steady process electrode (anode) gives the heat to the cathode through the material as well as to the environment. The heat exchange with the environment is via air which is a poor conductor of the heat. According to authors of [4, 5], for most materials in normal conditions, in the most effective modes from the perspective of quality indicators and process performance values of the gain are high, that is, the thickness of the formed layer on the cathode is limited. The limited thickness of the formed layers on the cathode (product) in many cases hinders the practical use of ESC [9, 10, 11]. Therefore, the analysis of the factors determining such nature of the process and opportunities for targeted management become highly important.

3. Process optimization

Numerous experimental studies of electrode erosion (the anode) and the metal transfer to the product (cathode) allows to manifest that in the process of ESC the erosion and the release material of the anode do not stop with time, rather settling of the anode on the cathode slows down due to evaporation to the environment [12, 13, 14]. The limitation of the layer thickness can only be explained by considering the entire set of physical, chemical and mechanical effects on the surface of working electrodes. As ESC is associated with multiple-pulse thermal and mechanical effects on the material of the electrodes of phase and structural transformations in the generated layers [15, 16, 17], we can assume that these processes should determine the nature of the growth and limitation of the thickness of the applied layer.

Phase and structural transformations, heating and cooling of the surface layers of the electrode's material under the influence of electric pulses lead to the formation of residual stresses of types I and II [15, 18, 19], characterising their ultimate irreversible stress state. Consequently, in real conditions of coating the whole complex of the processes occurring between the anode and cathode is determined by the law of energy conservation. Given the above-mentioned statements, it can be said with a small extent of error associated with the evaporation of materials into the environment (according to estimates of different authors [15, 20, 21], evaporation takes 3-5%) that the intensity of the heat flux of transferred energy is equal to

$$q_3 = \eta \cdot p \cdot \frac{W}{S_E} = m_E \cdot L_{fE} \tag{4}$$

This expression determines the melting rate (deposition) of the electrode material. The mass transferred per unit time of the electrode is determined by the product of the density and the volume:

$$m_E = \rho_E \cdot d \cdot h \cdot \upsilon \tag{5}$$

Therefore, expression (4) can be represented as follows:

$$\eta \cdot p \cdot \frac{4W}{\pi \cdot d^2} = \rho \cdot h \cdot d \cdot \upsilon \cdot L_{fE} \tag{6}$$

Hence,

$$W = \frac{\pi \cdot \rho \cdot L_{fE} \cdot d^3}{4 \cdot \eta \cdot p} \cdot h \cdot \upsilon \tag{7}$$

This expression links the power generator of impulses in specific process conditions with the coating thickness and the speed of the anode.

The thickness of the coating is selected from experience, applied for various materials and operating conditions, it may be different, but one should keep in mind that the greater thickness of the coating, the more fragile the coating becomes. Therefore, most often the coating thickness of the cutting tool is taken from the the range of 8-12 μm. The test tool in the production environment confirmed the validity of this statement. Coating thickness of 20 μm and more typically wears off due to chipping.

Given the equations above, the process of electro-spark alloying is described by the following system of equations:

$$W = \frac{2\pi \cdot \lambda \cdot S_E \cdot T_M}{\eta \cdot F(k)} \tag{8}$$

$$W = \frac{\pi \cdot \rho \cdot L_{fE} \cdot d^3}{4 \cdot \eta \cdot p} \cdot h \cdot \upsilon \tag{9}$$

$$W \le W_{max} \tag{10}$$

$$W \ge W_{min} \tag{11}$$

Let us consider the example of applying a carbide coating(ρ = 5.37 g/sm^3, λ = 0.21 W/(sm·°C), L_{fE} = 1.45·107 W/(g·sec)) on the tool steel (melting temperature T_M= 1550 °C) under the following conditions: a = 1, d = 2 mm.

Characteristics of the electrode material are presented in Table 1.

Table 1. Electrode properties.

The chemical composition of the electrode	Density, [g/sm3]	Hardness, [HRA]	σ, [kg/mm2]
TiC-Cr3C2-10Ni	5.37-5.38	92.5-93.0	90-100

For these conditions,*F(k)*is satisfactorily approximated by the exponential function

$$F(k) = \exp\left[-0.82\frac{\upsilon}{2\omega}\right] \tag{12}$$

Let us assume that the capacity of the plant is equal to W=150-400 Watts.
Therefore, the solution of equations (8)-(11) can be represented by the plot in Figure 1.

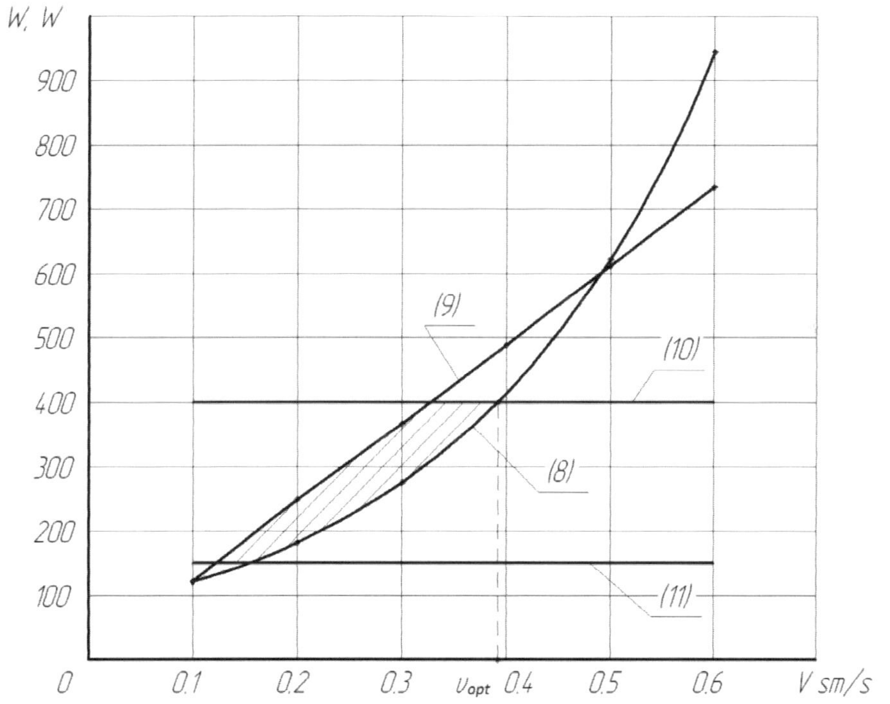

Figure 1. Optimisation of electro-spark alloying of high speed steel by electrode made of the solid alloy.

Fig 1 depicts that the function based on equation (8) shows the values of the generator power$W = F(\upsilon)$at which the temperature on a surface of the hardened product equals the melting temperature, that is, this function represents the lower limit in the optimisation system.

The function based on equation (9) shows that the maximum output power$W = F(\upsilon)$that provides the required coating thickness, that is, this function represents an upper limit in the optimisation system.

Functions represented by equations (10) and (11) are additional constraints.

In Fig. 1, functions (8)-(11) selected area searched possible solutions, the optimal solution corresponds to a point in the delineated area with υ_{max}. This point lies on a curve of equal temperature with a value of$\upsilon_{opt} \approx 0.39$ sm/sec.

Therefore, for these conditions the coating technological process should be conducted on the optimal mode for the following parameters:

$\upsilon_{opt} \approx 0.39$ sm/sec; $W = 400$ W; $h = 12$ μm.

Fig. 2 shows a photo of microsection of the sample after hardening at optimal conditions. The coating was applied using Elitron-22A. The figure shows that the coating has a relatively uniform structure without any breaks.

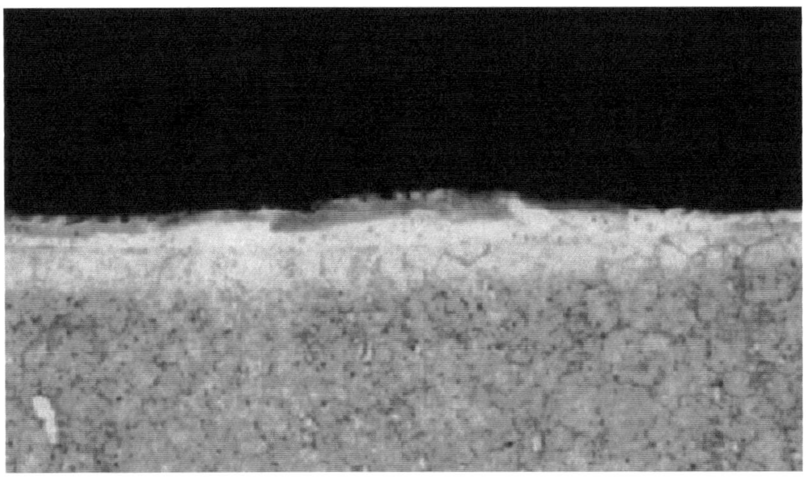

Figure 2. Image of microsection of a sample of high speed steel with coating of microsection of a sample of high speed steel with coating (magnification ×1000).

4. Conclusions

The paper presents the system of equations for the optimisation of the process where the objective function is the productivity. The lower limit is a function of equal temperatures ($\theta(x,y,\tau)=T_M$, the melting point of the electrode). The upper limit is the required thickness of the coating. Additional limitations are the capabilities of the equipment.

The optimisation system was pilot-tested and it showed the uniformity and reliability of the applied layer.

The introduction of electro-spark coating on a cutting tool increases its wear resistance, is embedded in the concept of the lean production and leads to a decrease in tool wear.

References

[1] Ketoeva N, Soldatova N and Ilyashenko S 2019 *E3S Web of Conferences***124** 04015
[2] Plugatyreva O A and Khusnutdinova E M 2019 *IOP Conf. Ser.: Mater. Sci. Eng.***570** 012073
[3] Korchagin A, Deniskina A and Fateeva I 2019 *E3S Web of Conferences***110** 02124
[4] Kolomeichenko A V, Titov N V and Baranov Y N 2020 *IOP Conf. Ser.: Mater. Sci. Eng.* **709** 044009
[5] Gertsriken D, Mazanko V, Qiao S and Zhang C 2009 *Mod. Phys. Lett.* B 23 pp 2369–76
[6] Loginov N Y 2017 *Hardening technologies and coatings*13 pp 67-70
[7] Loginov N Y, Gulyaev V A, Kozlov A A and Voronov D Y 2019 *IOP Conf. Ser.: Mater. Sci.*

Eng. **537** 022006
[8] Loginov N, Khudyakova O and Orlova E 2018 *MSF* **945** 919
[9] Prokopenko GI, Mordyuk BM, Volosevych PY, Vorona SP, Popova TV and Piskun NO 2017*Metallofizika i Noveishie Tekhnologii*39 pp 189–208
[10] Kreivaitis R, Žunda A, Kupčinskas A and Jankauskas V 2016*Tribol. Int.*103 pp 236–42
[11] Kuptsov KA, Sheveyko AN, Zamulaeva EI, Sidorenko DA and Shtansky DV 2019*Mater. Des.***167** 107645
[12] Korkmaz K, Ribalko FV and Sahin O 2013*Kovove Materialy*51 pp 283–9
[13] MashkovYK, KorotaevDN, Baibaratskaya MY and AlimbaevaBS 2015*Tech. Phys.*60 pp 1489–93
[14] MashkovYK, KorotaevDN, BaybaratskayaMY and Alimbaeva B S 2015*2014 Dynamics of Systems, Mechanisms and Machines, Dynamics 2014- Proceedings* 7005682
[15] VerhoturovAD, VlasenkoVD and KonevtsovLA 2018*J. Frict. Wear*39 pp 232–40
[16] PyachinSA, ErshovaTB, BurkovAA, VlasovaNM, KirichenkoEA, KulikMA and KrutikovaVO 2019*Letters on Materials*9 pp 191–6
[17] PadgurskasJ, KreivaitisR, RukuižaR, MihailovV, AgafiiV, KriūkienėR and BaltušnikasA 2017*Surf. Coat. Technol.*311 pp 90–7
[18] Kiryukhantsev-KorneevPV, SheveykoAN, ShvindinaNV, LevashovEA and ShtanskyDV 2018*Ceram. Int.*44 pp 7637–46
[19] PenyashkiT, KostadinovG and KandevaM 2018*Proceedings of 9th International Scientific Conference BALTTRIB 2017 - Dedicated to 100th Anniversary of Restitution of Lithuania* pp 34–40
[20] KozyrAV, KonevtsovLA, KonovalovSV, KovalenkoSV and IvashenkoVI 2018*Letters on Materials*8 pp 140–145
[21] AgeevEV, AltukhovAY, MalnevaYV and NovikovAN 2018*IOP Conf. Ser.: Mater. Sci. Eng.***327** 032001

IOP Conf. Series: Materials Science and Engineering **889** (2020) 012017 doi:10.1088/1757-899X/889/1/012017

The study of contact electrical resistance of coatings on the basis of refractory metal alloys Co-W, Co-Mo, Ni-W, Ni-Mo

G P Gololobov, D V Suvorov, M A Serpova and A S Arefiev

Ryazan State Radio Engineering University named after V.F. Utkin, 390005, Ryazan, Russia

E-mail: gololobov.gennady@yandex.ru

Abstract. The paper presents the research into contact electrical resistance of coatings on the basis of refractory metal alloys Co-W, Co-Mo, Ni-W, Ni-Mo. SEM-images of the tested samples' surface are presented, and the peculiarities of their microrelief are analyzed. The results of the research are presented as dependencies of contact electrical resistance on the sample electrode pressing force for each type of the surface. The comparative analysis of them is done.

1. Introduction

It is possible to single one of the most important working characteristics out of their wide complex of contact electrodeposits by metals and alloys, used in modern switching devices (magnetically operated sealed switches, MEMS-switchboards, relays, etc.); the most important characteristics are the size and the contact resistance stability. The contact resistance is the main operating parameter and in many respects it defines the operational resource of such type devices [1]. Despite the dependence of this parameter on a number of factors, such as the coating material, the method of its application, the application quality, and the state of the surface in contact; while developing these devices, the main attention is paid to the choice of the material.

During last decades a great amount of contact surfaces has been studied and tested. The contact surfaces were on the basis of rare-earth metals, such as gold and silver, their alloys with a number of alloying components, rhodium and ruthenium and their alloys [1, 2]. However, in many cases it was impossible to obtain the required parameters of the contact, the required erosion resistance, service life and the reliability of switching devices' work. The significant shortcomings of such type coatings are their mechanical characteristics (solidity, elasticity, melting temperature, etc.) which leave much to be desired, and as a consequence, their propensity to mechanical wear-and-tear, electrical erosion, and contacts' sticking. It is necessary to mention the high cost of rare-earth metals.

That is why recently contact coatings on the basis of refractory metal alloys W and Mo with the metals of iron group are becoming more and more spread [3-8]. These binary alloys have high wear resistance, corrosion resistance and solidity [9-11]. It is necessary to single out the resistance of these alloys to electrical erosion. The electrical characteristics, in particular contact electrical resistance of the coatings on the basis of these alloys, do not yield to the coatings on the basis of rare-earth metals.

Nowadays there is an opportunity to control the contact electrical resistance of the coatings due to the reasonable choice of the coating as well as due to the change of the technological conditions of their coating.

IOP Conf. Series: Materials Science and Engineering **889** (2020) 012017 doi:10.1088/1757-899X/889/1/012017

This scientific-research work is aimed at studying contact electrical resistance of the coatings, developed by us, on the basis of the whole range of refractory metal binary alloys. The data about similar comparative research in well-known sources are absent, thereby, the work is of great theoretical and practical significance.

2. Experiment details

For the research four batches of sample coatings on the basis of Co-W, Co-Mo, Ni-W and Ni-Mo alloys were got. The composition of the electrolytes and the deposition mode are presented in [4], [12], [8], [13] respectively.

The plates of oxygen-free copper of brand Cu-DHP with sizes 1×2 cm and thickness 1 mm were used as the bases. The bases underwent mechanical polishing by a felt circle without using abrasive, and then they were defatted in the solution containing NaOH and Na_2CO_3 in the concentration of 100 g/l for each component at the temperature of 70 °C. Straight before the deposition the bases were dipped into the 15% solution of HCl for 30 seconds and were washed thoroughly in bidistilled water. A platinum plate of 4 cm area served as an anode. The current density varied between 0.005–0.15 Acm^{-2}. The set values of electrodeposition current parameters were provided with the help of the potentiostat IPC Pro 3A by Volta (St. Petersburg, Russia). The electrolyte temperature during spreading the coatings in all the experiments was set and maintained constant with the help of the electromagnetic stirring with the heater C-MAG HS 7 digital by IKA (Staufen, Germany) with the system of feedback through the temperature sensor PT 1000 by Thermometrics (Northridge, CA, USA).

The thickness of coatings in each consignment was 6-7 μm. The content of the refractory metal component in all the alloys was maintained at the level of 18-19 atomic percent by using different deposition modes. The deposition conditions provided the minimum level of the crack and coating roughness formation.

For doing the test of contact electrical resistance R_c a stand was designed in order to measure R_c under the conditions of contact with the set value of contact pressure force of the sample electrode and the base with the coating. Its structural scheme is presented in Figure 1 a. The picture of the brass sample electrode (probe) is presented in Figure 1 b.

The stand consists of: 1 – the base; 2 – the load; 3 – the sample electrode; 4 – the measuring device (voltmeter); 5 – the source of constant current; 6 – the researched sample.

The electrical resistance was measured according to the four-conductors' scheme with the use of current and potential conductors. One pair of conductors (current and potential) was connected to the sample at different points; the second pair was connected to the sample electrode. A probe, made from brass with the working surface in the form of a sphere with the radius 1.5 mm (Fig. 1 b), was used as a sample electrode. With the help of the constant current source the required current of 50 mA was set. A voltmeter was used as a measuring device. The research was done using four different loads. Thus, the pressure force of the electrode (F_p) was 0.2 H, 0.45 H, 0.75 H and 1 H respectively. To do the experiment series of 5 sample surfaces on the basis of alloys Co-W, Ni-W and Co-Mo, Ni-Mo were prepared. For each sample the measurement of the voltage drop in the contact was done at 5 points, the measurement was done 5 times at each point. Further the obtained values were averaged.

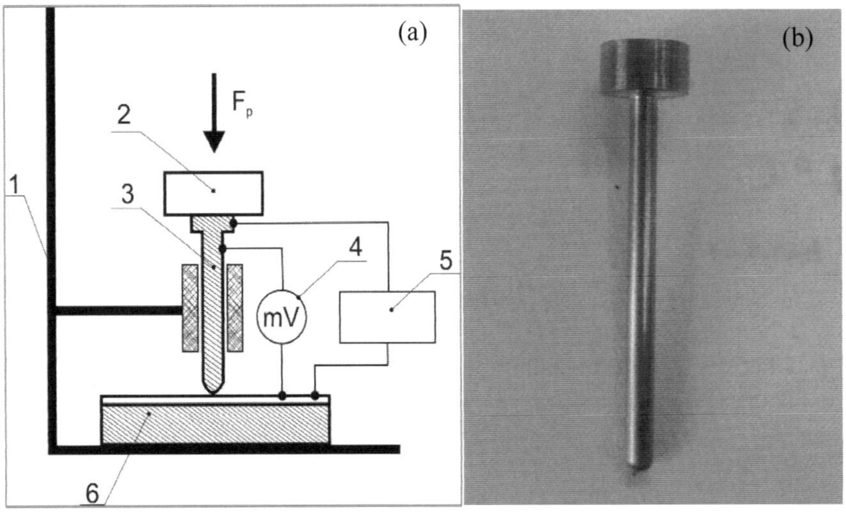

Figure 1. The structural scheme of the stand for testing contact resistance of coatings (a) and the image of the sample electrode (b).

Microstructural images were collected using a scanning electron microscope JSM-6610LV by JEOL (Tokyo, Japan) equipped with an energy-dispersive X-ray microanalyzer INCA X-MAX by Oxford Instruments (Abingdon, Oxfordshire, UK), which allowed for determining quantitative elemental composition of the coatings. We used a 30 keV acceleration potential in our experiments. The thickness of coatings was determined by means of cross-section microstructural measurements on the scanning electron microscopy (SEM) images. The relative accuracy of these measurements was estimated within the range of ±3.5%.

3. Results and discussion

It is known [1] that the state of the surface relief can influence significantly the value of the surface contact resistance. That is why at the first stage of the work the research of the surface structure and morphology the obtained samples' coatings was carried out. A comparative analysis was done.

The SEM-images of the typical surface areas of the researched sample coatings on the basis of alloys cobalt-tungsten, cobalt-molybdenum, nickel-tungsten and nickel-molybdenum are presented in Figure 2.

As it is seen in the SEM-images, the surface relief structure of the coatings of the four alloys under study is different. The surface relief structure of the coating by the alloy Co-W obtained under the given conditions is homogenous enough (Figure 2 a). The basis of the relief structure is the wedge-pyramidal formations with a strongly represented faceting, the lateral sizes of the formations are 0.5 – 1.5 microns. The structure elements arrangement is irregular. The analysis of the SEM-images of the typical surface areas of the sample coatings on the basis of alloys Co-Mo (Figure 2b) and Ni-Mo (Figure 2 d) do not allow revealing characteristic growing forms of the relief structure. In both cases the relief is composed of different in shape and size separate globule-formations. As it is in the first case, the lateral sizes of separate structure elements do not exceed 1.5 microns. The surface relief structure of the coating by the alloy Ni-W (Figure 2 c) is composed of the formations with the shape close to hemispherical. The arrangement order is also not observed in this case. The lateral sizes of the hemispheres are 0.5 – 1.5 microns.

It is necessary to point out that the roughness of the surface in the presented SEM-images is mainly determined by the sizes of separate elements constituting the relief structure and its value is commensurable in all four studied cases.

Figure 2. SEM-images of the typical surface areas of the sample coatings on the basis of alloys Co-W(a); Co-Mo(b); Ni-W(c) and Ni-Mo(d).

At the next stage of the work the research of contact electrical resistance of the sample coatings was carried out. The results of the research are presented in Figure 3 in the form of dependencies R_c on the pressure force of the sample electrode.

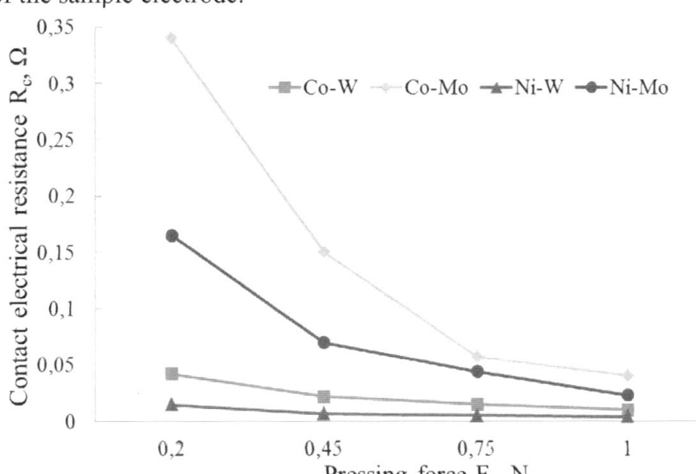

Figure 3. The dependencies of the contact electrical resistance of the coatings on the basis of alloys Co-W, Co-Mo, Ni-W, and Ni-Mo on the pressure force.

As the figure shows the dependencies in all the cases have a falling character along the whole length. At the range of the force 0.2 – 0.45 N the steepness is more represented. At the range 0.75 – 1 N there is a more sloping area; and at the value of pressure force 1 N the minimum value of the contact electrical resistance is reached.

In general, the dependence of contact resistance R_c on the pressure force corresponds to the conceptions, considered in the work [14], and is described by the following equation:

$$\lg R = \lg k - b \times \lg P, \qquad (1)$$

where P is the pressure force on the contact, [N]; b is the constant, depending on the type of contact (for point contacts – 0.5; for linear contacts – 0.6, and for planar contacts – 1.0).

Comparing the obtained results with the scientific research data [1], we may conclude that the decrease of R_c is connected with the increase of the common contact area (current transmission) as a result of surfaces modification and formation of the complementary relief on the contact surfaces. And partially it is connected with the forcing through the oxide films. In this case a steeper area of the curves is determined by physico-mechanical characteristics of this type alloys and by morphological peculiarities of the coating surfaces.

Comparison of the values R_c of the alloys under study shows considerable (for a degree of the value and more) differences of this parameter. The maximum values R_c at the level of 0.33 Ω and the maximum range of this parameter change (0.33 Ω – 0.02 Ω) with the increase of the electrode pressure force are typical of the coating on the basis of the alloy Co-Mo. The minimum value R_c at the level of 0.004 Ωwas fixed at the coating on the basis of the alloy Ni-W. The minimum range of this parameter change (0.015 Ω – 0.004 Ω) is also typical of this coating.

This great difference in the values of R_c of the researched alloys of one type with similar characteristics, such as electroconductivity, physic-mechanical characteristics (plasticity and elasticity), grain size of the alloy, presence of oxide films, surface roughness, can be explained by the heightened dispersion of the coatings structure and by the inclusion of nonmetal admixtures into the coatings in the process of electrochemical crystallization.

4. Conclusions

During the research we obtained the data on morphological peculiarities of the surface relief and contact resistance of some promising coatings on the basis of refractory metal alloys, such as Co-W, Co-Mo, Ni-W and Ni-Mo. The comparative evaluation and analysis of dependencies R_c on the pressure force of the sample electrode at the range of 0.2 N - 1 N allowed exposing the alloys with a maximum and a minimum value of this parameter as well as the range of its change for each alloy. The analysis and comparison of the obtained data shows that the coatings on the basis of binary tungsten alloys (Ni-W, Co-W) provide a considerably less value of the contact resistance in comparison to the alloys on the basis of molybdenum (Ni-Mo, Co-Mo) and can be recommended as a functional contact surface for switching devices on the basis of "dry contact" (magnetically operated sealed switches, relays, MEMS-switchboards).

The record low value Rc at the level of 0.004 Ω for the coating alloy Ni-W was obtained while using a standard methodology of its measuring.

Acknowledgment

The authors of the paper are grateful to the Regional Centre of Probe Microscopy of Collective Use and to the Candidate of Technical Sciences N. B. Rybin in person for the assistance in doing SEM-researches.

References

[1] S M Karabanov, N A Verlov, D V Suvorov, E V Slivkin 2015 *Technical Physics Letters* **41** 691–3

[2] L Shishkina, O Lokshtanova, S Karabanov 2012*Coatings* **21** 7

[3] A Bodaghi,J Hosseini 2012 *Int. J. Electrochem. Sci.* **7** 2584 – 95
[4] Ma Xi,Z Nie,T Dong, YMao2017. *Int. J. Electrochem. Sci.* **12** 1034 – 51
[5] MH Allahyarzadeh,M Aliofkhazraei, A R Rezvanian, V Torabinejad 2016 *Surf. Coat. Technol.* **307** 978–1010
[6] H B Lee2013 *Tribol. Lett.* **50** 407–419
[7] N Udompanit, P Wangyao,S Henpraserttae, Y Boonyongmaneerat2014 *Adv. Mater. Res.* **1025–1026** 302–9
[8] D V Suvorov,G P Gololobov, D Yu Tarabrin, E V Slivkin, S M Karabanov,A B Tolstoguzov2018 *Coatings* **8** 233
[9] N Tsyntsarua, H Cesiulis, E Pellicer, J-P Celis, J Sort 2013*Electrochim. Acta***104** 94–103
[10] N Eliaz, T M Sridhar, E Gileadi 2005*Electrochim. Acta* **50** 2893–2904
[11] Gomez E, Pellicer E, Valles E 2007 *Surface & Coatings Technology* **197** 238–46
[12] V S Kublanovsky, Yu SYapontseva 2014 *Electrocatalysis* **5** 372–378
[13] M Popczyk, B Łosiewicz2015 *SolidSt. Phen.* **228** 269-72
[14] Yu M Pyatin 1982 Materials in instrumentation and automation: reference (Moscow:Mashinostroenije) p 528

IOP Conf. Series: Materials Science and Engineering **889** (2020) 012018 doi:10.1088/1757-899X/889/1/012018

Quantitative phase analysis of Mg-Al-Zn and Mg-Zn-Zr alloys

Yu V Chernysheva, I A Grushin, I I Voskresenskaya and D S Savostin

MAI-Moscow Aviation Institute(National Research University) 125993, Moscow, Russia

E-mail: s.betsofen@gmail.com

Absract. A quantitative method has been developed for the phase analysis of magnesium alloys of the Mg-Zn-Al (MA2-1 and MA5) and Mg-Zn-Zr (MA14) systems based on the measurement of the lattice parameters of a solid solution, the use of Vegard's law, and balance equations of chemical and phase composition. Examples of the application of the technique for assessing changes in the content of alloying elements in a solid solution and the number of intermetallic phases as a result of heat treatment of MA2-1 alloy are presented.

1. Introduction Quantitative phase analysis makes it possible to objectively assess the structural-phase state of structural materials after various technological operations for the preparation of semi-finished products and welding. In our works, as applied to alloys of the Al-Mg-Li and Al-Cu-Li systems, a method for quantitative phase analysis based on measuring the lattice parameters of a solid solution, Vegard's law, and balance equations of chemical and phase composition is developed [1, 2]. This technique was used to assess the ratio of intermetallic phases during deformation, heat treatment and welding of industrial aluminum and lithium alloys [3-5], magnesium-based alloys [6-8]. In the present work, a similar technique is developed for the most widely used industrial alloys of the Mg-Al-Zn (MA2-1 and MA5) and Mg-Zn-Zr (MA14) systems.

2. Materials and research methods

The method of quantitative phase analysis was developed in relation to the most common magnesium alloys, the compositions of which are given in table. 1.

Table 1. The compositions of the alloys in weight%

Alloy grade	Al	Zn	Zr	Rest Mg
MA2-1	4.9	1.1	-	
MA5	8.5	0.9	-	
MA14	-	6	0.9	

The lattice parameters of MA2-1 alloy after quenching from 200 and 300^{0}C were determined on a DRON-3 diffractometer by recording the reflection (220) in CuKα radiation with a diffraction angle of ~ 75^0.

3. Results and discussion

Alloys Mg-Al-Zn Al alloys are characterized by the dominance of Al in comparison of Zn, therefore, it can be reasonably assumed that zinc is in solid solution. Therefore, it can be assumed that the alloy is two-phase and consists of a solid solution based on magnesium (α-phase) and β-phase ($Mg_{17}Al_{12}$). Moreover, in the solid solution contains a constant amount of zinc, equal to its concentration in the alloy and a variable aluminum content, which is determined from the lattice parameter of the solid solution. The mass percentages of both phases in the two-phase region of the double alloy can be determined from the balance equations of the chemical and phase compositions of the alloy:

$$X^{\alpha}_{Al(Zn)}W_{\alpha} + X^{\beta(MgZn_2)}_{Al(Zn)}W_{\beta((MgZn_2)} = 100X^{0}_{Al(Zn)} \quad (1)$$

$$W_{\alpha} + W_{\beta} = 100 \quad (2)$$

Fromhere:

$$W_{\beta} = \frac{X^{0}_{Al(Zn)} - X^{\alpha}_{Al(Zn)}}{X^{\beta(MgZn_2)}_{Al(Zn)} - X^{\alpha}_{Al(Zn)}} \quad (3)$$

The composition of the solid solution (X^{Al}_{α}) is related to the lattice parameters by Vegard's law, then:

$$a_{\alpha} = a_{Mg} + \left(\frac{\Delta a}{\Delta X}\right)_{Al} X^{Al}_{\alpha} + \left(\frac{\Delta a}{\Delta X}\right)_{Zn} X^{Zn}_{\alpha} \quad (4)$$

$$c_{\alpha} = c_{Mg} + \left(\frac{\Delta c}{\Delta X}\right)_{Al} X^{Al}_{\alpha} + \left(\frac{\Delta c}{\Delta X}\right)_{Zn} X^{Zn}_{\alpha} \quad (5)$$

$\left(\frac{\Delta a}{\Delta X}\right)_{Al}$ =-0,403x10^{-2} Å/wt.%; $\left(\frac{\Delta c}{\Delta X}\right)_{Al}$ =-0,540x10^{-2} Å/wt.%

$\left(\frac{\Delta a}{\Delta X}\right)_{Zn}$ =-0,186x10^{-2} Å/ wt.%; $\left(\frac{\Delta c}{\Delta X}\right)_{Zn}$ =-0,320x10^{-2} Å/ wt.% - a change in the lattice parameters "a" and "c" by 1 weight percent Al and Zn.

These values were calculated by the least squares method from the experimental values of the "a" and "c" parameters of the HCP lattice of a solid solution of aluminum and zinc in magnesium (figure 1), given in the Pearson monograph [9]. Thus, the Al content in the α phase is determined directly from (4)и (5), and then, knowing the chemical composition of the alloy, it is possible to calculate the amount of intermetallic phase from (1) и (2).

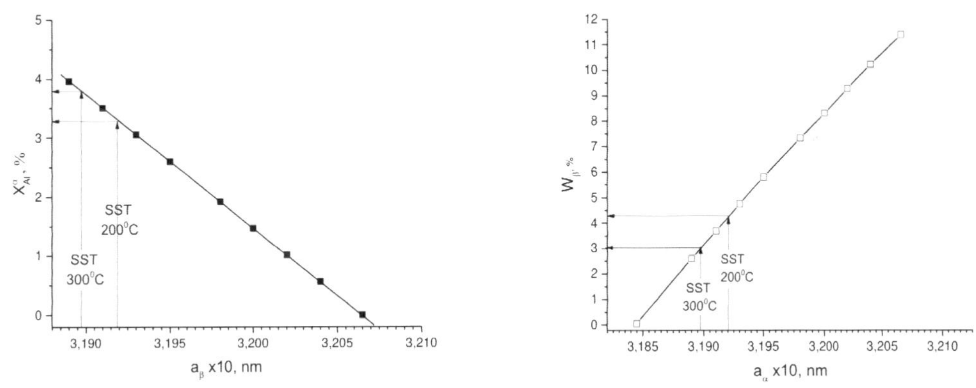

(a) (b)

Figure 1. The dependence of the aluminum content in the solid solution (X^{Al}_{α}) (a) and the amount of β-phase (W_{β}) (b) in the MA2-1 alloy (Mg-4.9Al-1,1Zn) on the value of the lattice parameter of the solid solution (a_{α}).

The accuracy of the determination depends on the accuracy of the determination of the lattice parameters and the slope of the "Vegard" dependences. In our case, with an error in determining the lattice parameter of 0.0001 nm, it amounts to 3-5% of the measured values, namely, the composition of the solid solution and the amount of the β phase. This is a good indicator, especially considering the fact that other methods for determining these quantities simply do not exist. Figure 1a shows the dependence of the α phase (a_{α}) on the aluminum content in the solid solution as a function of the lattice period "a" and the dependence of the amount of β phase in the MA2-1 (Mg-4.9Al-1,1Zn) alloy on figure 1b. An example of the application of the method for the quantitative phase analysis of MA2-1 alloy in the form of a pressed profile as the initial state and after quenching from 300^{0} C is shown, while the lattice parameter of the solid solution decreased from 0.3192 to 0.3187 nm, which corresponds to an increase in the aluminum content in the solid solution (X^{Al}_{α}) from 3.3 to 4.8% and a decrease in the amount of intermetallic phase (W_{β})) from 4.3 to 3%. Similar dependences are given for the MA5 alloy (Mg-8.5Al-0.5Zn), figure 2 and MA14 (Mg-6Zn-0.9Zr), figure 3.

IOP Conf. Series: Materials Science and Engineering **889** (2020) 012018 doi:10.1088/1757-899X/889/1/012018

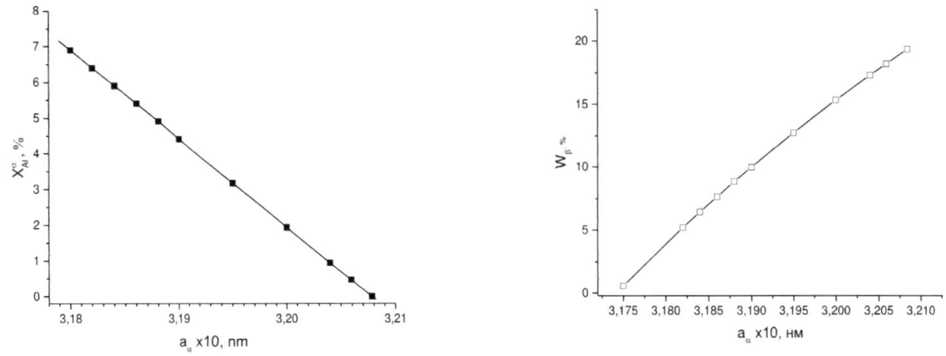

(a) (b)

Figure 2. The dependence of the aluminum content in the solid solution (X_α^{Al}) (a) and the amount of β-phase (W_β) in the MA5 alloy (Mg-8.5Al-0.5Zn) on the value of the lattice parameter of the solid solution (a_α)

.

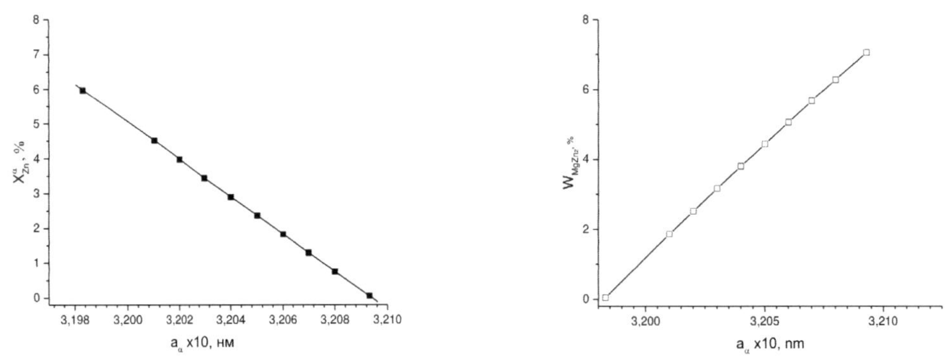

(a) (b)

Figure 3. The dependence of the aluminum content in the solid solution (X_α^{Al}) (a) and the amount of the β phase (W_β) (b) in the MA14 alloy (Mg-6Zn-0.9Zr) on the value of the lattice parameter of the solid solution (a_α).

Information on the quantitative ratio of phases in alloys allows reliable control of the state of alloys after deformation and heat treatment. In addition, knowing the phase composition, one can calculate the volumetric and linear dimensional changes accompanying the heat treatment. It must be taken into account that a change in the density and specific volume with a change in the composition of a magnesium-based solid solution is affected not only by a change in the lattice parameter, but also by the substitution of alloying elements with an atomic weight different from magnesium. Thus, when calculating the specific volume, it is necessary to take into account the average atomic weight as well as the lattice distance.The specific volume, V_α, of a single-phase magnesium-based solid solution can be calculated as follows [10]:

$$V_\alpha = \frac{N_A a_\alpha^2 c_\alpha \sqrt{3}}{2A_\alpha n} \tag{6}$$

Where: N_A is the Avogadro number 6.022×10^{23} (mol^{-1}); aα and c_α-parameters of the lattice of Mg solid solution at 25°C (cm) (calculated from (2)); n is the number of atoms per unit cell = 2. For pure Mg, we obtain $V_{Mg} = 0.5755$ cm^3/g ($A_{Mg} = 24.32$ g/mol).

The atomic weight (A_α) for a solid solution of aluminum in magnesium is calculated from the ratio:

$$A_{\alpha} = \frac{Y_{\alpha}^{Al} A_{Al} + (100 - Y_{\alpha}^{Al}) A_{Mg}}{100} \quad (7)$$

Where: Y_{α}^{Al} - atomic percentage of aluminum in the α phase; A_{Al} - atomic weight of aluminum = 26.97 g/mol

In accordance with the rule of mixtures, the specific volume of the Mg-Al alloy is calculated from the ratio:

$$V_{alloy} = \frac{W_{\alpha} V_{\alpha} + W_{\beta} V_{\beta}}{100}$$

Where: W_{α} and V_{α} wt. % and specific volume of α-solid solution, respectively; W_{β} and V_{β} are mass% and specific volume of the β phase, respectively. According to PDF # 73-1148, the density of the β phase ($Al_{12}Mg_{17}$) is 2.088 g cm^3, which corresponds to a specific volume of $V_{\beta} = 0.4789$ cm^3/g.

A change in the specific volume of a solid solution with a change in its concentration and a simultaneous change in the amount of the second phase have a major effect on the dissolution / precipitation processes. It's obvious that:

$$\Delta V = V_2 - V_1$$

The relative change in volume is:

$$\frac{\Delta V}{V_1} = \frac{V_2 - V_1}{V_1}$$

4. Conclusion

A method for quantitative phase analysis has been developed as applied to magnesium alloys of the Mg-Al-Zn and Mg-Zn-Zr systems, which include the most well-known industrial alloys MA2-1, MA5 and MA14.The method is based on the measurement of the lattice distances of a solid solution, Vegard's laws, and balance equations of chemical and phase composition.An example of the application of the developed technique for calculating the concentration of alloying elements in a solid solution and the amount of intermetallic phases in the MA2-1 alloy after heat treatment is shown.

5. References

[1] BetsofenS, AntipovV, GrushinI, KnyazevM, KhokhlatovaLand AlekseevA2015 *Russian Metallurgy (Metally)* **1** 51
[2] BetsofenS, AntipovVand KnyazevM 2016 *Russian Metallurgy (Metally)* **11** 929
[3] BetsofenS, AntipovV, KnyazevM andOglodkovM 2015*Russian Metallurgy (Metally)* **4** 326
[4] Betsofen S,LukinV, DolgovaM, Kabanova Y and Panteleev M 2018*Russian Metallurgy (Metally)* **4** 359
[5] Betsofen S,Sbitneva S, Panteleev M, Gordeeva M and Knyazeva Y2018*Russian Metallurgy (Metally)* **11** 1059
[6] Betsofen S,Osintsev O, Grushin I, Petrov A and Speranskii K 2019 *Russian Metallurgy (Metally)* **4** 361
[7] BetsofenS, Voskresenskaya I, Kolobov Yu, BozhkoS andVolkova E2015 *Russian Metallurgy (Metally)* **4** 257
[8] BetsofenS, RokhlinL, WuR,LozovanA and Voskresenskaya I2014 *Russian Metallurgy (Metally)* **11** 920
[9] Pearson W1958 *A Handbook of lattice spacing and structures of metals and alloys*(N.Y.: Pergamon Press)p 1044
[10] Hunsicker H1980 *Trans. of AIME* **VII** A 759

Effect of bias voltage on the texture of the TiN and ZrN coatings deposited by vacuum ion-plasma method

A A Ashmarin[1], S Ya Betsofen[2], L M Petrov[3]and M A Lebedev[2]

[1]BaikovInstitute of Metallurgy and Materials Science,Moscow, Russia
[2]MAI-Moscow Aviation Institute (National Research University), Moscow, Russia
[3]National Institute of Aviation Technologies, Moscow, Russia

E-mail: s.betsofen@gmail.com

Abstract. An X-ray diffraction method was used to study the formation of the crystallographic texture of vacuum ion-plasma TiN and ZrN coatings depend on the bias voltage on the substrate. It is shown that, at high values of the bias voltage (-100 V), a pronounced texture forms in TiN coatings and a two-component (111) + (113) texture in ZrN coatings. A decrease of the bias voltage to –10 V leads to a twofold increase in the hardness of the coatings, a non-texture state of TiN, and a weakening of the texture of ZrN coatings.

1. Introduction

It is well known that bias voltage plays a very important role in the deposition process of a coating. A negative bias voltage increases the energy of the ions that bombard the growing coating, which leads to a higher compaction of the coating, an increase in compressive stress and a change in the structure and texture of the coatings [1-6]. The temperature of the substrate is also important, which has a significant effect on the structure and properties of the coatings. In this paper, we investigated the effect of bias voltage on the texture and microhardness of TiN and ZrN coatings deposited by a vacuum ion-plasma method.

2. Materials and methods

Plasma flows of Ti-N and Zr-N systems were formed by electric arc evaporation during vacuum ion-plasma processing. Plasma flows of Ti and Zr were deposited in a 100% atmosphere N on a steel substrate (St1) in an NNV 6.6 I1 vacuum ion-plasma installation at a pressure of 0.13 Pa and an arc current of 110 A; the coating formation time was 60 minutes. The values of the accelerating reference voltage on the substrates were -100V and -10V. The interaction of the gas-metal plasma flow with the treated surface was carried out after preliminary electronic heating of the surface to a temperature of 320^{0}C and its bombardment by a flow of argon gas plasma for 5 min + 1 min. The heating temperature of parts in a vacuum chamber was controlled by an IP 140 digital infrared pyrometer. HV microhardness was measured on a Micromet 5101 microhardness meter in accordance with GOST RISO 6507-1-2007 at a load of 0.49 N. The texture was evaluated by the method of inverse pole figures (IPF) byDRON-3 diffractometerwith using CuK$_\alpha$ radiation. The pole densities (P_{hkl}) for each of the m-poles (hkl) at the IPF were estimated by the ratio of the integrated intensities of the reflexes of the textured (I_{hkl}) and non-textured reflection (R_{hkl},table 1) with Mueller normalization:

$$P_{hkl} = \frac{I_{hkl} / R_{hkl}}{\frac{1}{m}\sum_{i=1}^{m}\left(\frac{I_{hkl}}{R_{hkl}}\right)_i}$$

3. Resultsand discussion

In the process of forming coatings based on the TiN system, the bias voltage is an important technological parameter that determines the energy contribution of the gas-metal plasma stream to the acceleration and efficiency of the implementation of plasma-chemical processes of formation of interstitial phases. High values of the bias voltage (U_b=-100V) leads to an increase in temperature, an increase in the intensity of all processes on the surface of the substrate, which is accompanied by the formation of a pronounced texture (111) in TiN coatings (table 1).

Table 1. Intensities of non-textured (hkl) reflections (R_{hkl}) and pole densities (P_{hkl}) of TiN and ZrN coatings

		R_{hkl}		P_{hkl}			
				U_b=-100V		U_b=-10V	
No.	hkl	TiN	ZrN	TiN	ZrN	TiN	ZrN
1	(111)	793	1284	3.45	2.13	1.17	1.1
2	(200)	1260	866	0.07	0.29	0.8	0.93
3	(220)	677	540	0.08	0.04	0.92	0.52
4	(311)	274	375	0.4	1.58	1.11	1.44
5	(222)	194	168	3.45	2.13	1.17	1.1

When the bias voltage is reduced to U_b=-10 V, a fundamental change in the appearance of the X-ray diffraction pattern of the coating occurs, at which the intensities of all reflections approximately correspond to the non-textured state of the coating. The X-ray diffraction pattern of the ZrN coating at high values of the bias voltage (U_b=-100 V), as in the case of the TiN coating, leads to the formation of a more pronounced texture compared to the coating deposited at a lower value of the bias voltage (U_b=-10 V). However, unlike the TiN coating, the texture of the ZrN coating is two-component, in addition to the main component (111), similar to TiN, there is a texture (311), the pole density of this component is slightly inferior to the main (111) component (table 1).

A comparison of the pole densities of the main orientations (figures 1a and 2a) for TiN and ZrN coatings shows that the effect of lowering the reference voltage affects the texture of these coatings differently. For TiN coatings, a decrease in the reference voltage to -10 V leads to a textureless state, which is characterized by its deviation (P_{hkl}= 1) by no more than 20% (figure 1a, table 1), then for the ZrN coatings these deviations reach 50% (figure 2a, table 1). These differences, however, had almost no effect on the effect of increasing hardness, which increased approximately twice as a result of a decrease in the bias voltage.

The results obtained indicate that a decrease in the bias voltage on the substrate during the deposition of TiN and ZrN coatings eliminates the pronounced (111) texture and significantly increases the hardness of the coatings. A similar effect was obtained when replacing titanium atoms in TiN coatings of Al and Si, which was accompanied by an increase in hardness and corrosion resistance [7-9]. The presence of a pronounced (111) texture is usually associated with the presence of a columnar microstructure, as well as the enlargement of the subgrain structure, estimated by the width of the X-ray diffraction lines[9-12]. It can be assumed that the texture itself is not the reason for the change in properties, i.e. variations in properties are caused not by their anisotropy, but by the fact that, as a result of alloying or changing the energy parameters of the process, the coating formation mechanism changes, which is accompanied by a change in the texture and fine structure of the coating, which is reflected in the level of service properties. In those cases when, the composition or energy parameters change, the texture remains unchanged is an indication that these changes do not lead to a fundamental change in the mechanism of formation of the coating structure, i.e. essentially, this indicates the ineffectiveness of such changes in terms of optimizing the service properties of coatings.

Thus, the texture is a sensitive structural parameter for evaluating the effectiveness of the search for optimal compositions and technologies for producing coatings, while the texture itself apparently does not significantly affect the properties, since most coatings have highly symmetric crystal structures, in most cases a cubic NaCl lattice. For cubic crystals, even in the case of a sufficiently high single-crystal anisotropy, it is difficult to realize a substantial anisotropy of properties even in a highly textured polycrystal due to the high multiplicity factors of most crystallographic orientations (up to 48) and the presence of an obligatory region of texture spread, especially in the case of two or more texture components. In the latter case, the absence of anisotropy of properties is inevitable.

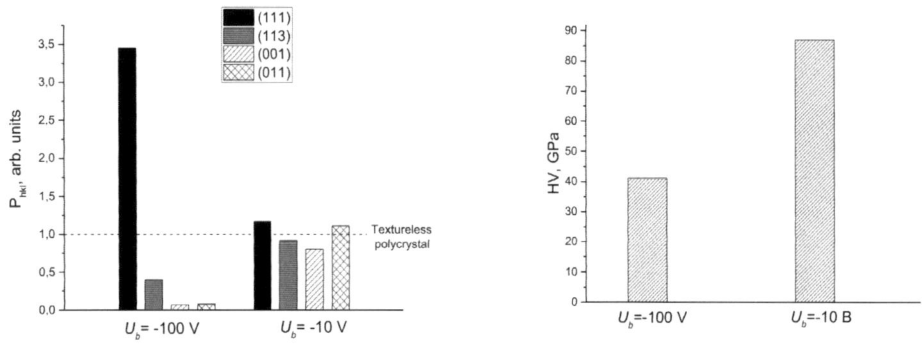

(a) (b)

Figure 1. Pole reflection densities (a) and hardness values for TiN coatings obtained by the vacuum-arc method at substrate temperatures of -100 and -10 V.

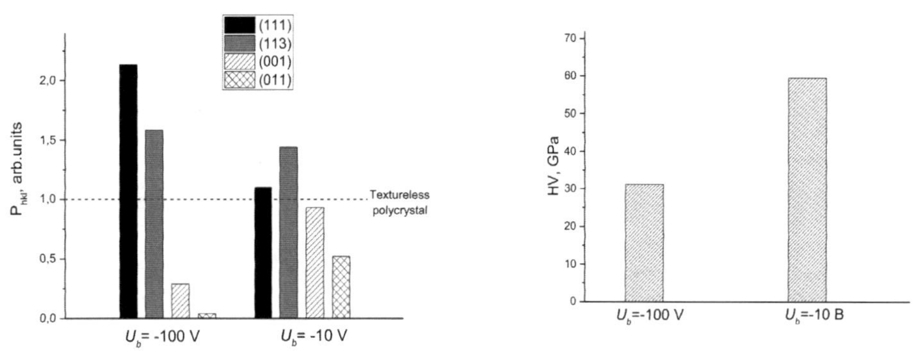

(a) (b)

Figure 2. Pole reflection densities (a) and hardness values (b) for ZrN coatings obtained by the vacuum-arc method at substrate voltages of -100 and -10 V.

4. Conclusions

1. A high value of the bias voltage (-100V) leads to the intensification of processes on the surface of the substrate, which is accompanied by the formation of a pronounced texture (111) in TiN coatings and a double texture (111) + (113) in the ZrN coating.

2. A decrease in the bias voltage leads to a twofold increase in the hardness of the coatings, a non-texture state of TiN, and a weakening of the texture of ZrN coatings.

5. References

[1] Lenkovets A, Lozovan A, Betsofen S, Bespalov A, Grushin I and Ivanov N 2019 *Journal of Physics: Conference Series* **1396** 012028

[2] Lozovan A, Betsofen S, Lenkovets A, Grushin I, Labutin A and Pavlov Yu2018 *Journal of Physics: Conf. Series* **1121** 012019
[3] Lenkovets A, Lozovan A, Betsofen S, Grushin I, Ivanov N and LebedevM2019*Journal of Physics: Conf. Series* **1281** 012047
[4] Lozovan A,Betsofen S, Alexandrova S, Lyakhovetskiy M and Kubatina E 2018*IOP Conference Series: Materials Science and Engineering,* **387** 012049
[5] KarlssonL, HultmanL andSundgrenJ 2000*Thin Solid Films* **371** 167
[6] BousserE, BenkahoulM, MartinuL and Klemberg-SapiehaM 2008J. *Surface & Coatings Technology* **203** 776
[7] KongaQ, LiH, LiuX, WangY, ChenJ and ZhouH 2011 *Applied Surface Science* **257** 2269
[8] Yoo Y, Le D, Kim J, Kim S andVinh P2008*Thin Solid Films* **516** 3544
[9] Chang C, Lee J and Tseng M 2009*Thin Solid Films* **517** 5231
[10] Zhang Z, Rapaud O,Bonasso N, MercsD, Dong C and Coddet C 2008*Vacuum* **82** 1332
[11] ChangCh and WuCh2009*Thin Solid Films* **517** 5219

IOP Conf. Series: Materials Science and Engineering **889** (2020) 012020 doi:10.1088/1757-899X/889/1/012020

Investigation of hydroabrasive resistance of internal anti-corrosion coatings used in the oil and gas industry

M Kovalev[1*], E Alekseeva[1] and N Shaposhnikov[1]

[1]Peter the Great St. Petersburg Polytechnic University, 195251 Saint Petersburg, Russia

E-mail: kovalev.ma@edu.spbstu.ru

Abstract. Anti-corrosion coatings are commonly used in the oil and gas industry to protect the inner surface of pipelines. Most of coatings have a polymer base and can effectively prevent the negative impact of the transported fluid. Existing standards allow to evaluate the quality of coatings and their durability in the anticipated operating conditions. The evaluation of the abrasive resistance of coatings usually limited to comparative tests on a Taber abraser. This paper presents the results of hydroabrasive resistance study of coatings using nonstandard technique. Comparative test results of various polymer (FBE), thermally sprayed coating and silicate coatings are obtained. The ways of modifying test method are proposed.

1. Introduction

In the oil and gas industry corrosion is one of the main problems. A large number of pipelines have been in operation for more than 30 years, which leads to an increased number of incidents related to a corrosion damage [1]. To reduce the number of incidents leading to economic and environmental losses, as well as to increase the operating life, various methods of corrosion protection are developed and applied. The main ones include inhibition and use of pipes with an internal anti-corrosion coating [2]. The use of inhibitors is an effective method of corrosion protection, although has several drawbacks such as the limited use in environments with a high concentration of abrasive particles and the presence of fixed costs throughout the entire period of operation. The use of pipes with an internal coating is one of the most effective methods of dealing with corrosion damage. The most common use is polymer-based coatings, such as fusion bonded epoxy (FBE). Such coatings protect the inner surface of the pipe from the influence of the transported medium, and due to the low surface roughness, increase the productivity of the pipeline [3].

Polymer coatings have been used for a long time, but in recent years there has been an increased interest in such products in the oil industry. The result of a large number of studies and the accumulation of experimental data led to the creation of the regulatory document GOST R 58346-2019 “Steel pipes and fittings for the oil industry. Protective coatings on the inner surface. General technical requirements” [4]. This document describes the certification procedure for internal anti-corrosion polymer coatings for use in oilfield pipelines and tubing. In the standard, test methods are collected and streamlined that allow to evaluate most of the properties of coatings and determine the possibility of their use in the selected conditions.

The least attention is paid to hydroabrasive wear of coatings, while the effect of solid mechanical particles in the fluid flow is one of the main reasons for the failure of oilfield equipment [5, 6, 7]. Material wear occurs as a result of cutting micro- particles from the surface or detachment of individual particles as a result of contact fatigue. The rate of polymer degradation under conditions of hydroabrasive wear is affected by the flow rate, the chemical composition of the operating medium, the concentration of abrasive, the temperature of the medium, the angle of interaction, coating hardness, elastic modulus, coating thickness, etc. [8].

The lack of standard methods for assessing hydroabrasive wear is due to the complexity of modeling wear processes. As a standard method for assessing the abrasion resistance of a coating, ASTM D4060-14 "Standard Test Method for Testing Organic Coatings for Abrasion Resistance Using the Taber Instrument" is generally used [9]. This method is comparative and does not simulate real operational (field) impacts, the inadmissibility of its use for evaluation was already written in [10], since the basis of this method is the process of dry friction, and not hydroabrasive wear, which is realized under operating conditions.

Therefore, to assess the hydroabrasive wear of oilfield anticorrosive coatings, a technique was applied based on the corrosion-mechanical test method to evaluate the metal of pipes [11] at the bench test, which was improved and used to assess the resistance of coatings.

The aim of the work is a comparative assessment of hydroabrasive wear of anticorrosive coatings (polymer, silicate-enamel, thermally sprayed Al-Mg) used to protect oilfield pipelines and oilfield equipment.

2. Experimental

The tests were carried out on a facility that allows testing materials in aqueous media with the addition of abrasive material. Hydroabrasive action occurs due to the introduction of abrasive into the solution in the form of silica sand with a concentration of 1.5% silica sand, the fraction size is 0.5 - 1.0 mm. A uniform distribution of abrasive particles in the solution is achieved due to the constant circulation of the test medium in the system. The flow rate is 8.5 m^3 / hour. The temperature of the test solution is 30 ± 5 ° C. The test duration is 8 hours. If the coating is worn to the substrate in less time, the test was terminated. Figure 1 shows the installation diagram and its general view in the assembly.

Figure 1. The scheme of test equipment.

Three samples were prepared for testing, with a size of 30 * 30 mm of each type of coating. The sample was installed at the angle of 95 ° relative to the direction of flow. Samples were weighed on an analytical balance with an accuracy of 0.0001 g before and after the tests, and also before the tests, the coating thickness was measured with an accuracy of 1 μm. Also, the wear time of the coating to the substrate was estimated.

3. Test results

The results of the tests for resistance to hydro abrasive wear are shown in table 1.

Table 1. Hydro Abrasion Test Results

Type of coating	№	Mass before, g	Mass after g r	Mass loss, g	Average mass, r	Thickness, μm	Time, h	Note
Thermally sprayed	1	39,97	39.70	0.27	0.27	105	2	Heavy wear
	2	42.91	42.64	0.27		101	2	
	3	41.56	41.30	0.26		109	2	
Epoxy 1-2	1	44.52	44.40	0.12	0.10	332	5	To substrate
	2	48.42	48.32	0.10		315	5	
	3	47.95	47.88	0.08		335	5	
Epoxy 2-1	1	36.09	35.99	0.10	0.10	502	8	No wear
	2	36.33	36.22	0.11		504	8	
	3	33.31	33.23	0.08		580	8	
Epoxy 2-2	1	40.80	40.55	0.26	0.28	540	8	To substrate
	2	43.82	43.51	0.31		521	8	
	3	41.57	41.29	0.28		563	8	
Epoxy Phenolic	1	57.93	57.49	0.44	0.42	495	4	Heavy wear
	2	55.32	54.91	0.41		498	4	
	3	54.63	54.21	0.42		492	4	
Silicate	1	36.27	35.80	0.46	0.5	1.28	4	Heavy wear
	2	31.40	30.83	0.58		1.83	4	
	3	30.37	29.90	0.47		1.54	4	

As a result of the tests. it was revealed that there were no patterns - the greater the thickness. the lower the wear out. Figure 2 shows a comparison of the state the samples surface after testing.

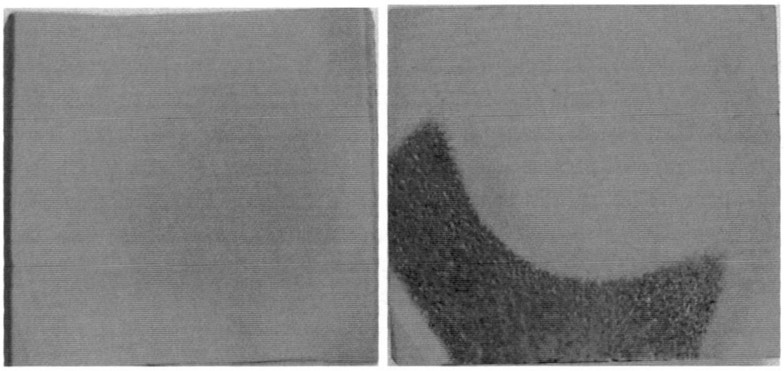

a) b)

Figure 2 - Samples after testing: a) epoxy №6-1. b) epoxy-phenolic № 8-1.

The coating marked 2-1 is a powder coating with a declared operating temperature of +110 ° C and a thickness of ≈ 500 μm. After 8 hours of testing. the coating did not lose its continuity. Epoxy phenolic coating with a maximum operating temperature of +60 ° C and a thickness of ≈ 500 μm was tested. After 4 hours of testing. the continuity of the coating was broken. Samples of epoxy phenolic coatings also have higher rates of hydro abrasive wear compared to powder polymer coatings of similar thickness. Such results can be explained by the lower hardness of the epoxy phenolic coating.

Figure 3 shows samples with a thermally sprayed coating and a silicate coating after testing. The lowest characteristics of resistance to hydroabrasive were determined for a thermally sprayed coating based on Al-Mg. Fig. 3. a. it wore to the substrate in 2 hours of testing. A high wear rate could be attributed to a roughness and a thickness of ≈ 100 μm. which is 3 to 10 times less than other coatings studied.

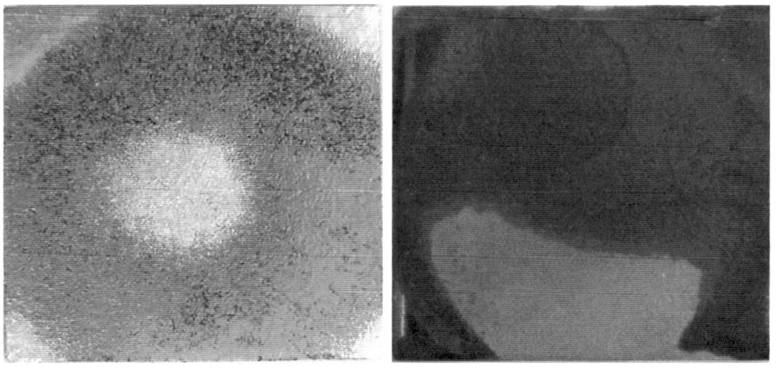

a) b)

Figure 3 - Samples after testing: a) metallization coating. b) silicate coating.

Silicate-enamel coating. despite a considerable thickness (more than 1 mm). was worn down to the substrate during 4 hours of testing. which may be due to the known fragility of coatings of this type. It is evident that wear occurs more likely to be chipped. and not due to uniform wear (figure 3. b).
Thus. comparative tests of various types of coatings were carried out. High test results were obtained when testing samples of one coating.

4.Conclusions

1. The technique used to assess the hydroabrasion resistance allows a comparative rapid assessment for choosing of various types of coatings both on plates and on samples from pipe segments.
2. The proposed method allows to assess the degradation of the hydro abrasive resistance of the coating after other laboratory tests.
3. According to the results of the tests. the epoxy coating turned out to be more resistant in comparison with the epoxy phenolic coating.
4. The thermally arc sprayed Al-Mg coating showed low hydroabrasive resistance. which is probably due to its thickness and application features.
5. The silicate coating showed low hydroabrasive resistance due to its brittleness.
6. Verification of the obtained values in experimental field trials would allow us to assess the degree of applicability of the methodology.

5. References

[1] Laptev A B and Bugay D E 2010 Fighting against corrosion in oil and gas complex of Russia: problems and the ways of their solving *European Corrosion Congress EUROCORR* 191
[2] Heidersbach R 2018 Metallurgy and corrosion control in oil and gas production *Wiley Series in Corrosion John Wiley & Sons Second* 368
[3] Protasov V N 2007 *The theory and practice in application of polymeric coatings to equipment and constructions in the oil and gas industry* (Moscow: Nedra) 373
[4] GOST R 58346 2019 *Steel pipes and fittings for the oil industry Protective coatings on the inner surface General technical requirements*
[5] Bulat A V Degovtsov A V Ivanovskiy V N Korolkov D O and Sabirov A A 2018 Features of the work of well separators mechanical impurities in the composition of sucker-rod pumping unit *Territorija NEFTEGAS Oil and Gas Territory* **7(8)** 42-52
[6] Rodionova I G and Novoselova O N 2012 *Modern approaches to improving the corrosion resistance and operational reliability of steels for oil field pipeline* (Moscow: metallurg izdat) 172
[7] Alkhimenko A Shaposhnikov N Shemyakinsky B and Tsvetkov A 2019 Several erosion test results of means of sand control *E3S Web of Conferences EDP Sciences* **121**

[8] Protasov V N 2011 Physical and chemical mechanics of materials of equipment and facilities of the oil and gas industry. *Russian State University of Oil and Gas named by Gubkina* 204
[9] ASTM D4060 2014 *Standard test method for abrasion resistance of organic coatings by the taber abraser*
[10] Shtyrev O O 2017 The method of the resistance control of the material of the Inner polymer coating of oil country tubular goods to hydro abrasive depreciation in the certification and periodic laboratory testing *Territorija NEFTEGAS Oil and Gas Territory* **3** 28-33
[11] Mitrofanov A V Alkhimenko A A Kolesov S S Shemyakinsky B A Tikhonov S M Kuznetsov D V Komissarov and A A Sidorova 2018 *Patent №2697030 Method for piping testing for oil pipes for corrosion-abrasive wear E P* 10122018

IOP Conf. Series: Materials Science and Engineering **889** (2020) 012021 doi:10.1088/1757-899X/889/1/012021

Operational properties of drill pipes made of coated aluminum alloy 2024

O V Shvetsov[1*], S Y Kondrat'ev[1], V A Yakhimovich[1] and M K Kurakin

[1]Peter the Great St. Petersburg Polytechnic University, 195251 Saint Petersburg, Russia

* shvec_off@mail.ru

Abstract. The main aim of this article is to provide for the comprehensive study that has been carried out. We tested the performance characteristics of coated aluminum alloy 2024 under conditions simulating the operation of drill pipes in the oil industry. In the course of our study we experimentally determined electrochemical potential and corrosion rate of the coated aluminum alloy 2024. Moreover, a comparative analysis of adhesion and wear resistance of coatings during the use of the alloy has been carried out. The authors have also studied structures of various coatings and the nature of damage to aluminum alloy drill pipes under operation.

1. Introduction

The increasingly complex conditions for oil and gas wells drilling require the use of new materials to manufacture drill pipes and the reasonable choice of such materials in order to comply with specific operational conditions [1-6]. Aluminum alloys (in comparison with steel) can be considered as the most promising materials for the production of drill pipes that are used in oilfield development [7-11]. Despite demonstrating a higher specific strength and corrosion resistance [12-13], aluminum drill pipes are inferior to steel ones in terms of durability. Thermal treatment, cold deformation, surface toughening, etc [14-17] are used to increase the wear resistance of aluminum pipes. However, it is most economical to apply a protective coating on the metal of the pipe. Such coating should have the necessary adhesion with the main metal and the continuity and the technology of its application should not change the service properties of the substrate. There is little information in the literature on how the protective coatings affect aluminum drill pipes, so our study is obviously seen as a relevant one.

The purpose of the study is to investigate the performance of aluminum drill pipes covered with various protective coatings.

2. Research material and methodology

We tested drill pipes (147 mm and 126 mm in diameter) made of aluminum alloy 2024 under tempering (hardening and natural ageing). Chemical composition of the pipe material (table 1) is consistent with the standard aluminum alloy 2024 (GOST 4784-97). We used various mixtures provided by three manufacturers to apply coatings with high-speed flame spraying. P and K coatings were applied to the tubular blank of the aluminum alloy 2024 in the after quenching and natural aging state, and G coating was applied to the blank subsequent to annealing.

We carried out metallographic analysis of the alloy longitudinal and transverse grinding with «Reichert-Jung Mef3a» optical microscope (x 100-500 zoom). SUPRA 55/55VP autoemission raster electron microscope was used to determine chemical composition of the coatings. We used equipment of the «Buehler» company to manufacture and prepare all the metallographic grinders according to the ASTM E 3-95.

Table 1. Real chemical composition of the aluminum alloy 2024 samples covered with various coatings.

Alloy type	Coating material	Basic chemical element content, mass. %								
		Al	Mg	Zn	Mn	Cu	Cr	Ti	Fe	Si
Aluminum alloy 2024 (Base P)	P 1	Base	1.59	0.24	0.49	4.39	0.02	0.04	0.23	0.20
	P 3	Base	1.50	0.05	0.67	4.89	0.01	0.04	0.44	0.24
Aluminum alloy 2024 (Base K)	K 1	Base	1.50	0.15	0.78	4.79	0.03	0.06	0.08	0.06
	K 2	Base	1.57	0.16	0.70	4.75	0.03	0.06	0.08	0.03
	K 3	Base	1.57	0.12	0.75	4.68	0.01	0.05	0.06	0.07
Aluminum alloy 2024 (Base G)	G 1	Base	1.26	0.14	0.64	4.71	0.04	0.09	0.14	0.08
	G 2	Base	1.25	0.13	0.68	4.20	0.03	0.09	0.15	0.05
	G 4	Base	1.32	0.14	0.73	4.81	0.03	0.03	0.05	1.99

The electrode potential of the metal was measured in a 5% sodium chloride solution using three samples per each coating type. The potential was measured with a high-resolution voltmeter using a standard silver chloride comparison electrode that features its own potential relative to a standard hydrogen electrode of -0.200 V. The experiment lasted 120 hours. The electrode potentials of the samples were measured during 60-minute time period every day.

The rate of corrosion of the «metal-coating» pair by mass loss was determined in accordance with the requirements of ASTM G71-81 standard and calculated according to GOST 9.908-85 formula (1):

$$V_{cor.} = (m_K - m_0)/St. \qquad (1)$$

Where Vcor is the corrosion rate, g/m^2 per hour; m_k is the mass of the sample after the test, gram; m_0 is the mass of the sample before the test, gram; S is the surface area of the sample, m^2; t is the testing period, hour.

The microhardness was determined on the basis of the Vickers method with the use of «Reichert-Jung Micro-Duromat 4000E» device following ASTM E 92 procedure. Mechanical properties of coated alloys were determined while performing the three-point bending tests [18]. Taking the geometry of each sample into account, we used a finite-element method and Cosmos Works package to calculate stress during testing

We carried out durability tests of the coatings using a test bench. The test bench is equipped with three engines that allow rotating the shaft with the testing body (TB), rotating the counter face (CF) by means of a cogged pinion and chain gear and moving the TB against the CF simultaneously. The test conditions for coating durability are as follows: TB pressurization force to CF: 100 kg-wt; TB rotation speed: 100 rpm; TB displacement speed along the shaft: 3 m/h; CF rotation speed: 0.1 rpm; washing fluid: water. Control panel is used to ensure control over the measured parameters (frictional moment, clamping force and shaft speed) and to control engines that provide three types of movement (rotation and longitudinal movement of the sample, and rotation of the CF). Dacell TMS Ver 1.1 software is used to process sensor incoming data. The data are stored in files in tables and graphs.

3. Research findings and discussion

Figure 1 and table 2 show the micro-structure and the microhardness values of the alloy covered with different coatings. The alloy structure consists of α-solid solution grains and inclusions of intermetallides. Anisotropy of the deformed alloy is more apparent in the tempered and aged state. Intermetallic phases are more evenly distributed over the thickness and length of the heated pipe blanks. Hence, the microhardness of the heated alloy covered with G coatings is significantly lower than the tempered and aged alloy covered with P and K coatings.

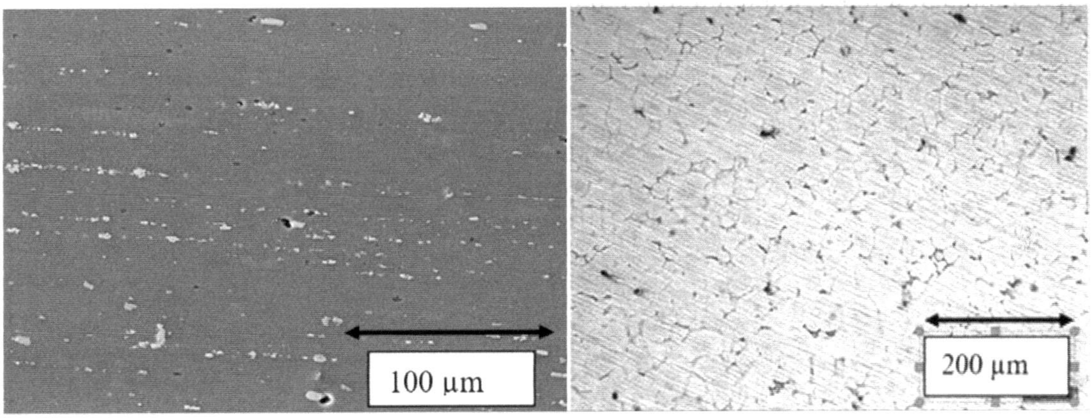

(a) (b)

Figure 1. Micro-structure of aluminum alloy 2024 covered with P coating (a) after quenching and natural ageing and covered with G coating after annealing (b): longitudinal section (a) and cross-section (b).

Table 2. Microhardness of aluminum alloy 2024 tested samples covered with different coatings.

Coating	Microhardness HV, Pa
P 1	1400-1550
P 3	1350-1450
K 1	1350-1500
K 2	1400-1550
K 3	1400-1500
G 1	750-850
G 2	730-780
G 4	710-770

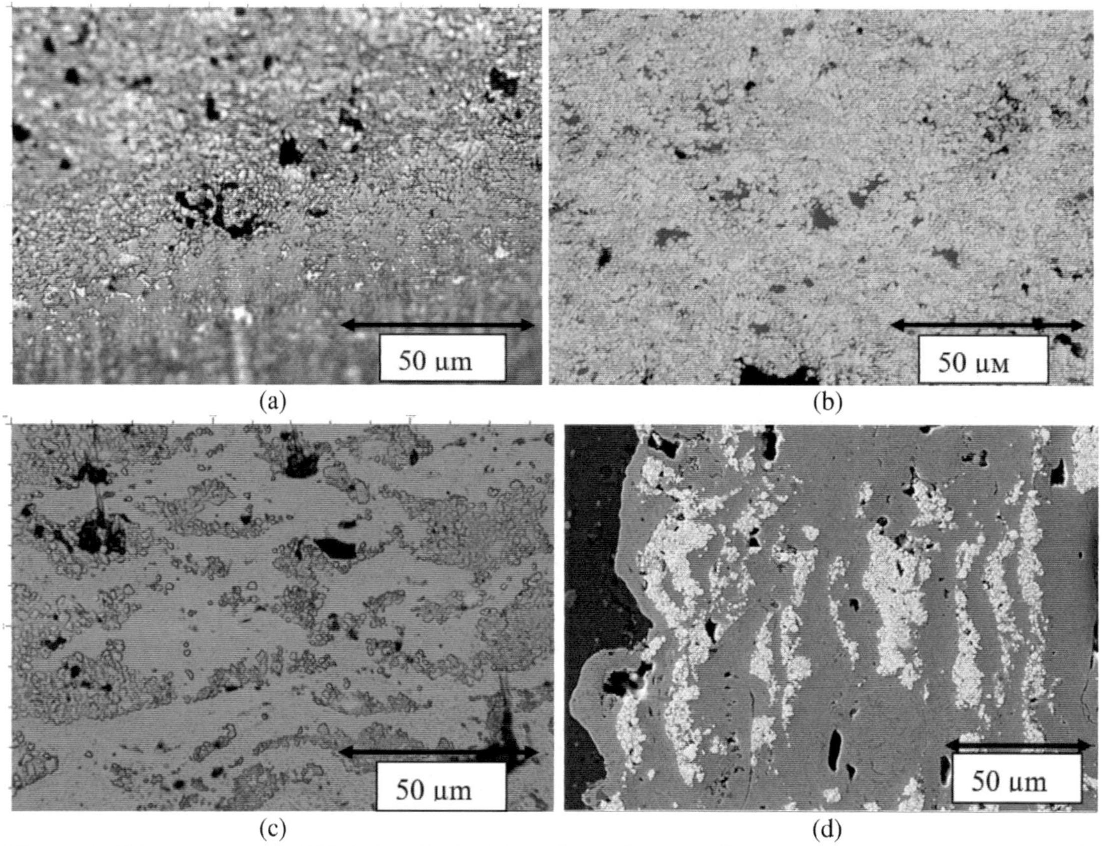

(a) (b) (c) (d)

Figure 2. Microstructure of P 1 (a . b) P 3 (c . d) coatings: optical microscope (a . c); electronic microscope (b . d).

The microstructure of the P 1 coating (Figure 2, a-b) with a thickness of 120 μm is composed entirely of WC (tungsten carbide) representing its high microhardness of 11.300 HV. The coating is homogeneous over the entire surface of the sample and a small number of pores are evenly distributed over the section and the length of the coating; the amount of porosity is 1.94%. The microstructure of P 3 coating (Figure 2, c-d) contains: ~ 50 vol. % of WC (tungsten carbide) with hardness of 12.000 HV and ~ 50 vol. % of Ni-Cr-Si connection with hardness of 8700 HV. The porosity of this coating (3.25 %) is greater in comparison with P 1coating, and its thickness is 160 μm.

Figure 3 presents K 1, 2 and 3 coatings. K 1 coating structure is homogeneous; porosity is 0.038%; thickness is 80 μm. This coating consists entirely of WC (tungsten carbide); its hardness is 12300 HV. The structure of K 2 coating (thickness is 160 μm) consists of two components: ~ 53 vol. % of WC (tungsten carbide) with hardness of 12,000 HV and ~ 47 vol. % of Ni-Cr-Si mixture with hardness of 8700 HV. The porosity of the coating increases from 0.032% in the center to 0.21% on the surface. K 3 coating structure consists mainly of the Ni-Cr-Si mixture with hardness of 6300 HV, and a small amount (~ 3.98 vol. %) of Al2o3 oxide, which forms conglomerates. K 3 coating porosity is 4.45%, and its thickness is 170 μm.

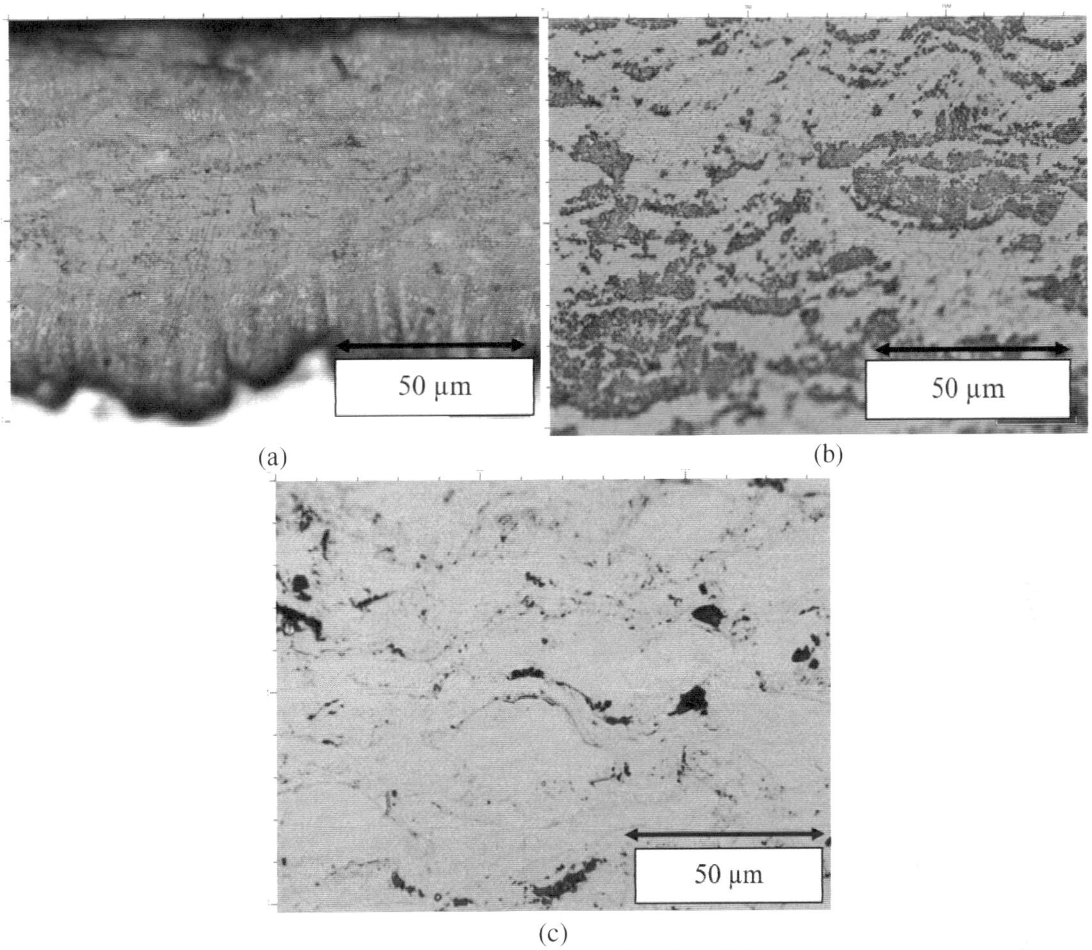

Figure 3. Micro-structure of K 1 (a) . K 2 (b) . K 3 (c) coatings.

Figure 4 shows the micro-structure of the G 1, 2 and 4 coatings. The structure of the G 1 coating consists of two layers: an outer layer (with a thickness of up to 30 μm) consists of WC (tungsten carbide) with a hardness of 11.600 HV, and an inner layer consists of a mixture of the Ni-Cr-Mo system with a hardness of 5.500 HV. The average surface porosity is 0.54% and is concentrated mainly in the inner layer. The total thickness of the coating is 120 μm; the thickness of the outer layer is 25-30 μm. The continuity of this layer is low. The microstructure of G 2 coating (140 μm thick) is homogeneous and consists of a Fe-Cr-Ni mixture (microhardness is 7900 HV; porosity is 0.083%). The microstructure of G 4 coating (326 μm thick) is also relatively homogeneous and consists of WC (tungsten carbide) (hardness is12.900 HV; porosity is 0.76%).

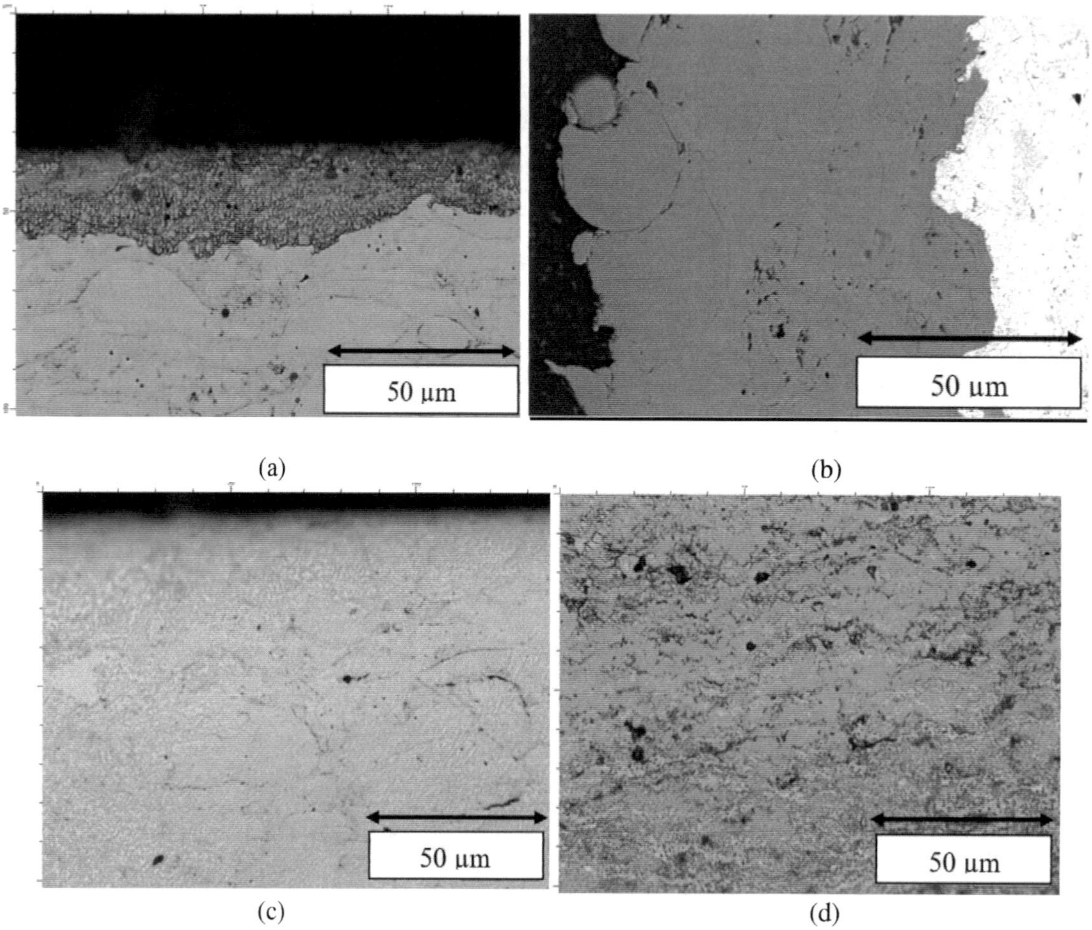

Figure 4. Microstructure of G 1 (a . b) . G 2 (c) . G 4 (d) coatings: optical microscope (a . c . d); electronic microscope (b).

Having compared the results of metallographic analysis (table 3), we can conclude that K 1 coating is the best one according to the studied characteristics. Performance of P 1, K 2, G 2 and G 4 coatings is also sufficient. Due to the low continuity of the outer layer and the low hardness of the inner layer, G 1 coating is not recommended to use in practice.

Table 3. Qualities of the surfaces examined

Coating title	**Coating composition**	**Structural homogenity**	**Porosity, %**	**Microhardness, HV, Pa**
P 1	WC	+	1.94	11300
P 3	WC Ni-Cr-Si	-	3.25	12000 8700
K 1	WC	+	0.04	12300
K 2	WC Ni-Cr-Si	-	0.03	12000 8700
K 3	Ni-Cr-Si	-	4.45	6300
G 1	WC Ni-Cr-Ni-Mo	+	0.54	11600 5500
G 2	Fe-Cr-Ni	+	0.08	7900
G 4	WC	+	0.76	12900

Figure 5 shows the values of electrode potentials of aluminum alloy 2024with and without different coatings. If the aluminum alloy 2024 is not covered with a coating, its electrochemical potential is 580-612 mV, which is more than 100 mV negative than its values for coated alloy 2024. Consequently, the aluminum alloy will be corrosion-soluble at the contact point with the coating. Therefore, in order to determine the contact corrosion rate of the samples mass loss, we only considered the surface area of the aluminum sample not covered with a wear-resistant coating.

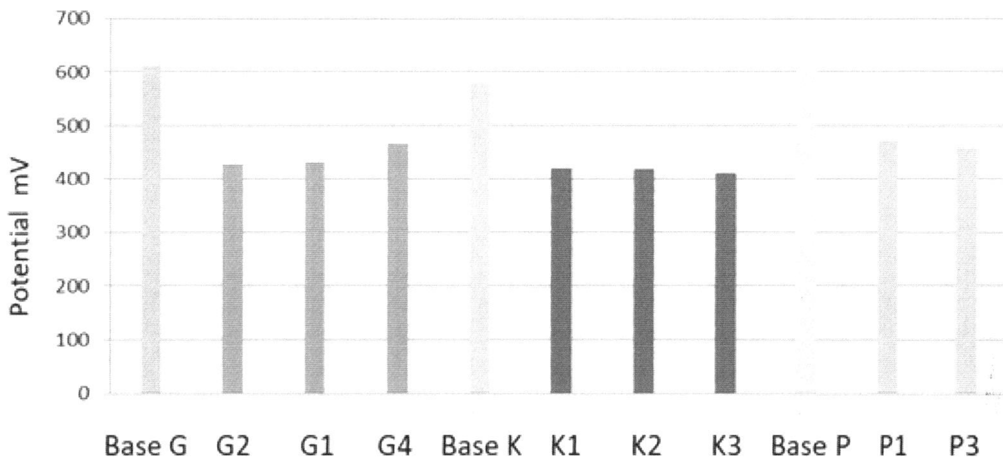

Figure 5. Values of electrode potentials of aluminum alloy 2024 (table 1) with and without different coatings.

The results of the bi-metallic corrosion tests showed that the coatings contribute (at the point of contact with the main metal) to some increase in the corrosion rate of the aluminum alloy 2024 from 0.06 mm/year to 0.08-0.12 mm/year. We can come to a conclusion that a groove up to 1 mm deep may appear at the point of contact with the aluminum alloy coating in 8-12 years.

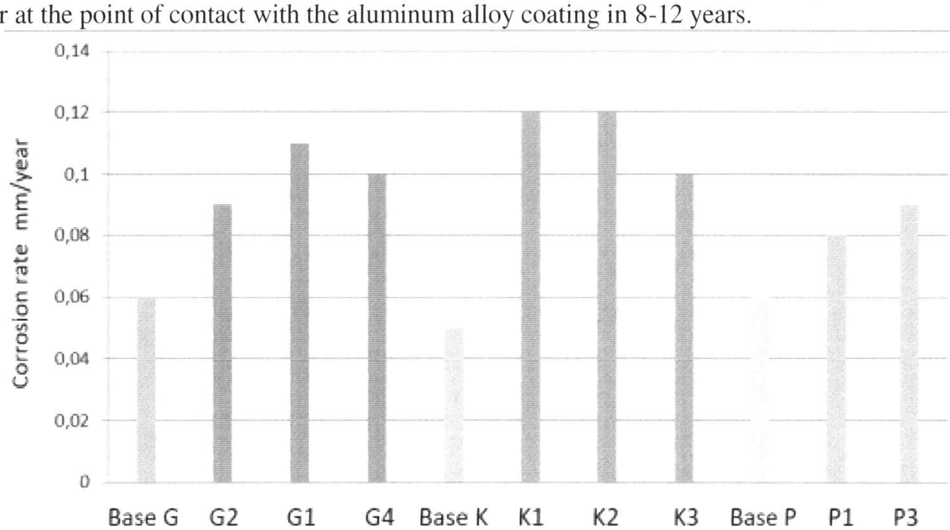

Figure 6. Corrosion rate of aluminum alloy 2024 (table1) with and without different coatings.

Table 4 shows the results of three-point bend tests of coated aluminum alloy 2024. G1 and G2 coatings have the maximum deflection to the fracture formation and the highest breaking stress level of 1040 MPa. P 1, G 4, K 1 and K 3 coatings have average breaking stress level; P 3and K 2 coatings feature the lowest breaking stress level.

Table 4. Results of three-point bend tests of coated aluminum alloy 2024

Coating	Maximum breaking stress level of the coated area, MPa
P1	1000
P3	800
G1	1040
G2	1040
G4	1000
K1	1020
K2	980
K3	1020

Table 5 represents the results of depreciation tests of coated aluminum alloy 2024. It is obvious that P 1, K 1 and G 4 coatings are the most wear-resistant ones.

Table 5. Results of aluminum alloy 2024 wear-resisting properties tests

Coating	Marking	Weight loss, g	Test period, min
P1	P1.1	2.4	240
	P1.2	0	480
	P1.3	0.3	480
P3	P3.1	6.2	120
	P3.2	8.7	480
	P3.3	6.9	480
G1	G1.1	0.2	30
	G1.2	0.7	30
G2	G2.1	12.5	30
	G2.2	19.5	30
G4	G4.1	0	480
	G4.2	0	480
	G4.3	0	480
K1	K1.1	0.1	480
	K1.2	0	480
	K1.3	0	480
K2	K2.1	31.5	30
	K2.2	10.6	30
	K2.3	9.8	30
K3	K3.1	2.2	480
	K3.2	2.2	480
	K3.3	2.2	480

We measured surface roughness with MarSurf 1 profilometer in order to compare specific features of worn out surfaces of coatings. Table 6 shows the results of roughness measurements.

IOP Conf. Series: Materials Science and Engineering **889** (2020) 012021 doi:10.1088/1757-899X/889/1/012021

Table 6. Assessment of surface roughness of aluminum alloy 2024 coatings (after wear tests)

Coating type	Profile chart	Ra . µm
P1	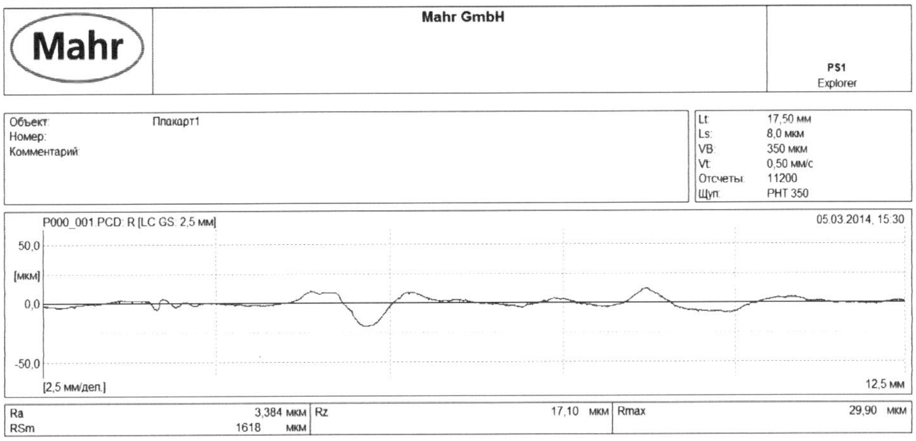	3 .4
K1	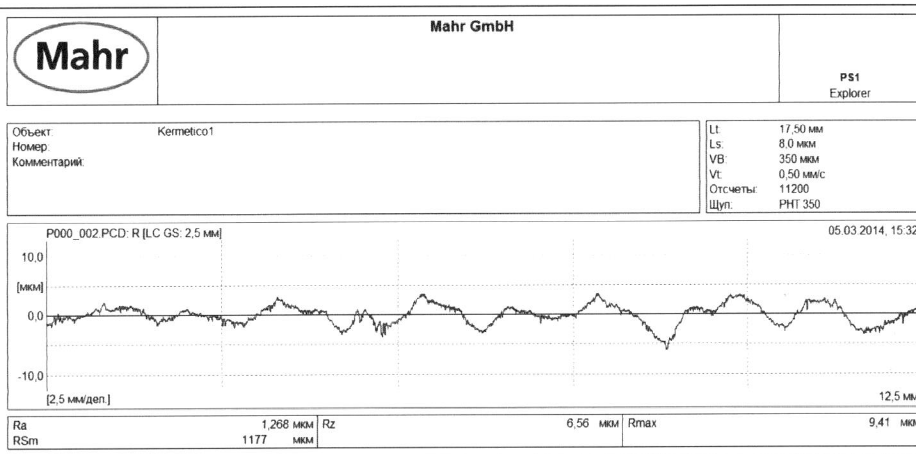	1 .3
G4	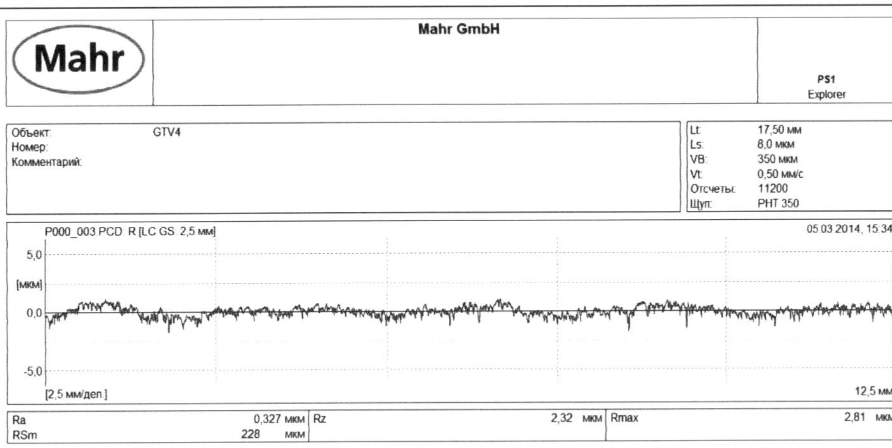	0 .3

G 4 coated alloy samples have the smallest roughness value (Ra). P 1 and K 1 coated alloy samples appeared to have significant local depreciation areas of up to 20 µm.

To summarize the results of the complex study of aluminum alloy 2024 coatings, it is possible to rank their performance on a scale of one to ten (Table. 7).

Table 7. Totals for the operational characteristics study of coated aluminum alloy 2024

Coating	Corrosion resistance	Stress-related properties	Wear-resisting properties	Total
P1	9	9	8	26
P3	9	8	5	22
G1	9	10	1	20
G2	9	10	1	20
G4	9	9	10	28
K1	9	9	9	27
K2	9	8	1	18
K3	9	9	7	25

4. Conclusion

Having carried out a study of the operational characteristics of coated aluminum alloy 2024 to be used as drill pipe material, the authors may conclude the following. All the coatings examined increase the properties of aluminum drill pipes, but the effectiveness of the influence depends on their composition and structure. The corrosion and mechanical properties of the alloy are comparable across all coatings, but their impact on wear resistance varies considerably. To use WC (tungsten carbide) coated aluminum alloy 2024 as a drill pipe material would be the most beneficial for petroleum production. In contrast, WC+Ni-Cr-Si coating is the least recommended one.

5. References

[1] Gelfgat M, Alkhimenko A and Kolesov S 2019 Corrosion and the role of structural aluminum alloys in the construction of oil and gas wells *E3S Web of Conferences* 7

[2] Basovich V S, Buyanovskiy I N and Petunkin I V 2015 Prospects aluminum drill pipe application while drilling horizontal wells great length *Drilling and oil* 81-5

[3] Tan M, Gan L 2009 Equilibrium equations for nonlinear buckling analysis of drill-strings in 3D curved well-bores *Science in China Series E: Technological Sciences* pp. 590–5

[4] Fuks M D, Zelenin Y V and Kondrat'ev S Y 2012 Study of thick-walled pipes of metal quality made of corrosion-resistant steels *Finished products in mechanical engineering* pp. 36-8

[5] Gelfgat M, Chizhikov V, Kolesov S, Alkhimenko A and Shaposhnikov V 2013 Application of aluminum alloy tube semis: Problems and solutions in the development of exploration production and transportation business of hydrocarbons in the Arctic *Society of Petroleum Engineers* - SPE Arctic and Extreme Environments Conference and Exhibition Moscow SPE-**166919**-RU. 20 p

[6] Shakirova A I, Nazarova M N, Ismakov R A, Khafizov A R and Tsenev N K 2015 On promising directions of development of technologies for the manufacture of aluminum drilling pipes for work in the Arctic *Materials for technical devices and constructions used in the Arctic Moscow FSUE ARSRIAM* 1-20

[7] Zakharov V V, Yelagin V I, Rostova T D and Samarina M V 2008 Ways of evolution an improvement of high-strength Al-Zn-Mg-Cu system alloys *Technology of light alloys* 7-13

[8] Kolesov S S, Kondrat'ev S Yu, Chizhikov V V and Shvetsov O V 2011 Investigation of structure and properties of drilling pipes made of aluminum alloy 2024 after exploitation under conditions of oil extraction *Finished production in mechanical engineering* 39-43
[9] Hong L J, Scaruffi P G, Brewster T W and Balka S M 2010 HZ25-4 ERD challenging drilling project–new strategy and new technology implementation are the keys to improved drilling performance *SPE ATCE. 19-22 September*, Florence, Paper SPE **134949** 3604-19
[10] Kondrat'ev S Yu and Shvetsov O V 2013 *Effect of high-temperature heating on the structure and properties of aluminum alloys in the production of drill pipes Metal Science and Heat Treatment* 91-6
[11] Kondrat'ev S Yu, Zotov O G and Shvetsov O V 2014 Structural stability and variation of properties of aluminum alloys 2024 and 1953 in production and operation of drill pipes *Metal Science and Heat Treatment* pp.526-32
[12] Kondrat'ev S Yu, Shvetsov O V and Alhimenko A A 2014 The Changing of mechanical properties of aluminum alloys D16T and 1953T1 in the corrosion environment of oil wells *SPbPU Journal of Engineering Science and Technology* 93-9
[13] Shvetsov O V, Alhimenko A A and Kondratiyev S Yu 2014 Corrosion damages of boring pipes from aluminum alloys D16 and 1953 when operating in the conditions of oil wells *SPbPU Journal of Engineering Science and Technology* 180-91
[14] Kolachev B A, Yelagin V I and Livanov V A *Metal Science and Thermal Processing of Non-ferrous Metals and Alloys* (MISIS Moscow 1999)
[15] Friedlyander I N 2003 Regularity of changes in properties of aluminum alloys at aging *Metal Science and Heat Treatment* pp. 8-11
[16] Gorynin V, Kondrat'ev S and Popov V 2010 Laser modification of the tribological properties of steels and nonferrous alloys *Photonics* 26-33
[17] Kondrat'ev S Yu, Gorynin V I and Popov V O 2011 Optimization of parameters of surface-hardened layer at laser hardening of components *Welding production* 11-5
[18] Kondrat'ev S Yu *Mechanical properties of metals: training manual* SPbPU Saint Petersburg 2011

IOP Conf. Series: Materials Science and Engineering **889** (2020) 012022 doi:10.1088/1757-899X/889/1/012022

Polymeric nanomaterialsbased on iron (III) tris-acetylacetonate: synthesis, structure and properties

N P Shapkin[1*], I G Khalchenko[1] and E K Papynov[1, 2]

[1]Far Eastern Federal University, Vladivostok, Russia
[2]Institute of Chemistry, Far-Eastern Branch Russian Academy of Sciences, Vladivostok, Russia

*e-mail: npshapkin@gmail.com

Abstract. The formation of iron (III) polychelates under different conditions has been investigated: at mechanochemical activation of iron (III) tris-acetylacetonate with subsequent condensation in boiling toluene (method 1) and through chelate heating in air at 100, 150, and 200°C (method 2).The obtained compounds have been studied by the methods of gel permeation chromatography, diffractometry and infrared spectrometry, scanning electron microscopy, and gamma resonance spectroscopy.Based on the obtained data, the maximal particle size has been calculated. The fractal structure of spherical particles was presented.

1. Introduction

Dielectric paramagnetic films based on metal acetylacetonates keep finding an extensive application in electronics [1]. They are used in surface passivation of semiconductor devices parameters stabilization, as shielding coatings, and in creation of film condensers and insulators in multilayer structures [1, 2]. Oxides in the form of homogeneous solid films are deposited from the vapor phase through decomposition of metal acetylacetonates [1, 2].To obtain polymeric nanocomplexesof metals we usedthe interaction of magnesium acetylacetonate containing the hydrate water in synthesis of polymetalorganosiloxanes, a low yield of the target product is caused by the formation of magnesium polyacetylacetonate upon heating[3].Similarly, at the interaction of vanadylacetylacetonate with polyphenylsiloxane [4] with boiling in toluene, a low yield of the final product is observed as well. It depends on the formation of vanadylpolyacetylacetonate ($[(Acac)_xVO]_n$) participating in the reaction only under strict conditions.

To sum up, there arises the necessity to study the behavior of metal acetylacetonates under different conditions: heating in air, mechanochemical impact, and boiling in solvent.

2. Results and Discussion

Iron (III) tris-acetylacetonate polymerization was carried out by two methods. In the first one, the initial chelate was treated mechanochemically, dissolved in toluene, boiled until complete solvent evaporation, and heated in vacuum up to 80°C (10 mm Hg – synthesis 1). In the second one, the chelate was heated in air at 100, 150, and 200°C (syntheses 2, 3, and 4, respectively).

The data of element analysis and gel permeation chromatography (GPC) are shown in Table 1.

One observes essential differences in iron polyacetylacetyonates (IPA) treated mechanochemically with isolation from solution and by heating in air. Here, from the GPC data, the mass of soluble trimers and tetramers somewhat increases at heating up to 100°C for 1 h (Fig. 1c). Further heating up to 150°C under similar conditions results in producing an insoluble fraction with an output of 30 %, in which 1.5 of acetylacetonate group and 0.5 of oxygen atom are per 1 iron atom.

Table 1. Element analysis and gel permeation chromatography data for the obtained materials.

Synthesis	Elements contents, %						
	Found		Calculated		Molecularweight	Generalformula	
	C	M	C	M			
Initial chelate	51.2	16.1	50.9	15.8	≈1090 – 700	$(C_5H_7O_2)_3Fe$	a
№1: mechanochemistry, solution	43.5	20.9	44.4	20.7	mixture ≈ 4000 – 12% ≈ 2100 – 64%	$[Fe(C_5H_7O_2)_2OH]_n$	b
№2: 1 h, 100°C	51.7	14.9	50.9	15.8	mixture ≈ 1500 – 55% ≈ 1000 – 38%	$[(C_5H_7O_2)_3Fe]_n$	c
№3 1 h, 150°C	51.3	14.8	51.0	15.0	mixture ≈ 700 – 55% ≈ 1000 – 38%	$[(C_5H_7O_2)_xFe]_n$	d
	42.3	25.8	42.5	24.5	insoluble fraction, output 30 %	$[Fe(C_5H_7O_2)_{1.5}O_{0.5}]_n$	
№4 1 h, 200°C	51.7	14.3 5	50.5	14.4	mixture ≈ 3200 – 30% ≈ 1000 – 66%	$[(C_5H_7O_2)_xFe]_n$	e
	36.1	29.6	35.3	29.3	insoluble fraction, output 57%	$[Fe(C_5H_7O_2)O]_n$* $2(C_5H_7O_2)_n$	

The data of gel permeation chromatography for oligomers from the synthesis 1 are shown in Figure 1b.

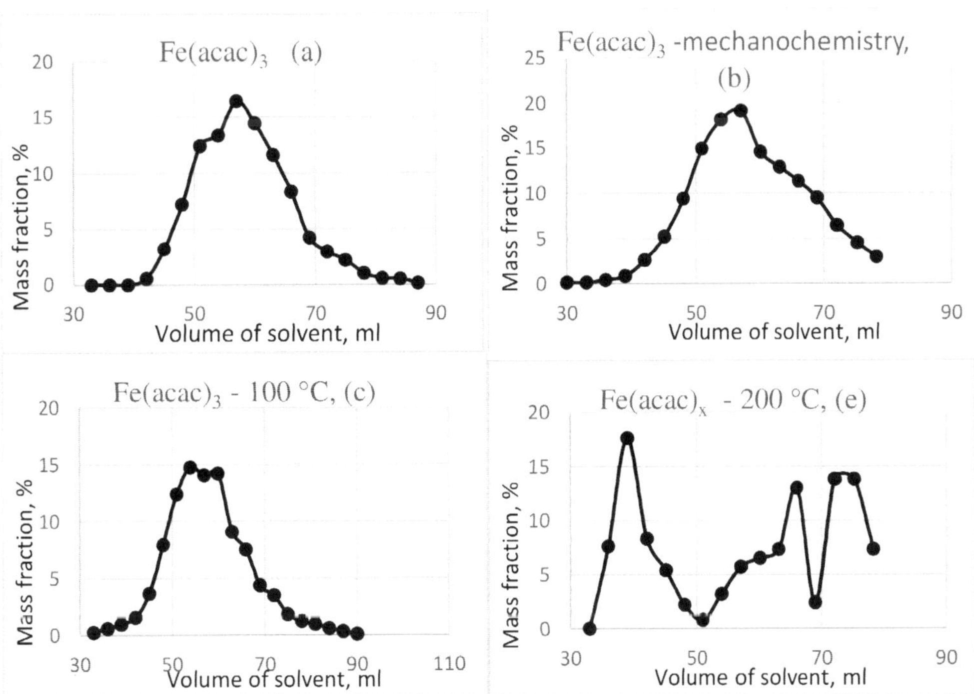

Figure 1. Gel permeation chromatograms: a – initial chelate; b – synthesis 1; c – synthesis 2; d – synthesis 4.

Heating up to 200°C (synthesis 4) under similar conditions results in the increase of the quantity of the insoluble fraction up to 57 %, where 1 iron atom corresponds to 1 acetylacetonate group and 1 oxygen atom. There are also end groups: hydroxyl groups or water molecules. The soluble part (synthesis 4) comprises a mixture of trimers, dimers, and oligomers with n of 9–10 (Fig. 1d).

The IR spectroscopy data corroborate the suggested general formulas. All the spectra contain absorption bands at 1570–1571 cm^{-1} and 1524–1526 cm^{-1} corresponding to C=O and C=C bonds vibrations in the acetylacetonate ring, respectively. The absorption bands at 2922–2924 cm^{-1} correspond to stretching vibrations of the C–H bond of the methyl group. The most informative is the band at 435–436 cm^{-1} corresponding to stretching vibrations of the Fe–O bond [5].

The absorption band corresponding to vibrations of the Fe–O bond in the initial chelate is at 435 cm^{-1}, while in the course of polymerization it shifts to higher frequencies, for the oligomer (synthesis 1) it is at 435 cm^{-1}, and for insoluble fractions – at 441 cm^{-1}.

The obtained oligomers were studied by X-ray diffraction analysis (Figure 2).

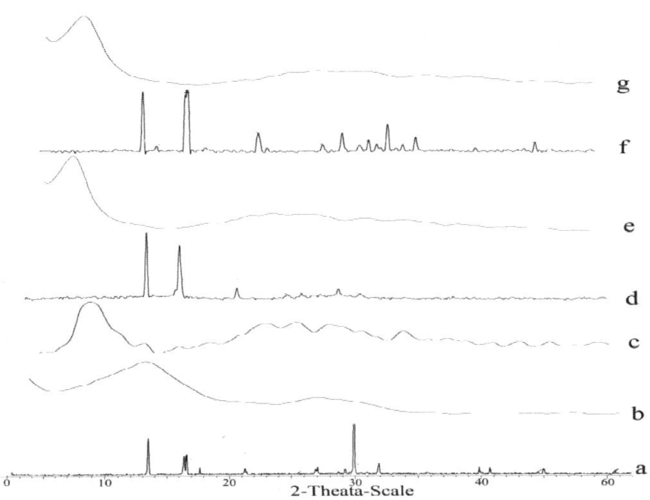

Figure 2. Diffractograms of: a – iron tris-acetylacetonate; b – oligomers obtained in solution (synthesis 1); c – product of the synthesis 2; d – soluble fraction of the synthesis 3; e – insoluble fraction of the synthesis 3; f – soluble fraction of the synthesis 4; g – insoluble fraction of the synthesis 4.

The diffractograms of the oligomer from the synthesis 1 (Fig. 2b) corresponds to the structure of a mesomorphous polymer. The soluble fractions of the syntheses 2–4 comprise mixtures of crystalline forms of dimers, trimers, and tetramers of iron acetylacetonates (Figs. 2d,f). The diffractograms of the insoluble fractions of the syntheses 3 and 4 (Figs. 2e,g) correspond to amorphous products. The diffractogram of the soluble polychelate at 200°C in air corresponds to the structure of a mesomorphous polymer.

According to [6,7], in case of mesomorphous polymers, the first reflection corresponds to the distance between adjacent chains. Using the calculation methods suggested in [7], the volumes of coherent scattering regions (CSR) and the cross-section areas were calculated on the Selyakov–Shearer formula[8]. The calculated data are shown in Table 2.

Table 2. Calculated physical-chemical characteristics from the data of X-ray diffraction analysis.

Compounds	d, 100 Å	L_{CSR}, Å	S, Å^2	V_{CSR}, Å^3
Initialchelate	8.31	1150.0	58.8	67620.0
Oligomer, synthesis 1	8.48	10.9	60.9	663.8
Oligomer, synthesis 4	9.26	13.25	70.5	930.6

The coherent scattering region size for the initial crystalline chelate is two orders of magnitude larger than that for the oligomer with n = 8–9. One can assume that in the case of the chelate crystalline form the CSR size is that of the crystallite (0,12 μm).

To obtain more accurate notions on the structure of polymeric forms of $[(C_5H_7O_2)_xFe]_n$, geometric parameters and theoretical cross-section areas of its tris-chelate and polychelate were calculated.

The theoretical calculations were carried out using the FireGly 8.1.0 software package [9] partially based on the initial GAMESS code (US) [10]. Calculations were carried out by the method of density functional theory (DFT), whereas to describe the electron interaction the hybrid exchange-correlation B3LYP5 functional was used (basis 6-31G) [11]. The initial complexes geometries were taken from the experimental data, prior to the optimization procedure.

The spectra were processed using the CASA XPS program. The energy scale was calibrated according to the internal standard technique – the C1s level was selected (285,0 eV).

In comparison with $Fe(C_5H_7O_2)_2$, in the $[(C_5H_7O_2)_xFe]_n$ complex, one observes the increase of interatom distances and angles. The largest cross-section area of the initial $Fe(C_5H_7O_2)_2$ complex is 57.93 Å^2, while that of the $Fe(C_5H_7O_2)_2^+$ complex – 56.6 Å, according Figure 3.

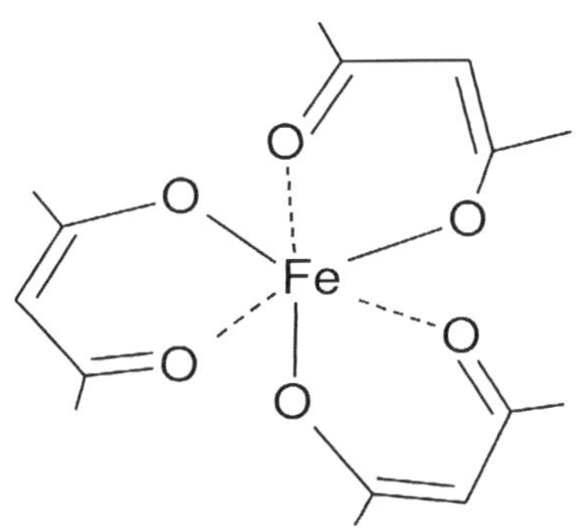

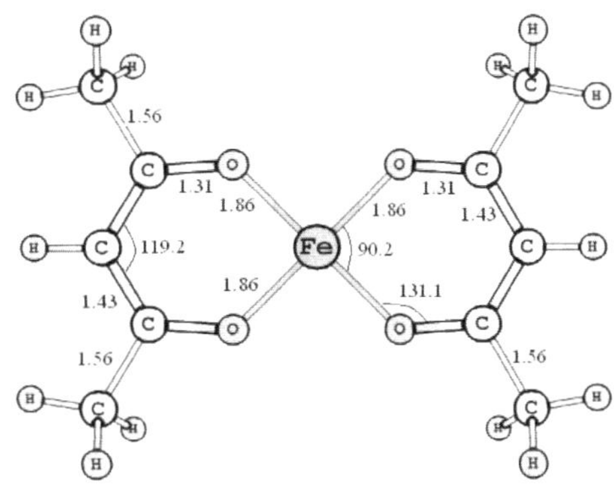

a b

Figure 3. Thestructures:a) $Fe(C_5H_7O_2)_2$; b) $[(C_5H_7O_2)_xFe]_n$ complex.

Based on the element analysis (Table 1), GPC, diffractometry, and calculated data (Table 2), the structure of the $[(C_5H_7O_2)_{16}Fe_8{*}2OH]_n$ oligomer with MW 2100–2200 Da was revealed (Figure 4).

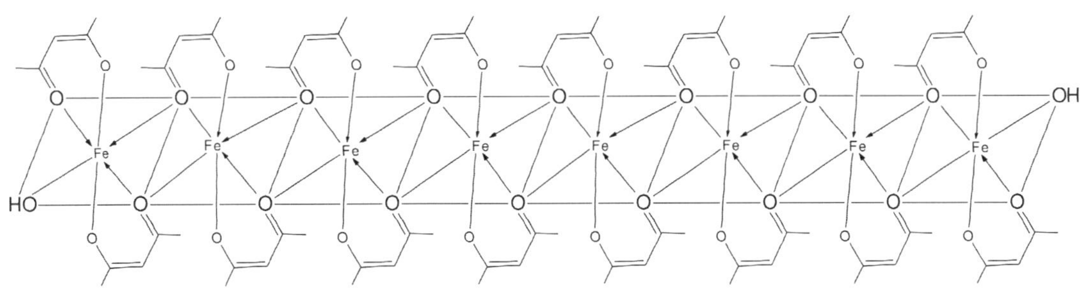

Figure4. Oligomer structure (synthesis 1).

The oligomer with the molecular weight of 2066 (calculated data) and 2100 (found value) is well soluble in organic solvents. Unlike the polychelate obtained through heating in air, in which a significant loss of acetylacetonate groups takes place, in the former case, the iron chelate transforms into a spatial insoluble polymer due to formation of oxygen bridges (150-200 °C).

Upon heating up to 200°C, further loss of acetylacetonate groups takes place and the number of oxygen bridges increases dramatically. The element analysis confirms the formula $[Fe_8(C_5H_7O_2)_8O_{18}]_n$.

Since the structure of solid-state samples of iron polychelates is of practical importance for their further application as magnetic materials, to study magnetic properties, the following samples were

IOP Conf. Series: Materials Science and Engineering **889** (2020) 012022 doi:10.1088/1757-899X/889/1/012022

taken: initial chelate, polychelate obtained in solution (sample 1, Table 1), polychelate obtained at heating up to 200°C in air (sample 4). Samples were studied using an American vibrating type magnetometer LakeShore VSM 7410.

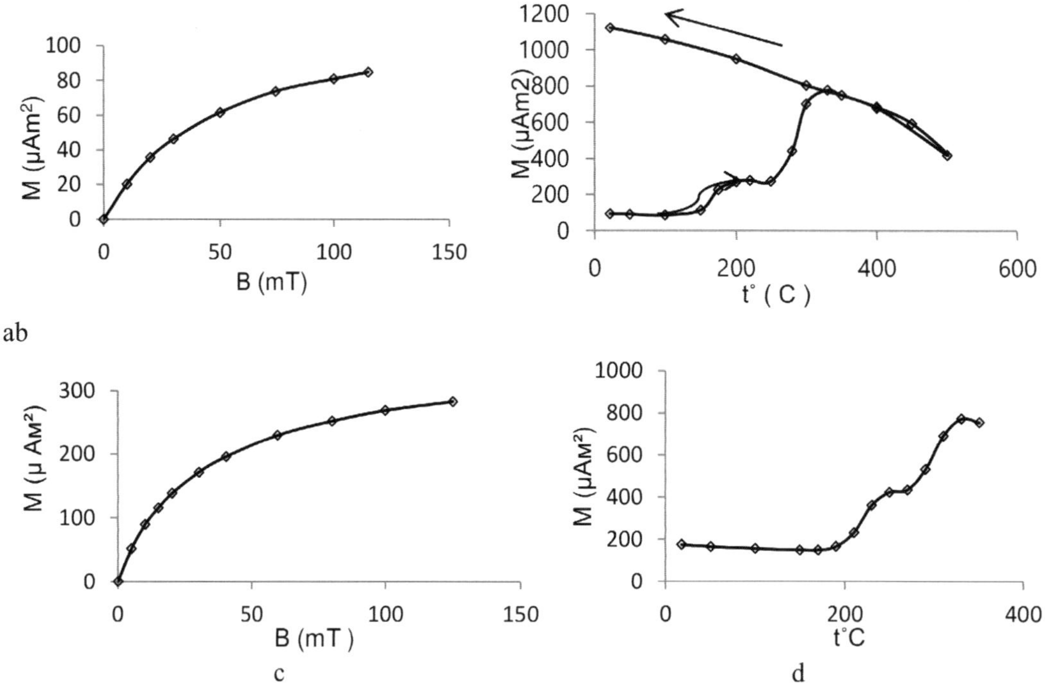

Figure5. a) Dependence of the sample inductive magnetic moment on the induction of the applied magnetic field of the sample 1 in the initial state; b) Dependence of the saturation magnetic moment on temperature for the sample 1 (→ heating, ← cooling); c) Dependence of the inductive moment on induction of the applied field for the sample 4 ($[Fe(C_5H_7O_2)_x]_n$); d)Dependence of the inductive moment of $[Fe(C_5H_7O_2)_x]_n$ (sample 4) in the field of 100 mT on temperature.

The curve is characteristic of ferrimagnetics (Figure 5 a-d): at low field intensities – the area of fast moment changes related to shifting boundaries between domains; at high intensities (from 40 to 100 mT) – slow changes in the magnetic moment related to rotation of spin moments of iron atoms, closeness to saturation. From the moment close to saturation, one can estimate the magnetic moment per iron atom: at the iron content of 21.0 % and the density of 1.46 g/cm^3.The decrease of the Bohr magneton number for the samples 1 and 4 (Table 3) is determined by the polymeric structure – formation of oxygen bridges, whose role belongs to oxygen atoms of the polyacetylacetonate group.The authors of [12] observed similar changes in the magnetic moment for bridge groups of the type below.

Table3. Magnetic studies data.

Substance, sampleno.	Initialchelate	Iron polyacetylacetonate (sample 1)	Iron polyacetylacetonate (sample 4)
Bohrmagnetonnumber	6.0	4.32	0.33

The data of Table 4 were obtained from the Mössbauer spectra of the initial chelate and the samples 1 and 4.

Table 4. Physico-chemical characteristics of polymers calculated using with the help of Mössbauer spectra.

Substance	δ	ΔE, mm/s
Initialchelate	0.36	0.365
Sample1	0.40	0.395
Sample 4	0.44	0.55
	1.38	2.77

The data of GR spectroscopy (Table 4) show that polyacetylacetonate obtained in solution (sample 1) is characterized with smaller isomeric shift and lower ΔE value in comparison with the sample 4 heated in air up to 200°C. Moreover, for the sample 4, there emerges a doublet corresponding to the divalent iron atom with the isomeric shift of 1.38. Thus, according to the above data (Table 4), an oligomer of more regular structure is formed in the solution.

The sample 1 morphology was studied by the method of electron microscopy (Figure 6).

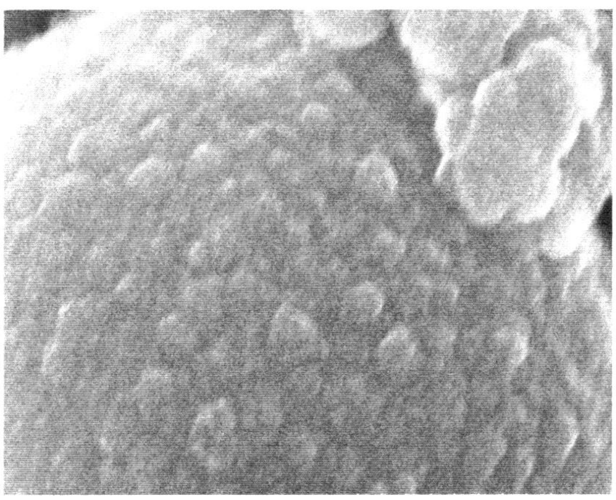

200 nm

200 μm

Figure 6. Scanning electron microscopy images of polychelate (sample 1).

Iron polyacetylacetonate (sample 1) in solid phase comprises a pack of spherical particles of various diameters – sizes vary from 200 to 800 nm.

Physical-chemical characteristics of polyacetylacetonates (samples 1, 4) (S_{spec}, V_{spec}) were determined from low-temperature adsorption. The obtained data are shown in Table 5.

Table 5. Physical-chemical characteristics of polyacetylacetonates.

Sample№	S_{spec}, m²/g	V_{spec}, cm³/g	D_{pore}, nm	ρ, g/cm³
1	0.92	0.003	3.0–4.0	1.46
4	1.25	0.0022	3.0–4.0	1.32

The found values of density and specific surface area (Table 5) allow determination of the layer of oligomer spherical particles (sample 1) by method [13].

First, the substance specific volume is calculated, then the obtained value is divided to the specific surface area:

$$L(layer) = \frac{1/\rho}{S} \quad (1)$$

where: L_{layer} – the layer thickness (nm, Å); ρ – the substance density (g/cm³); S – the specific surface area measured on nitrogen (nm).

The layer thicknesses calculated in this way are equal to 740 nm (sample 1), 600 nm (sample 4). The found values are in agreement with the scanning electron microscopy data (Fig. 6).

The values L_{layer} = 740 nm (sample 1) and 600 nm (sample 4) correspond to the maximal size of a spherical particles (Fig. 6) while, as can been at sufficient magnification, these large particles consist of those of smaller sizes. The final smallest size of spheres corresponds to a chain of the oligomer formed by three CSR volumes linked to each other, analogy in work [14] this is fractal structure.

3. Conclusions

Nanomaterials based on iron (III) tris-acetylacetonate have been fabricated by two methods. The first one included mechanochemistry and condensation in boiling toluene solution, the second one – heating up to 200°C. The difference between them consisting in more ordered structure of the oligomer obtained in solution (method 1) was demonstrated by the methods of XRD, gel permeation chromatography and gamma resonance spectroscopy. The cross-section values for oligomer molecules were found from the XRD data (60 Å), calculated from the X-ray structural analysis data (59 Å). The volume CSR data for the initial chelate found from the XRD (67620) data enabled us to conclude that the suggested approach to studies of solid oligomeric iron oxides was rather efficient. The first method is more promising in deposition of iron (III) oxides on the substrate surface.

References

[1] RazuvaevGA, GribovBG, DomrachevGA, SalamatinBA1972*Organometallic Compounds in Electronics*(Moscow: Nauka)p 479

[2] SiddiqiMA, SiddiqiRA, AtahanB2007*Surf. and Coat. Techn.***201** 9055-9059.

[3] ShapkinNP, TokarEA, RazovVI, GardionovSV, KuryavyiVG, MayorovVY, PapynovEK2019*Silicon***1**12283-2292

[4] ShapkinNP, KapustinaA A, GardionovS V, Khal'chenkoI G, LibanovV V, TokarE A2019*Silicon* **11** 2261-2266

[5] Diaz-AcostaJ, BakerJ, HintonJF, PulayP2003*Spectrochim. Acta* Part A **59**363–377

[6] JaberM, Mieche-BrendleJ, PouxM, DentzerJ, Le DredR, Louis GuthJ2002*New. J. Chem.***26** 1597–1600

[7] GardionovSV, Shapkin N P,Razov V I,BalanovMI, VasilyevaVV, TrukhinVO2016*J.Struct. Chem.***57** 507–511

[8] MillerRL, BoyerRF1984*Polymer. Sci. Polym. Phys. Edition.* **22** 2043–2050

[9] GranovskyAA2012*J. Chem. Theory Comput.* **8** 4069–4080

[10] SchmidtMW, BaldridgeKK, BoatzJA1993*J. Comput. Chem.***14** 1347–1363.

[11] StabnikovPA, PervukhinaNV, BaidinaIA, SheludyakovaLA, BorisovSV2007*J. Struct. Chem.* **48** 186–192

[12] WuChS, RossmanGR, GrayHB, HammondGS, SchugarH1972*J. Inorg. Chem.***11** 990–994

[13] Shapkin N P,Razov V I,Khal'chenkoI G,MayorovVY, ShkuratovAL, TrukhinVO 2015*Appl. Mechan. and Mater.* **752** 257–262

[14] Shapkin N P,Razov V I,KorochentsevVV, TokarE A, GardionovSV, PanasenkoAE, Khal'chenkoI G,BalanovMI, SlobodyukAB2017*J. Molec. Struct.***1145** 300–308

Effects of «simplest» post-treatment techniques on the composition of porous silicon surfacc

A S Lenshin[1], O V Chernousova[2], P V Seredin[1], K A Barkov[1] and D A Minakov[1]

[1] Voronezh State University, 1 UniversitetskayaSq., Voronezh, 394006, Russia,
[2] Voronezh State Technical University, 84 20-letiya Oktyabrya Str., Voronezh, Russia, 394006

lenshinas@mail.ru

Abstract. The article is concerned with the investigation of techniques of post-treatment of porous silicon. It was shown that the applied techniques have considerable effect on por-Si nanowires shell without a considerable influencing on their core. To study the features of changing the phase composition and the composition of chemical bonds of the surface of porous silicon, we used infrared spectroscopy, Ultrasoft X-Ray Spectroscopy and photoluminescence spectroscopy.

1. Introduction

Complicated composition of porous silicon surface including both crystalline and amorphous silicon, silicon dioxide and sub-oxides as well as adsorbed hydrogen and hydroxyl groups is an advantage and at the same time disadvantage of this material [1,2]. On the one hand, a wide set of possible morphology and composition characteristics opens up great possibilities for utilization of porous silicon in a lot of applicable problems, but on the other hand, it provides problems in a choice of the suitable technique for a certain practical application. In addition, this puts on limitations for its storage conditions and the following technological operations with this material [2].

For example, now a number of works are actively performed in the field of materials design in order to obtain a targeted delivery of medications where researchers try to introduce porous silicon nanoparticles into the living organisms or separate cells. It is important to note that the cells and organs do not experience toxic effect of silicon particles [3-8].The surface of these particles can be inoculated with the molecules named bio-markers which are used in the diagnostics of the analyzed cells, as well as the medications applied for the recovery of the faulted cell functions. However, a high surface activity of porous silicon that allows adsorption of certain active molecules to its surface at the same time stipulates instability of the surface properties for this material[3-8].

Therefore, an important task in the work with porous silicon and nanopowders on the basis of this material is the controllable stabilization and further functionalization of its surface. So, in order to solve these problems several techniques were proposed: thermal annealing of porous silicon in the oxygen medium, carbonization, treatment in certain acids. However, these techniques are not applicable in the cases when the results of post-treatment lead to the formation of the phases that are not characteristic for the source material or to the appearance of the new morphological features or either to the adsorbed products of reactions on their surface. For example, surface treatment under increased temperature is known to result in the appearance of additional mechanical stresses in the porous layer and to its cracking. In its turn, this can result in the degradation of the quality for the films grown on the structure at the further technological operations or in a decrease of the size of the synthesized nanoparticles [2,9-12]. This work is intended to demonstrate that even the simplest post-treatments can provide a significant impact on the composition of porous silicon surface and it luminescence characteristics.

2. Materialsandmethods

Porous silicon samples were obtained by electrochemical etching of single-crystalline silicon of KEF (P- doped) (100) type with a resistivity of 0,2 Ohm*cm. Solution on the basis of fluoric acid with the addition of isopropyl alcohol and hydrogen peroxide was used as an etchant for the ratio of the components of 2:2:1, respectively. Standard post-treatment of the samples applied for removal of the reaction products and a partial passivation of the surface included rinsing in the distilled water and soaking in isopropyl alcohol for 5 minutes in both procedures.

Below, we present investigations of the effect of the simplest, or, in other words, "basic" techniques for the post-treatment of porous silicon on the composition of its surface without any temperature exposure or reactants that were not used for producing of porous silicon. Technique for post-treatment of the surface was chosen on the basis of the general ideas concerning the etching mechanism while obtaining of the porous silicon.

Thus, trying to perform additional oxidation of porous silicon surface according to the reaction $Si + 2H_2O \rightarrow SiO_2 + 2H\uparrow$ additional treatments were implemented by soaking in the distilled water and 30% solution of hydrogen peroxide, while in order to decrease the fraction of surface oxides according to the reaction of $SiO_2 + 6HF \rightarrow H_2SiF_6 + 2H_2O$, chemical sub-etching in the alcoholic solution of fluoric acid was selected as $HF:C_3H_7OH$ 1:5. Technique of production and the steps of post-treatment are presented in details in Table 1.

Table 1. Technique for obtaining the samples and post-treatment parameters.

Series, designation	Type of the plate of single-crystalline silicon, current density, mA/cm^2	Post-treatment, order of the process, solution	Time of post-treatment over the steps, min.
stand.		1)H_2O 2) C_3H_8O	5+5
stand. +H_2O		1) H_2O 2) C_3H_8O	15+5
stand.+H_2O_2	KEF-0.2 (100), 25	1)H_2O 2) H_2O_2 3) C_3H_8O	5+10+5
stand.+HF		1)H_2O 2)$HF:C_3H_8O$ 3)C_3H_8O	5+10+5

To study the influence of post-treatments on the composition of porous silicon the following analytical techniques were applied: IR, photoluminescence (PL) spectroscopy as well as ultrasoft X-ray emission spectroscopy (USXES). Chemical bonds in porous silicon were analyzed with the use of infrared Fourier-spectrometer Vertex 70 (Bruker) applying an attachment for attenuated total reflection spectroscopy. Using X-ray spectrometer-monochromator RSM-500 by USXES technique phase composition of the samples in the near-surface area of the samples was studied (depth of analysis was of 60 nm). Luminescence spectra were registered with fiber-optics spectrometer USB4000-VIS-NIR (Ocean Optics), connected to the computer. Laser diode radiating at the wavelength of 405 nm was utilized as a source of excitation.

3. Results and discussion

According to SEM data the investigated post-treatment processes did not result in the noticeable changes in morphology of the samples. SEM images of the cleavage plane and surface of the samples are presented in figure. 1.

a) b)

Figure 1. SEM images of: a) cleavage plane and b) surface of mesoporous silicon obtained accordingto the "standard" technique (sample Standard, analogue of the sample №3).

IR-transmission spectra for the samples of porous silicon are presented in figures 2-4. They were obtained after 3 and 45 days after preparation and post-treatment of the samples (functionalization). All this time the samples were stored in the laboratory conditions within the closed plastic packets.

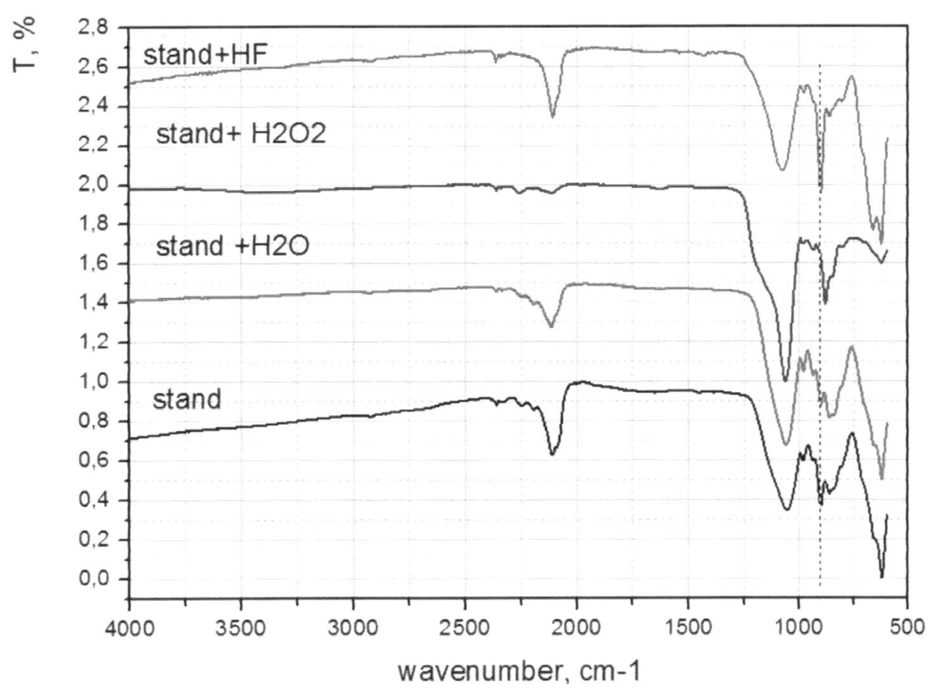

Figure 2. IR-transmission spectra of the porous silicon samples (three days after their obtaining) with different kinds of post-treatment.

IOP Conf. Series: Materials Science and Engineering **889** (2020) 012023 doi:10.1088/1757-899X/889/1/012023

According to the IR-spectroscopy data after 3 days of storage in the atmosphere in the spectrum of "standard" sample spectral bands can be observed characteristic for porous silicon, attributed to Si–Si and Si-H bonds (616 and 634 cm^{-1}), the band at 750-1000 cm^{-1}, corresponding to various kinds of Si-H and O_xSi_y bonds (see figure 3); one can also see oxide composite peak Si-O-Si (1060-1250 cm^{-1}) and the band at 2150-2250 cm^{-1}, corresponding to different kinds of vibrations attributed to the bonds of adsorbed hydrogen and oxygen-containing groups (2084 cm^{-1} Si-H stretching longitudinal, 2110 cm^{-1} Si-H_2 stretching longitudinal, 2200 cm^{-1},SiH-SiO_2 structural vibrations, 2245 cm^{-1} O_3SiH valence vibrations) [13].

Post-treatment of the samples in water and in hydrogen peroxide resulted in the additional oxidation of the samples as compared with the standard technique of obtaining the samples. This is observed as an increase of intensity of the absorption band corresponding to Si-O-Si bonds as compared with that one attributed to Si-Si bonds. Moreover, post-treatment of the samples in water and hydrogen peroxide, as can be expected, resulted in a decrease of the percentage of Si-H_x bonds on the surface of pores and a corresponding increase of the number of oxygen-containing O_xSiH_y bonds. The treatment in hydrogen peroxide undoubtedly resulted in stronger oxidation of porous layer. Post-treatment in the fluoric acid solution, on the contrary, diminished the amount of oxide on the surface. This decrease was accompanied by an increase of intensity for the bands corresponding to the different types of Si-H bonds figure 3.

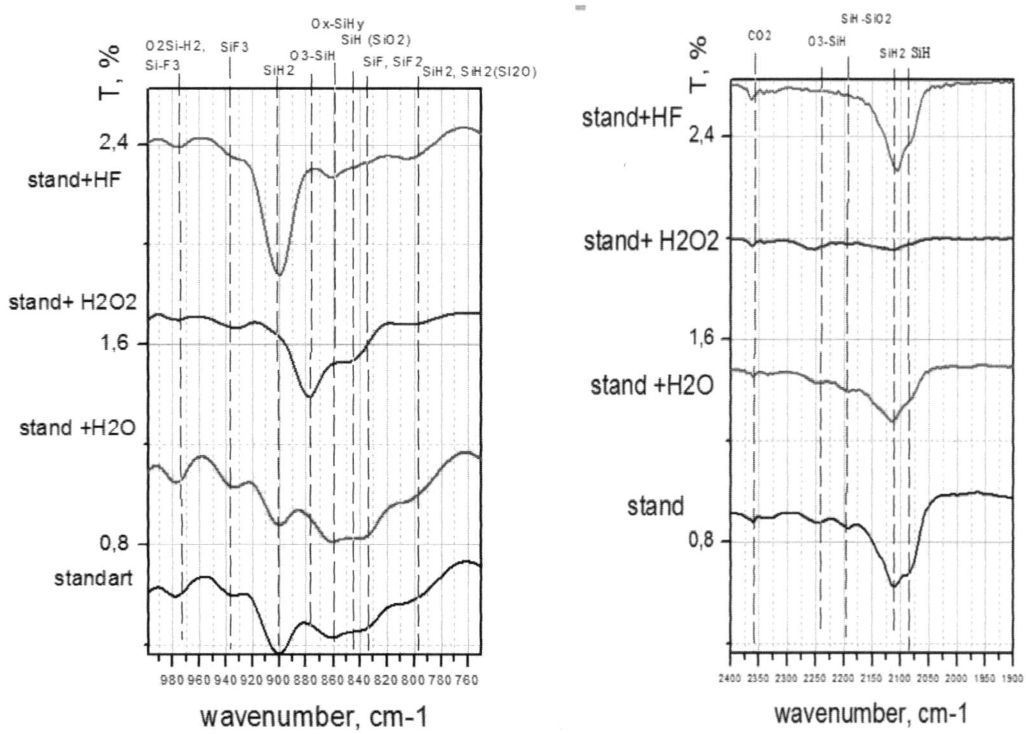

Figure 3. IR-transmission spectra of porous silicon samples with different kinds of post-treatment (3 days after obtaining of the samples)

1.5 months later under exposure in the laboratory conditions a common tendency to oxidation was observed as for the standard sample, as for the samples after post-treatment procedures (figure 4). It is manifested as a decrease of intensity for the absorption band attributed to Si-Si and Si-H bonds (600-650 cm^{-1}) and increase of intensity for the band corresponding to Si-O- Si bonds (1000 -1300 cm^{-1}). Moreover, for all of the samples during their natural ageing a decrease of different variants of Si-Hx bonds and increase of the number of oxygen-silicon bonds of O_xSiH_y type can be seen. It is observed as a considerable change in the shape of absorption bands in the spectral intervals of 750-1000 cm^{-1}

and 2050-2300 cm^{-1}. It should be noted that the shape of the spectrum for the samples treated in hydrogen peroxide changed less than those ones for other samples after 1.5 months of exposure in the atmosphere.

Electron structure of the valence band in porous semiconductors was studied using ultrasoft X-ray emission spectroscopy (USXES) technique. It enables to obtain information on the occupied electron states in the materials with rather high energy resolution.

In order to simulate phase composition of the investigated porous silicon samples X-ray emission spectra were represented as a sum of Si $L_{2,3}$ spectra of the reference phases involving silicon: single-crystalline silicon c-Si, amorphous hydrogenated silicon a-Si (a-Si:H), and two kinds of silicon oxides – silicon sub-oxide $SiO_{1.3}$ and silicon dioxide SiO_2. All of the spectra for the reference substances were retrieved from our previous work [13]. X-ray spectra of the investigated structures were obtained 3.5 months after preparation of the samples and their post treatment under accelerating voltages at the X-ray tube anode of 3 kV that provided the depth of analysis of about 60 nm. The data obtained in these experiments are presented in Table 2 and in the figure 5.

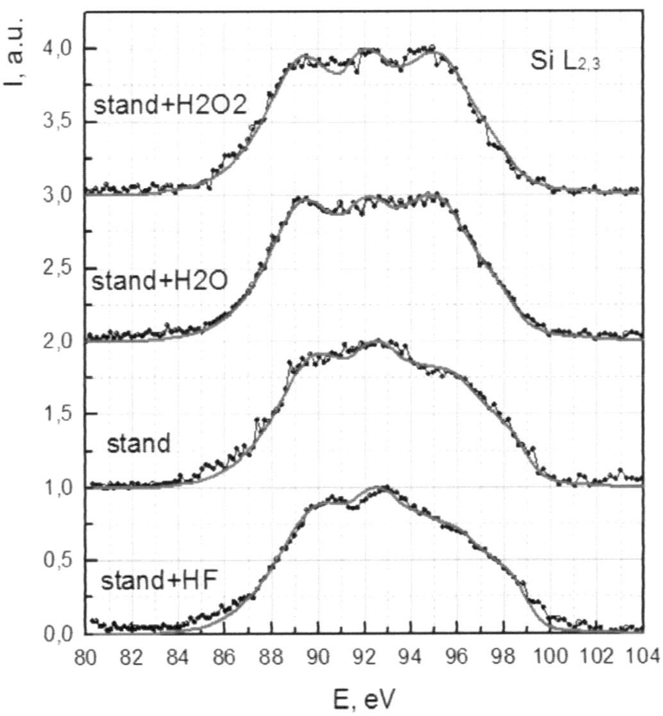

Figure 5. USXES Si $L_{2,3}$spectra of porous silicon samples with different kinds of post-treatment (3.5 months after obtaining of the samples). Black line – experiment, red line – fitting according to [14,15].

Table 2. Distribution of the components in the phase composition as their percentage in the porous silicon samples with different kinds of post-treatment (functionalization).

Sample	**c-Si, %**	**a-Si:H, %**	**SiO_x, %**	**SiO_2, %**	**Error, %**
stand.	37	32	9	22	9
stand. +H_2O	37	11	18	34	8
stand.+H_2O_2	30	20	0	50	8
stand.+HF	32	45	0	23	10

USXES results are in good agreement with IR spectroscopy data. Basing on the problem definition including the change of surface layer composition utilizing post-treatment procedures, namely, to attain an additional oxidation or, on the contrary, to remove a part of oxide layer from the surface of porous layer, after analysis of the USXES data (depth of analysis is about 60 nm) one can conclude the following. Soaking of the samples in the distilled water and in the solution of hydrogen peroxide results in the change of the ratio of the oxidized (SiOx +SiO_2) and non-oxidized phases (c-Si +a-Si:H) in the porous layer from ~ 30%:70 % to 50%:50%. Note, that the treatment of the sample in hydrogen peroxide leads to the formation of the phase of silicon oxide with a greater degree of oxidation (SiO_2) on the surface of silicon wires in por-Si. Treatment of por-Si surface in the solution of fluoric acid, on the contrary, leads to a decrease of oxides contribution into composition of the porous layer up to ~ 20:80, as a result of the dissolution of defect silicon oxide with the formation of a-Si:H. It should be specially noted that the use of post-treatments in our investigations does not in fact change (with the account of errors of simulation) the percentage of crystalline silicon within the porous layer of the samples. Basing on the quite well-established ideas on the phase composition of por-Si, where the fragments (wires, columns, arrays of nanoparticles) are distinguished by the presence of crystalline core inside, covered by the layer of amorphous silicon with a gradual transfer to sub-oxide and dioxide of silicon in the ratios depending on the technique of production. Applying the presented post-treatment procedures, we can control the composition of the core "shell" of the nanostructures without a noticeable effect on the "core". This result can be of a great importance for different practical applications of porous silicon. This behavior was also observed in the porous layer for the greater depth of analysis (120 nm). In all of the samples the contribution of non-oxidized phases into the phase composition was expected to be enhanced with depth of analysis [14].

Photoluminescence spectra (PL) of the porous silicon samples were measured with the use of fiber-optical spectrometer USB4000-VIS-NIR (350-1000 nm) produced by OceanOptics company. Photoluminescence spectra excited by the radiation with λ = 405 nm are presented in figure 7 were obtained 1.5 months later after preparation of the samples.

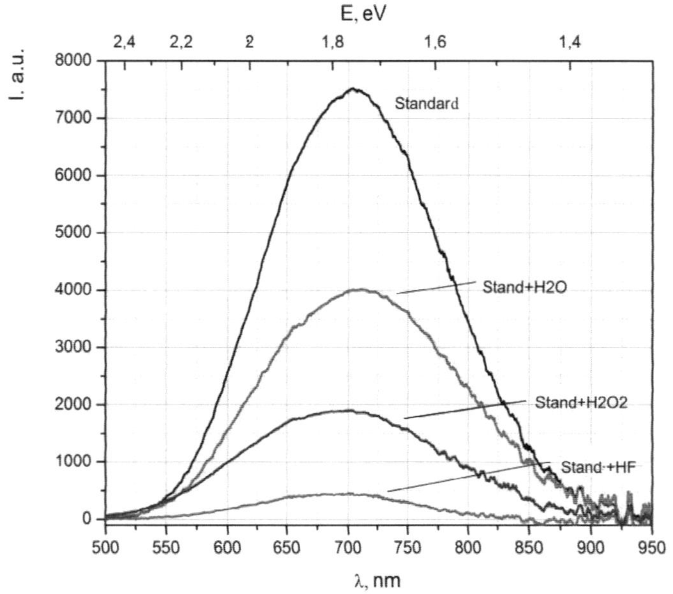

Figure 6. Photoluminescence spectra of porous silicon samples before and after post-treatment.

According to PL data the utilized post-treatments resulted in a decrease of intensity of the photoluminescence band in the primary porous silicon samples while its position and bandwidth is kept invariable. In addition, basing on the conventional ideas concerning phase composition of porous silicon and quantum-size model of its PL it is also possible to make a conclusion that the applied experimental techniques have no serious effect on the mean size of silicon nanocrystallites forming the "core", responsible for the bandwidth and position of PL peak. However, they have a significant impact on the "shell" consisting of amorphous and oxide phases. An increase of the "shell" contribution to the phase composition of the investigated silicon nanostructures, presented in the form of amorphous (hydrogenated) silicon and in the form of silicon sub-oxides and dioxide, leads to a considerable decrease of PL intensity. This result is in a good agreement with the ideas on the negative impact of the defects in silicon sub-oxide and hydrogen bonds on PL properties of por-Si, resulting in the formation of a great number of non-radiative recombination centers. Thus, one can conclude that in order to improve PL properties of porous silicon in the visible and near-IR spectral range it is necessary to form silicon nanostructures consisting mainly of the (nano)crystalline core with a thin shell of the perfect oxide SiO_2.

4. Conclusions

Freshly obtained porous silicon demonstrates instability of the chemical composition: surface of the pores is covered with hydrogen weakly bound to silicon atoms. At the same time Si-OH complexes are also formed in the while etching of crystalline silicon. Later on weakly bound hydrogen starts to release from the pores while the dangling unstable bonds of silicon are passivated by the atmospheric oxygen. The interaction of the surface silicon atoms with water vapour from the atmosphere is also a common reaction when keeping of por-Si samples in the atmosphere. The processes occurring in porous silicon layer is considered as ageing. Ageing of por-Si is difficult to control and regulate. That is why post-treatment procedures are highly recommended after producing of porous silicon.

It was found that even simplest post-treatment techniques can significantly change the chemical composition of silicon nanowire shell. Increase of silicon oxide and amorphous phase content in the por-Si nanowire shell results in degradation of its PL intensity.

5. References

[1] Lenshin A, Kashkarov V, Domashevskaya E, Bel'tyukov A and Gil'mutdinov F 2015 *Appl. Surf. Sci.***359** 550
[2] Canham L 2014 *Handbook of Porous Silicon* (Basel, Switzerland : Springer)
[3] Park J H,Gu L, Maltzahn G, Ruoslahti E, Bhatia S and Sailor M 2009 *Nat. mat* .**8** 331
[4] Wang G., Yau S T, Mantey K And Nayfeh M 2008 *Opt. Comm.* **281** 1765
[5] Salcedo W, Fernandez F and Rubim J 2004 *Phys. stat. sol. (c)* **1** 26
[6] Polkovnikova Yu, Lenshin A, Seredin P and Minakov D 2017 *Inorg.Mat.* **53**(5) 477
[7] KostishkoB., AtazhanovSh. andMikovS. 1998*Tech. phys.lett.***24** 633
[8] Heikkinen J.,Heiskanen J. andHormi O. 2006*Pol.Adv.techn.* **17(6)** 426
[9] Foll H., Christophersen M., Carstensen J. and Hasse G. 2002*Mat. Sci. Eng. R.***280**1
[10] Sreseli O, Goryachev D, OsipovV, Belyakov L. Voul'S., Serenkov I., Sakharov V and Voul' A 2002 *Semicond.* **36** 574
[11] Nagornov Yu., Kostishko B., Mikov S., Atazhanov Sh., Zolotov A and Pchelintseva E. 2007 *Tech.Phys* **52** 1093
[12] Lenshin A., Kashkarov V, Tsipenyuk V, Seredin P, Agapov B, Minakov D and Domashevskaya E 2013 *Tech. phys.* **58(2)** 284
[13] TolstoyV.P., ChernyshovaI.V. andSkryshevskyV.A 2003.*Handbook of infrared spectroscopy of ultrathin films* (New York: Wiley–Interscience)
[14] Len'shinA., KashkarovV., TurishchevS., SmirnovM. and DomashevskayaE.2011 *Tech. Phys. Lett.* **37**(9) 789
[15] Terekhov V, Kashkarov V, Manukovskii E, Schukarev A and Domashevskaya E 2001 *J. El.Spectr. Rel. Phen.* **114** 895

Acknowledgments

The study was funded by RFBR and Voronezh State Government according to the research project № 19-42-363004.

IOP Conf. Series: Materials Science and Engineering **889** (2020) 012024 doi:10.1088/1757-899X/889/1/012024

Technologies of reusing materials obtained during step-by-step demolition of buildings

S N Zolotukhin[1], V P Volokitin[1] and N M Kovalev[1]

[1]Federal government-financed educational establishment of higher education "Voronezh State Technical University", the Russian Federation, 394006, Voronezh, 84 20-letiya Oktyabrya Street

E-mail: n.kovalv@list.ru, ser6812@yandex.ru

Abstract. Nowadays scientists are required to create closed-cycle technologies for handling the waste; they have to reduce the waste amounts and the anthropogenic carbonic trace on the Earth, and they are supposed to use the recycled materials. The main amounts of the construction waste at the landfills of solid domestic waste appear after using the explosion method and the mechanical way of destruction, which are used for the demolition of buildings. Smart demolition, used during the renovation in Moscow, assumes crushing the obtained stone materials for their reuse. However, technologically crushing is not always possible, and economically it is cost-based. The research of the durability characteristics of load-bearing elements of the demolished buildings and existing regulatory documents allow reusing of these materials.

1. Introduction

The growth of land prices in the city centres, the modernization of the industrial factories, obsolescence of buildings require the renovation of the city territories and reallocation of the areas for new construction. The technologies, mostly used nowadays while demolishing different objects, lead to the obtaining of construction waste, which is one of the reasons for the solid domestic waste landfills constant growth, and also lead to the ecological problems.

The elimination of the negative consequences of the mentioned above problems in the EU is achieved by creating the closed-cycle economy for waste management, which requires the prevention of waste formation and material reuse [German law on the closed-cycle economy for waste management dated February 24, 2012]. In 2012, 30th April, in the Russian Federation there was adopted a programme "The Fundamentals of the Government Policy in the Sphere of the Russian Federation Ecological Development in the period till 2030" (approved by the President of the Russian Federation) where it is indicated that "the main tendency of managing the wastes is the prevention and reduction of waste generation, and step-by-step introduction of imposing a ban on waste burial". That is why the development of the technologies aimed at reusing materials, items and constructions, obtained during the building demolition, is an urgent task set to the scientists in the construction sphere.

2. Aims and objectives of the research

The development of the technology for reusing materials, items, and constructions, obtained during step-by-step building demolition:

The assessment of the possibility to reuse materials, items and constructions, obtained during step-by-step dismantling of the industrial buildings.

The practice of adopting technology of step-by-step building dismantling with the further reuse of the obtained materials in building low-rise constructions.

3. Materials and the research methods

The following methods were used: systematization, comparative analysis, theoretical generalization of the data, obtained during the detailed analysis of the scientific resources.

The research of the concrete durability was carried out according to the requirements of State Standard 28570-2019 Concretes. Methods of determining durability by the samples, chosen from the constructions; State Standard 22690-2015 Concretes. Determining the durability by the non-destructive control methods; State Standard 17624-2012 Concretes. The ultrasound method of durability determining; State Standards P 53231-2008 Concretes. Rules of control and durability evaluation.

The research of brick durability was done according to State Standard P 57349-2016 Brick and blocks. Method of determining the durability for compression.

The research of the reinforcement durability was carried out according to State Standard 12004-81. Reinforcing steel. Method of tension testing (with Amendments N 1, 2), State Standard 1497-84. Metals. Methods of tension testing (with Amendments N 1, 2, 3).

4. Results and discussions

The main methods of dismantling buildings and constructions in the world practice are explosion and mechanized methods. The special ways of crushing objects and their constructions are hydroexplosive, thermal, electrohydraulic and hydrosplitting, which are used rarely [7]. The explosion method is used to demolish multistoried buildings in America, China, Great Britain, Germany [1-7].

The explosion method has the following disadvantages:

1. Dynamic loads of the neighboring buildings.
2. Formation of a huge amount of dust during the explosion.
3. Heterogeneity of the obtained materials; this complicated the loading of the materials onto vehicles and labour-intensiveness of further processing.
4. Ecological problems of the obtained construction waste utilization.

The mechanized method of demolishing is used while dismantling building up to 9 stories. The advantage of this method is high efficiency, achieved by using special equipment.

Obvious advantages of these methods are high speed of work execution on dismantling and its low cost.

The disadvantages of the explosion method and mechanized method are difficulties during loading, low efficiency of transportation, as it is difficult to provide a full loading of broken ferroconcrete constructions; this increases the cost price of work execution, causes ecological problems or recycling and utilizing the obtained waste.

Nowadays during renovation in Moscow they use the technology of “smart demolition”. “Smart demolition” is the works, executed at several stages [8, 9]. At the first stage such elements as faience, ceramics, glass and woodwork are taken off. Then they are transported to specialized landfills with the aim of reuse.

At the second stage the mechanized method of demolition is used, the remnants of the construction are crushed and recycled. After crushing metal scrap is separated and sent for remelting.

The main material, obtained after recycling, is crushed stone which is used for filling the technological roads, filling the foundation pit pockets, making drainage in boggy areas [10-16].

The advantages of the “smart demolition” recycling are:

- reduction of the construction waste amounts which require burial at the solid domestic waste landfills;
- reuse of stone materials;
- solution to ecological problems due to production of the initial materials.

The disadvantage of this method is high cost of the obtained stone materials, their low homogeneity and durability [17-24].

The analysis of the maintenance and real longevity term of the load-bearing constructions of the buildings which are being demolished shows that they were exploited only 40-70 years while the regulatory terms of maintaining buildings from brick and ferroconcrete is 150-175 years according to MPP-3.2.23-97 [Methodological recommendations on economic basis for using constructive elements and technologies, providing the improvement of the investments efficiency due to reduction of operational costs, improvement of longevity of buildings and constructions, reduction of construction duration, and other effective decisions at increase of one-time costs while projecting and constructing and simultaneous growth of the estimate costs. / Economical efficiency of project solutions./ Part 1, item 1.3. Longevity of buildings, constructions and constructive elements and the estimation of longevity influence

IOP Conf. Series: Materials Science and Engineering **889** (2020) 012024 doi:10.1088/1757-899X/889/1/012024

on maintenance costs. Approved by the first Vice-Prime Minister of Moscow Government V. I. Resin, Moscow 1997].

Our research showed that durability characteristics of brick, concrete (figure 1, figure 2), reinforcing steel (figure 3), being in the body of the ferroconcrete item, and being exploited under favourable temperature-humidity conditions (plus temperatures and the presence of humid surrounding) and the absence of factors, which contribute to the destruction of these materials (chemically aggressive environment, washing-out of the cement stone from the concrete structure by precipitation, or freezing and thawing of the materials in the water-saturated state) do not sink, the durability characteristics of the reinforcing steel is higher than the ones guaranteed by the producing factories by 3-5%, and concrete durability even increases by 20-200 %.

According to [25, 26] an average durability limit at compression of brick batch aged more than 150 years equals to 6.8 MPa , the brick brand is M50. According to the scheme of projecting the development of concrete durability in time [26], with time the durability of concrete will grow up to 112.7 years.

Table 1 shows the results of testing durability characteristics of the main construction materials, obtained by the staff of VSTU by destructive and non-destructive methods of control.

Table 1. Durability characteristics of the construction materials after their maintenance during 42 years and more

	Concrete	**Silicate brick**	**Ceramic brick**
Durability under compression (MPa)	20-100	7.5-10	3.5-10

Figure 1. Determining the concrete durability by non-destructive methods.

Figure 2. Determining the concrete durability by destructive methods.

Figure 3. Determining the durability of reinforcing steel by destructive methods.

The analysis of the following documents [State Standard P 57678-2017 Resource-saving, waste management, construction waste destruction. Moscow, Standartinform, 2017, STO NOSTRA 2.33.53.2011. Demolition (dismantling) of buildings and constructions. The National Association of Builders. Moscow 2012, TCP 45-01.03-186-2009 (02250) Products, materials, equipment. The regulations of reuse. RUE "Stroytechnorm". Ministry of architecture and building of the Republic of

IOP Conf. Series: Materials Science and Engineering **889** (2020) 012024 doi:10.1088/1757-899X/889/1/012024

Belarus, Minsk 2010] shows that almost all the materials, products and constructions, obtained during demolishment have to be reused according to their intended purpose.

From all the mentioned above, we may conclude that the temporal factor of exploiting construction materials, products and constructions often does not lead to the decrease of their durability, and they may be reused and have to be reused, and crushing of these constructions, having huge durability resources, is economically unreasonable.

The practice of step-by-step dismantling technology, supervised by the professionals of the chairs of the VSTU construction departments, shows that reuse of brick, foundation blocks, timber and breakage of building stone does not lead to psychological problems of private customers, who are the main consumers of these items and materials during low-rise construction, and these items are widely used during constructing low-rise buildings. However, the disassembling of ferroconcrete carcasses due to getting defects in the areas of concreting during step-by-step dismantling causes significant problems during their reuse for their intended purposes. That is why the scientists of VSTU have elaborated the technologies of reusing ferroconcrete columns, crossbars (figure 4), plates of coverings and bridgings, both cavitated (figure 5) and ribbed (figure 6), in the foundation constructions of the low-rise buildings.

The reuse, especially of the ribbed plates, obtained during step-by-step dismantling, allows reducing by 4-5 times the weight of foundation constructions, the loads onto foundation soils and the cost of strip foundation constructions in comparison with the new ones from monolithic concrete with an equal support area. And construction of box-like section foundations from these plates is possible while constructing under complicated ground conditions, and allows using them even while constructing on gliding slopes (figure 7) [27].

Figure 4. Construction of strip foundation using ferroconcrete crossbars and columns.

Figure 5. Construction of foundation using cavitated plates of bridging.

Figure 6. Construction of foundation reusing ribbed plates of bridging, obtained at step-by-step buildings dismantling.

Figure 7. Construction of breastwalls from ribbed ferroconcrete plates.

Figure 8. The use of stone construction materials breakage for strengthening the soils of foundations while building on the impound territories.

In 2019, according to the National project "Dwelling and urban environment" (approved on 24th December, 2018, by the decision of the presidium of the Russian Federation President Council for Strategic Development and National Projects), the amount of settled unfit available housing apart was 0.14 mln m2, and in 2021-2024 this figure is supposed to reach 2.14 and 9.54 mln m2 respectively. A great amount of two-storied buildings from slag stone have to be demolished; during dismantling such buildings a huge amount of construction waste is obtained, which was reused to strengthen clayey soils on the impound territories (figure 8) [28].

The novelty of the developed technologies is confirmed by patents [29, 30] of the Russian Federation and implemented during constructing dozens of buildings and constructions in Voronezh and Voronezh region. The implementation of the developed technologies allowed reducing by 40% the amount of the construction waste transported to the solid domestic waste landfills in Voronezh. The amount of the implementation of the reuse technology is constantly growing and improving along with the training of architects, designers, planners, organizations involved into step-by-step dismantling, engineers who control the quality of the materials obtained during the demolition, builders, constructing low-rise buildings and constructions. Even without the support of the government bodies it is planned that in 5-7 years the construction waste in Voronezh will not be transported to the solid domestic waste landfills.

The technologies of adamantine cutting [7] during step-by-step buildings and constructions dismantling allow avoiding the breach of ferroconcrete constructions integrity; this will allow using the technologies with the reuse of ferroconcrete carcasses wider in the process of constructing low-rise buildings and constructions.

5. Conclusions

The main reason for the formation of the construction waste is the demolishing technologies used by the builders nowadays.

The researches prove that construction materials and reusable constructions under favourable exploiting conditions do not yield to new materials by their durability characteristics; that is why such materials may be reused and have to be reused.

The technologies of step-by-step dismantling of building constructions allow obtaining construction materials, items and constructions for their reuse.

The implementation of step-by-step building dismantling technologies with the reuse of the obtained materials in the low-rise building showed high efficiency of the trend of these technologies development.

The reuse, especially of ribbed plates in the construction of breastwall foundations, in low-rise construction has a high economical reasonability.

Russia has a significant potential for the rational use of construction materials, items and constructions, obtained during demolition of buildings and constructions, due to the development of the closed-cycle innovative technologies implemented in building.

6. References

[1] Huang B, Wang X, Kua H, Geng Y, Bleischwitz R and Ren J 2018 Construction and demolition waste management in China through the 3R principle *Resources, Conservation and Recycling* **129** 36-44, DOI: 10.1016/j.resconrec.2017.09.029

[2] Galaeva N L 2019 The use of explosion method for demolition of buildings and constructions under urban building conditions *The perspectives of science* **5** (116) 54-6

[3] Velikanov N L, Naumov V A and Tarasov D A 2011 The use of impact fracture while demolishing building constructions *The proceedings of KSTU* **20** 48-53

[4] Kleemann F, Lederer J, Aschenbrenner P, Rechberger H and Fellner J A 2016 Method for determining buildings' material composition prior to demolition *Building Research and Information* **44** (1) 51-62. DOI: 10.1080/09613218.2014.979029

[5] Mihai F C 2019 Construction and demolition waste in romania: The route from illegal dumping to building materials *Sustainability (Switzerland)* **11**, DOI: 10.3390/su11113179

[6] Dobrosotskikh M G, Potekhin I A, Kim T S and Kostina D P 2018 The organization of step-by-step dismantling of a building with the further reuse of the building constructions and materials // *Construction and real estate* **1-1**(2) 123-8

[7] Zolotukhin S N, Abramenko A A, Artemova Y A and Kolodyazhny S A Destruction of buildings and use of materials formed when renovating urban territories// *Bulletin of MSBU* 273-83

[8] Fakhratov M A,Suleymanov K A and Bolotin O A 2018 The peculiarities of concrete casting and building dismantling in the frames of step-by-step system *Innovations and investments* **4** 341-4

[9] Jesus S, Maia C, BrazãoFarinha C, de Brito J and Veiga R 2019 Rendering mortars with incorporation of very fine aggregates from construction and demolition waste *Construction and Building Materials* **229**, DOI: 10.1016/j.conbuildmat.2019.116844

[10] Jiménez J R, Ayuso J, López M, Fernández, J M and De Brito J 2013 Use of fine recycled aggregates from ceramic waste in masonry mortar manufacturing *Construction and Building Materials* **40** 679-90, DOI: 10.1016/j.conbuildmat.2012.11.036

[11] Solís-Guzmán J, Marrero M, Montes-Delgado M V and Ramírez-de-Arellano A A 2009 Spanish model for quantification and management of construction waste *Waste Management* **29** 2542-8, DOI: 10.1016/j.wasman.2009.05.009

[12] Grigoriadis K, Whittaker M, Soutsos M, Sha W, Napolano L, Klinge A and et. al. 2019 Improving the recycling rate of the construction industry *Sustainable Construction Materials and Technologies: 5th Int. Conf. on Sustainable Constr. Mat. and Techn.* **1**

[13] Chen J, Su Y, Si H and Chen J 2018 Managerial areas of construction and demolition waste: A scientometric review *Int. J. of Environmental Research and Public Health* **15**, DOI: 10.3390/ijerph15112350

[14] Iacovidou E, Purnell P and Lim M 2018 The use of smart technologies in enabling construction components reuse: A viable method or a problem creating solution? *J. of Env. Management* **216** 214-23, DOI: 10.1016/j.jenvman.2017.04.093

[15] Rose T M, Manley K and Agdas D A 2017 Conceptual framework to investigate the adoption of on-site waste management innovation in Australian building projects// *PICMET 2016 (Portland International Conference on Management of Engineering and Technology: Technology Management For Social Innovation)* 1830-7, DOI: 10.1109/PICMET.2016.7806745

[16] Guignot S, Touzé S, Von der Weid F, Ménard Y and Villeneuve J 2015 Recycling Construction and Demolition Wastes as Building Materials: A Life Cycle Assessment *J. of Ind. Ecology* **19** 1030-43, DOI: 10.1111/jiec.12262

[17] Pittau F, Amato C, Cuffari S, Iannaccone G and Malighetti L E 2019 Environmental consequences of refurbishment vs. demolition and reconstruction: A comparative life cycle assessment of an Italian case study *Earth and Environmental Science. International Sustainable Built Environment Conference on Resilient Built Environment for Sustainable Mediterranean Countries* **296**, DOI: 10.1088/1755-1315/296/1/012037.

[18] Islam R, Nazifa T H, Yuniarto A, Shanawaz Uddin A S M, Salmiati S and Shahid S 2019 An empirical study of construction and demolition waste generation and implication of recycling *Waste Management* **95** 10-21, DOI: 10.1016/j.wasman.2019.05.049

[19] Letelier V, Henríquez-Jara B I, Manosalva M and Moriconi G 2019 Combined use of waste concrete and glass as a replacement for mortar raw materials *Waste Management* **94** 107-19, DOI: 10.1016/j.wasman.2019.05.041

[20] Pavlu T, Pesta J, Volf M and Lupísek A 2019 Catalogue of Construction Products with Recycled Content from Construction and Demolition Waste *IOP Conf. Ser.: Earth and Environmental Sci.* **290**, DOI: 10.1088/1755-1315/290/1/012025

[21] Kianimehr M, Shourijeh P T, Binesh S M, Mohammadinia A and Arulrajah A 2019 Utilization of recycled concrete aggregates for light-stabilization of clay soils. *Constr. and Bui. Mat.* **227** 11, DOI: 10.1016/j.conbuildmat.2019.116792

[22] Akhtar A and Sarmah A K 2018 Construction and demolition waste generation and properties of recycled aggregate concrete: a global perspective *J.of Cleaner Production* **186** 262-81, DOI: 10.1016/j.jclepro.2018.03.085.

[23] Cardoso R,Silva R V,De Brito J and Dhir R 2016 Use of recycled aggregates from construction and demolition waste in geotechnical applications: a literature review *Waste Management* **49** 131-45

[24] Zolotukhin S N 2018 The possibilities of creating a regional cluster for construction waste reuse *Resource effective technologies in the construction complex of the region* **10** 209-12

[25] Samokhina M N, Ishkov A N and Shmelev G D 2019 The peculiarities of prediction of concrete durability characteristics changes under the conditions of the restricted information. *Housing economy and municipal infrastructure* **1** (8) 9-17

[26] Shmelev G D, Kryuchkova E A and Episheva A K 2019 The basis for the possibility of building materials and constructions reuse *Housing economy and municipal infrastructure* **1** (8) 25-35

[27] Zolotukhin S N, ChigarevA G and Abdulloev A R 2018 Designing and Constructing Foundations on a Landslide Slope *IOP Conf. Series: Mat. Sci.and Eng.* **463** 032088 DOI:10.1088/1757-899X/463/3/032088

[28] Lobosok A S, Abramenko A A, Vyazov A U, Kukina O B and Zolotukhin S N 2018 *Method of bulk cementation of soils* (RF patent 2656656 C2, 06.06.2018. Bul. 15)

[29] Lobosok A S, Abramenko A A, Vyazov A U, Kukina O B, Zolotukhin S N, Milovanova V I and Kolodyazhny S A 2018 *Method for manufacturing solid box-section slab foundations from ribbed floor slabs* (RF patent 26447521 C1, 16.03.2018, Bul. 8)

[30] Lobosok A S, Abramenko A A, Vyazov A U, Kukina O B, Zolotukhin S N, Milovanova V I and Kolodyazhny S A 2018 *Method for building a wall Foundation using ribbed floor slabs* (coatings) (RF patent 2671019 C1, 29.10.2018, Bul. 31)

IOP Conf. Series: Materials Science and Engineering **889** (2020) 012025 doi:10.1088/1757-899X/889/1/012025

Thermal transformations in polymineral technogenic cement raw

O A Miryuk[1]

[1]Rudny Industrial Institute, Kazakhstan

E-mail: psm58@mail.ru

Abstract. The results of physicochemical studies of composition and thermal transformations of large-tonnage wastes of skarn-magnetite ore dressing in Kazakhstan are presented. To determine materials' composition of the following methods were used: X-ray, differential and thermal, Mössbauer and infrared spectroscopy. The stepwise nature of transformation of anthropogenic material during firing, due to polymineral composition of ore dressing wastes is revealed. Provides information on the chemical and mineral composition, thermal transformations of ore dressing wastes. It has been established that the chemical-mineral characteristics of natural silicates determine formation processes and properties of clinker phases. The features of the formation of clinker phases with the participation of the technogenic component of the raw mix, which consist in the activity of low-temperature interactions, the variety of intermediate phases are revealed. It has been established that the staging of belite formation with the participation of actinolite, albite, andradite, grossular, diopside, chlorite, epidote is due to gradual transformations of minerals into intermediate phases C_2AS, C_3MS_2, of which C_2S is formed at temperatures of 1000 – 1300°C. Alite formation is intensified due to high-temperature belite formation and doping impurities. Skarn-magnetite ores wastes are an effective cement raw material, providing high-quality portland cement clinker using low-energy and resource-saving technology.

1. Introduction

Intensive development of metallurgy is accompanied by an increase in production and processing of natural raw materials. During extraction and processing of most minerals, 50 – 90% of mineral raw materials are sent to dumps. On the other hand, mineral and raw material base of the construction industry is depleted, which often leads to excessive consumption of fuel and energy, and a decrease in quality of finished products. Reducing reserves of natural raw materials for building materials requires search for its substitutes. Features of building materials technology provide the opportunity for wide involvement inproduction of waste from various industries.

A large consumer of industrial raw materials is the cement industry. Resource conservation is ends and means of cement technology improving. The further development of the industry is facilitated by saving of material, fuel and energy resources. Depletion of natural reserves determines a fundamental change in the raw material base due to the development of technogenic sources. Logical continuation of the development of recent decades to replace scarce raw materials with waste is optimization of technological solutions with the rational use of technogenic materials [1 – 20]. Experience of a number of industry enterprises indicates that the use of waste is associated not only with expansion of raw material base, but with an increase in technological processes efficiency.

Waste management efficiency is often low due to the lack of an integrated approach to assessment of technogenic sources. Selection of new raw materials is often limited to replacing the scarce traditional component with waste that is similar in composition and condition. For example, the valuable properties of large-capacity man-made materials of complex chemical and mineral composition remain unclaimed. Strict regulation of dispersion of new raw materials virtually eliminates the use of coarse reactive materials. Wastes of enrichment plants vary significantly depending on the type of a deposit, methods of raw materials processing and require individual research [7, 9, 16, 17].

The world's largest deposits of skarn-magnetite ores are located in Kazakhstan. At the first stage of skarn-magnetite ores dressing, more than 4 million tons of tailings of dry magnetic separation (DMS) are formed annually. Information on the use of crushed-stone tailings of DMS ores of various deposits in cement production is scarce.

To justify the possibility of using skarn-magnetite ore dressing waste in cement clinker production, it is necessary to understand thermal transformations of minerals, the nature of technogenic component mixture influence on clinker phases formation.

A comprehensive study of technogenic material was the objective of this work

2. Object and research methods

Dry magnetic separation waste is a disintegrated mass (particle size up to 25 mm) with low humidity (1 – 2%), exceeding cement clinker and quartz sand in grinding ability. Therefore, the introduction of a technogenic component will not complicate the grinding of the raw material charge.

Low values of the specific effective activity of radio nuclides (53 – 55 Bq kg^{-1}) indicate radiation and environmental safety of ore dressing wastes.

Mineral base of DMS tails is composed of silicates that differ in genesis, composition, structure, physical properties, chemical activity and thermal stability wt. %: pyroxenes (diopside) 20 – 25; epidote 10 – 13; feldspars 8 – 12; chlorites 7 – 10; scapolite 8 – 11; grenades (andradite, grossular) 7 – 12; amphiboles(actinolite) 7 – 14 (figure 1). In the waste is present, wt. %: calcite 4 – 7; pyrite 4 – 8; quartz 2 – 4; magnetite 3 – 4.

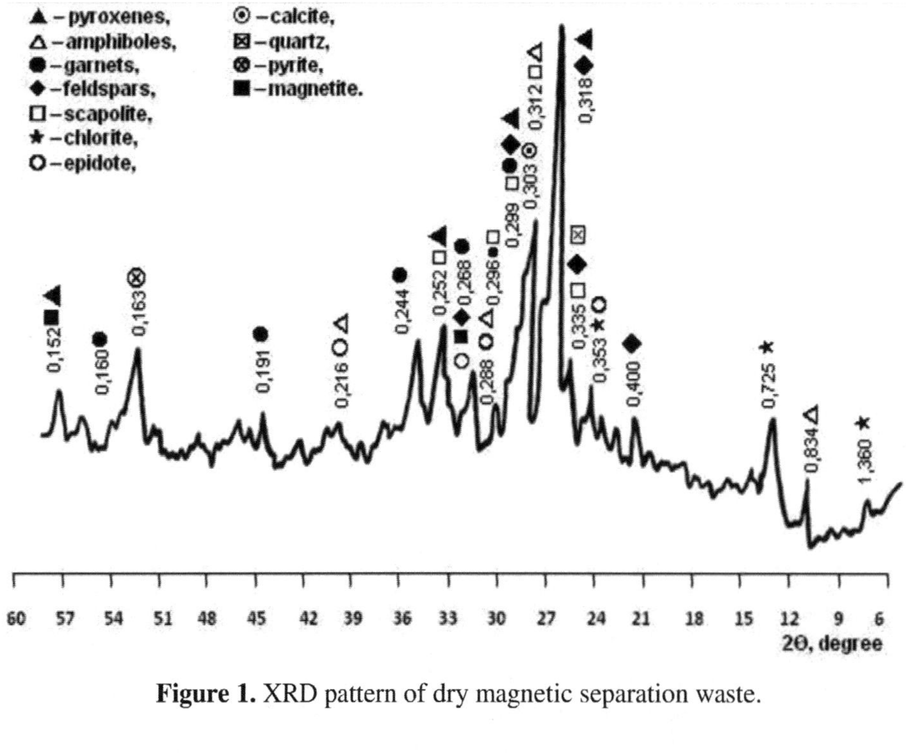

Figure 1. XRD pattern of dry magnetic separation waste.

Chemical composition of dry magnetic separation tailings is presented, wt. %: SiO_2 40 – 45; Al_2O_3 10 – 12; Fe_2O_3 16 – 18; FeO 6 – 8; CaO 12 – 13; MgO 5 – 6; in the order 3 – 6. A number of catalytic and modifying elements, wt. %: S 2 – 5; R_2O 2 – 4; TiO_2 0.50 – 0.53; P_2O_5 0.25 – 0.30; MnO 0.35 – 0.40; V_2O_5 0.04 – 0.06; Cl 0.09 – 0.12; Cu 0.04 – 0.05; Ni 0.007 – 0.008.

Presence of clinker-forming oxides, an increased concentration of alloying and modifying impurities of sulfur, alkalis, magnesium, titanium, phosphorus, manganese, chlorine, copper, vanadium, and nickel are capable of ensuring the efficient use of waste in the composition of cement raw material mixture. The silicate base of a technogenic material, presence of non carbonate calcium oxide determines material and energy advantages of raw mixtures with a technogenic component.

Materials were calcined in a laboratory furnace with silicon heaters. Exposure at maximum temperature is 15 min. Mineral composition of a technogenic material and firing products were identified by X-ray diffraction, differential thermal microscopic methods, as well as by infrared spectroscopy.

The Mössbauer method was used to determine the composition of iron-containing compounds. X-ray phase analysis was carried out on a DRON-3 general-purpose X-ray diffractometer with a copper cathode and a nickel filter according to the powder method in the double-angle range $2\Theta = 4 - 65^0$. Thermo chemical processes were studied by the differential-thermal analysis method, which was performed on a "MOM" derivatograph by F. Paulik, I. Paulik and L. Erdey. Infrared absorption spectra were recorded on "Specord – 75JR" automated spectrometer. To determine a degree of oxidation, a coordination number of iron atoms and phase composition of iron-containing minerals, the nuclear gamma resonance spectroscopy method (Mössbauer) was used. The studies were carried out on "YaGRS – 4M" spectrometer in the speed range up to 10 mm s^{-1} with Co^{57} source in a Cr matrix.

3. Results and discussion

3.1 Thermal transformations of ore dressing wastes

Investigations of transformations during the firing of wastes from skarn-magnetite ore dressing in the temperature range of cement clinker formation were made. Variegated chemical and mineral composition of skarn-magnetite ore dressing waste led to the multi-step nature of thermal transformations (figure 2).

Firing of DMS tailings is accompanied by complex transformations and interactions of separate minerals (figure 3). Low-temperature transformations (exothermal effects at 300 - 600°C) are associated with magnetite and pyrite oxidation. Decomposition of pyrite seems to be multi-stage [14]. A noticeable removal of sulfur from pyrite occurs at a temperature of about 500°C, then the reactions begin:

$$S_2 + 2O_2 \rightarrow 2SO_2 \quad (1)$$

$$FeS + 1{,}5O_2 \rightarrow FeO + SO_2 \quad (2)$$

The general scheme of pyrite oxidation has the form:

$$4FeS_2 + 11O_2 \rightarrow 2F_2O_3 + 8SO_2 \quad (3)$$

Under the catalytic effect of iron oxides at temperatures up to 600°C, SO_2 is oxidized into SO_3. Intensity of thermal effects at 300 – 600°C indicates a significant heat release during oxidative processes. Calculations showed that the amount of released heat is 831 kJ mol^{-1} of FeS_2 according to the reaction (3); 147 kJ mol^{-1} FeO due to its transition into Fe_2O_3 and 98 kJ mol^{-1} of SO_2 during the

IOP Conf. Series: Materials Science and Engineering **889** (2020) 012025 doi:10.1088/1757-899X/889/1/012025

formation of SO_3. Ambiguous nature of the change in the test sample mass upon heating (figure 2) confirms oxidation of magnetite (increase in mass at 300 – 400°C) and decomposition of pyrite, accompanied by partial removal of gaseous products (decrease in weight at 400 – 500°C).

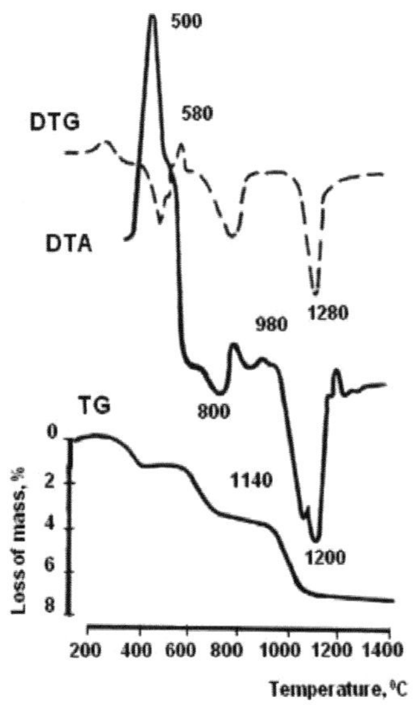

Figure 2. Derivatogram of dry magnetic separation waste.

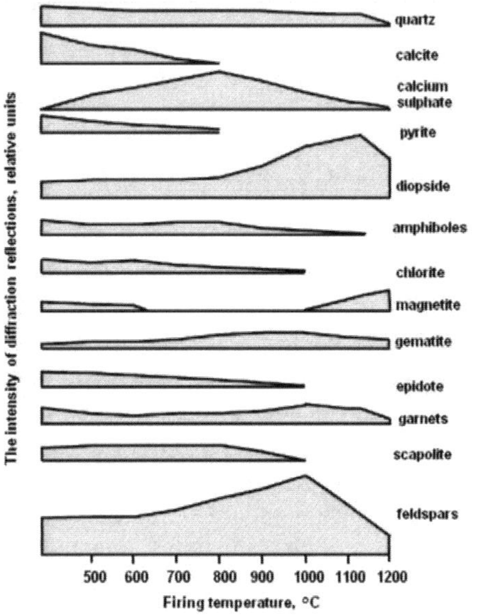

Figure 3. Thermal changes in mineral composition of dry magnetic separation waste.

When DMS tailings are heated up to 580°C, mass growth is observed due to oxidation of residual Fe_3O_4 and FeO released in the reaction (2).Presence of pyrite accelerates thermal processes: under the catalytic influence of FeO, calcite decarbonization is intensified. Released CaO interacts with SO_2 and SO_3 to form $CaSO_4$. Firing of DMS tailings at temperatures above 800°C is accompanied by decomposition of $CaSO_4$ anhydrite and a decrease in concentration of sulfur-containing compounds. In the temperature range of 600 – 800°C, amphiboles, epidote dehydration, removal of bound water from talc-like layers of chlorites occurs [7]. This is confirmed by intense loss of material mass (figure 2). Significant changes are observed when the temperature rises up to 1000°C: minerals of a late geological period (scapolite, epidote, chlorites) are absent; amphibole content decreases (figure 3). Disappearance of aluminosilicate reflections on the DMS tailings X-ray under a simultaneous increase in the intensity of pyroxenes maxima, garnets and feldspars indicates a close interconnection of processes. Epidote decomposes with the formation of anorthite, garnets; amphiboles when heated pass into pyroxenes. A decrease in the content of calcium sulfate and quartz indicates participation of the latter in these transformations. An additional amount of pyroxenes (diopside) is formed by interaction of the decomposition products of chlorite with calcium oxide. Melting of DMS tailings takes place stepwise (figure 2): at first, the crystal structure of garnets and feldspars is destroyed (end effect at 1140°C). When heated to a temperature of 1200°C, the content of pyroxenes decreases. It is known [15], that the initial temperature of melting of clays – traditional cement raw materials varies in the range of 1135 – 1300°C, and the end of melting occurs at temperatures of 1400 – 1500°C. DMS tailings are completely melted at a temperature of 200 – 300°C lower than clay.

The heat treatment of DMS tailings is accompanied by structural changes in silicates (figure 4). After firing at a temperature of 800°C, layered minerals (950 cm^{-1}) are not diagnosed. An absorption band of 1050 cm^{-1} indicates the presence of garnets, and its decrease confirms the destruction of minerals at temperatures above 1000°C. The spectra of wastes calcined at temperatures of 800 – 1000°C are characterized by simultaneous amplification and expansion of the absorption region of 850 – 1000 cm^{-1}. This indicates an increase in the proportion of chain and frame structures. Heating of DMS tailings to a temperature of 1200°C is accompanied by strong amorphization and disordering of silicon-oxygen tetrahedra: absorption intensity of the band of 850 – 1050 cm^{-1} decreases. Consequently, pyrogenic changes in the constituent tailings of DMS mirror the formation of minerals. First of all, late formations are destroyed, their transformation into primary minerals is observed. Composition of new formations during firing of DMS tailings is not limited to the listed compounds. However, a very complex nature of pyrogenic transformations complicates their detailing.

Iron is concentrated in ore minerals (33 – 35%), sulfides (18 – 22%), silicates (45 – 47%). Among iron-containing silicates the following dominate; they are andradite, actinolite, epidote, and chlorite. Isomorphic substitutions in minerals determine the presence of iron ions in diopside, grossular. The nuclear-gamma-resonance spectrum of the initial DMS tailings (figure 5) can be interpreted as overlapping lines characterizing iron ions in various oxidized and coordination states [18].

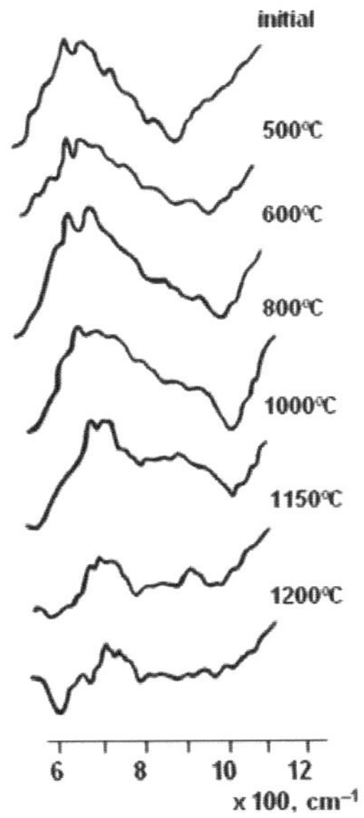

Figure 4. Infrared spectra of dry magnetic separation waste.

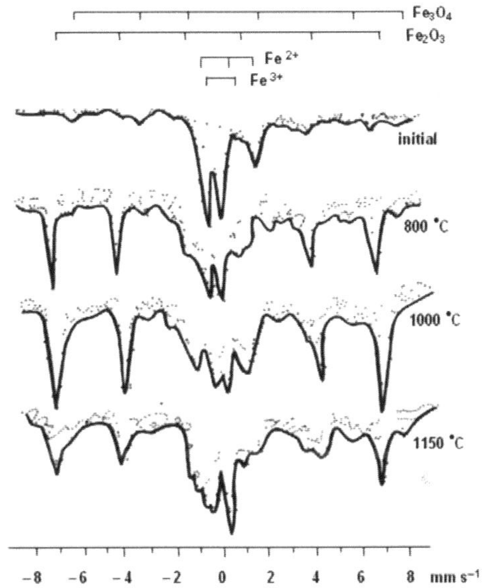

Figure 5. Mössbauer spectra of dry magnetic separation waste.

Multivalent iron ions are predominantly represented by octahedral coordination. Oxidation of pyrite and magnetite upon heating the DMS tailings to a temperature of 800°C is marked by appearance of on the spectrum of hematite sextet. A subsequent increase in hematite content is associated with its release during the decomposition of silicates (andradite, actinolite). The nitrous form of iron oxide persists at elevated temperatures: at 1150°C, transformations in the tailings of DMS are accompanied by the formation of magnetite and an increase in the content of pyroxenes. As a result, the area of the central doublet grows, while the size of hematite sextet decreases, and magnetite lines appear. The advantage of octahedral coordination of iron ions when heating the DMS tailings is retained. This provides high mobility of the melt. In DMS tailings calcined at temperatures of 800, 1000 and 1150°C, the oxidation coefficient of $Fe^{3+} / (Fe^{3+} + Fe^{2+})$ is 0.78; 0.85; 0.76; in initial – 0.48; this corresponds to the nature of transformations of iron-containing minerals during waste firing.

3.2 The formation of clinker phases with the participation of the wastes

The study of phase transformations during firing of mixtures of calcium carbonate with minerals predominant in the composition of skarn-magnetite wastes allows determining the influence of technogenic minerals on the main phases formation of the cement clinker (belite C_2S and alite C_3S). Activity of reactions depends on the composition and structure of natural minerals (table 1).

It has been established that the staging of belite formation with the participation of natural silicates is due to gradual minerals transformation into intermediate phases C_2AS, C_3MS_2, from which C_2S is formed.

Table 1. Influence of the structure of natural silicates on phase formationin mixtures of "calcium carbonate – mineral".

Minerals-silicates		Structural formulae of intermediate silicates	Temperature of cement clinker phase formation, ^0C	
mineral's name	structure tetrahedron		C_2S	C_3S
Grossular (insular)	$[SiO_4]^{4-}$	$Ca[Al_2Si_2O_8]$ – frame $Ca_2Al[AlSiO_7]$ – ring	950 – 1300	1300 – 1350
Andradite (insular)	$[SiO_4]^{4-}$	$Ca_2(Al,Fe)[AlSiO_7]$ – ring	950 – 1300	1250 – 1350
Epidot (insular)	$[SiO_4]^{4-}\times[Si_2O_7]^{6-}$	$Ca_2(Al,Fe)[AlSiO_7]$ – ring $Ca_2Al[AlSiO_7]$ – ring	950 – 1300	1250 – 1350
Diopside (chained)	$[Si_2O_6]^{4-}$	$Ca_2Mg[Si_2O_7]$ – ring $Ca_3Mg[SiO_4]_2$ – insular	1000 – 1300	1350 – 1400
Actinolite (belt)	$[Si_4O_{11}]^{6-}$	$CaMg[Si_2O_6]$ – chained $Ca_3Mg[SiO_4]_2$ – insular	1000 – 1300	1300 – 1400
Chlorite (layered)	$[AlSi_3O_{10}]^{4-}$	$Ca_2Al[AlSiO_7]$ – ring $Ca_3Mg[SiO_4]_2$ – insular	1000 – 1300	1250 – 1400
Albite (frame)	$[AlSi_3O_8]^-$	$Na_2[Al_2Si_2O_8]$ – frame $Ca[Al_2Si_2O_8]$ – frame $Ca_2Al[AlSiO_7]$ – ring	1100 – 1300	Not revealed
Scapolite (frame)	$[Al_{1+x}Si_{3-x}O_8]^-$	$Na_2[Al_2Si_2O_8]$ – frame	950 – 1300	1300 – 1350

The temperature of the onset of belite formation is determined by transformation complexity of silica oxygen motifs of skarn minerals into an orthosilicate structure and rises in the series: isular → chain → layered → frame. The rate of belite formation depends on the stability of intermediate silicate phases. Formation of the highly basic C_3S phase occurs in a very narrow range, often bordering the melting point (fracture) of minerals structure or intermediate phases formed from them.

The regularities of individual minerals influence are revealed in synthesis of the clinker based on skarn-magnetite ore dressing waste during the firing of the raw mixture "limestone – DMS tailings" containing 35% of a technogenic component. Exothermic oxidative processes intensify decarbonization reactions and decomposition of unstable minerals (figure 6).

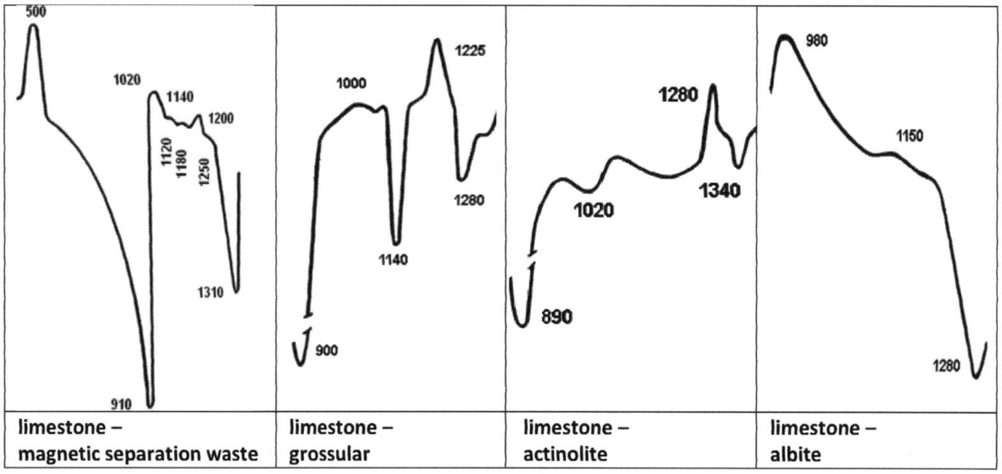

Figure 6. Thermograms of the raw material mixtures.

Interactions of components begin at low temperatures due to the Hadwall effect during the period of structural transformations of minerals. Low-temperature melting of the technogenic component until a clinker molten appearing ensures formation of a liquid low-viscosity phase and promotes the activation of phase transformations. Difference in crystalline structures of natural silicates and intermediate silica-containing phases predetermined the staging of belite formation with the predominant formation of C_2S in the high-temperature region (1050 – 1250^{0}C). Formation of alite is accelerated during the melt enrichment with magnesium oxide and is intensified due to high-temperature whitening and alloying impurities.

The increased temperature of belite formation, deformation of the crystal lattice, and doping of clinker phases provide increased hydration activity of clinkers with a limited content of the high-temperature phase C_3S. This allow to synthesize low-energy clinker.

4. Conclusions

Stepwise nature of thermal transformations of poly-mineral wastes from skarn-magnetite ore dressing waste, accompanied by interactions of minerals with the formation of numerous intermediate compounds is established.

Regularities of technogenic minerals influence on the formation of clinker phases were revealed. Relationship of clinker formation with the crystalline structure of minerals is noted. It is shown that due to the clinker phases inheriting the crystalline structures of natural minerals, close contact of the main and isomorphic elements, dicalcium silicate (belite) of increased reactivity is formed.

Possibility of obtaining cement clinker based on skarn-magnetite ore dressing waste is proved. Preference for the synthesis of active cement clinkers with a high content of belite, characterized by low energy intensity is substantiated.

The need for a comprehensive study of technogenic raw materials is shown. An integrated approach to the study of technogenic raw materials made it possible not only to substantiate the possibility and expediency of using skarn-magnetite ore dressing wastes in cement production, but also to determine preference for the phase composition of cement clinker. The research results are the basis for development of low-energy cement clinkers with a high content of belite phase.

References

[1] Wu Q, Wu Y, Tong W and Ma H 2018 *Constr. Build. Mater.* **93** 426 – 434
[2] Garcia-Lodeiro I, Carcelen-Taboada V, Fernández-Jiménez A and Palomo A 2016 *Constr. Build. Mater.* **105** 218 – 226
[3] Radayev S, Seleznyova O, Ilyukhin K, Ivanov K and Forosevich N 2016 *Mater. Scien. Forum* **871** 90
[4] Gartner E and Sui T 2018 *Cem. Concr. Res.* **114** 27 – 39
[5] Shettima AU, Hussin MW, Ahmad Y and Mirza J 2016 *Constr. Build. Mater.* **120** 72 –79
[6] Torkittkul P and Chaipanich A 2010 *Cem. Concr. Comp.* **32** 440 – 449
[7] Miryuk O A 2019 *IOP Conf. Ser.: Mater. Scien. Eng.* **510** 012012
[8] Kavas T, Angelopoulos G N and Iacobescu RI 2015 *Cement Wapno Beton* **5** 328 –334
[9] Miryuk OA 2019 *ARPN J. Eng. Appl. Scien.* **14** 2605 – 13
[10] Vinnichenko V and Ryazanov A 2017 *MATEC Web Conf.* **116** 01020
[11] Wang Yu, Yong C, Kim J, Choi O, Lee K M and Lachemi M 2009 *Constr. Build. Mater.* **23** 2481 – 86
[12] Ahmedzeki NS, Abbas MN, Joodee AM and Jaed YM 2018 *J. Chem. Technol. Metall.* **53** 239 – 244
[13] Wu F, Cao Z, Wang Sh and Zhong H 2018 *J. Clean. Prod.* **10** 831 – 843
[14] Zhang X, Kou J and Sun C 2019 *J. Analyt. Appl. Pyrol.* **138** 41 – 53
[15] Laita E and Bauluz B 2018 *Appl. Clay Scien.* **152** 284 – 294
[16] Wu K, Shi H, Schutter GD, Guo X and Ye G 2012 *Cem. Concr. Comp.* **3** 322 – 327
[17] Hlystov AI, Shirokov VA and Vlasov AV 2015 *Proc. Eng.* **111** 290 – 296

[18] Kalska-Szostko B and Kropiewnicka K 2012 *Mössb. stud., Cur. Appl. Ph.* **12** 896 – 902
[19] Rezaee F, Danesh S, Tavakkolizadeh M and Mohammadi-Khatami M 2019 *J. Clean.Prod.* **214** 749
[20] Buruberri LH, Seabra MP and Labrincha JA 2015 *J. Hazard. Mater.* **286** 252 – 260

IOP Conf. Series: Materials Science and Engineering **889** (2020) 012026 doi:10.1088/1757-899X/889/1/012026

Calculation update of strength of expanded clay fiber-reinforced concrete elements with mixed reinforcement

D R Mailyan[1*], A M Blyagoz[2] and K M Kretinin[2]

[1]Don State Technical University, 162 Sotsialisticheskayastr, Rostov-on-Don, 344022, Russia
[2]Federal State Budgetary Educational Institution of Higher Education «Kuban State Agrarian University named after I.T. Trubilin» 13, Kalininastr, Krasnodar, 350044, Russia

*alfa-maikop@yandex.ru

Abstract. The paper deals with the methodology for calculating the strength of the expanded clay fiber-reinforced concrete elements with mixed reinforcement with the most complete view of the post-yield properties of prestressed and non-stressed rebars, as well as the influence of prestressing on the properties of the expanded clay fiber-reinforced concrete.

1. Introduction
The developed method of calculation is based on a complete account of the work of prestressed and non-stressed rebars beyond the conditional yield point, as well as the effect of prestressing on the properties of the expanded clay fiber-reinforced concrete.

For these purposes we need to use analytical dependencies that describe the complete actual diagrams of the tensile strength of high-strength reinforcement and the effect of prestress on them.

Currently, there are many proposals for an analytical description of steel deformation diagrams [1-5], from which there were selected the dependencies obtained in the Russian State Social University [4,5].

2. Experimental program and research results
These dependencies are flexible, because they are suitable for all types of steel, simple and extra-strong.

According to [5], the steel deformation diagram is divided into two sections: elastic (from 0 to σ_{el}) and non-elastic (from σ_{el} to σ_u) work. Within the section from the elastic limit σ_{el} to the temporary resistance of steel σ_u, the analytical description of the deformation diagram has the following form:

$$\sigma_s = S - K\frac{l_n 10(\varepsilon_s 10^3 + L)}{\varepsilon_s 10^3 + L}, \quad (1)$$

where $S = \sigma_{0.2}$ (1,475 $\sigma_u / \sigma_{0.2}$ -0,475)-30; (2)

$$K = 3700\left[\frac{\sigma_{0.2}(\sigma_u/\sigma_{0.2}-1)+150}{1000}\right]^2 - 96, \quad (3)$$

$$L = \sigma_{0.2}\left[\frac{12.5(\sigma_u/\sigma_{0.2}-1)}{1000} - \frac{1000}{E_s}\right] + 0.96, \quad (4)$$

The relationship between deformations and stresses within the section of the elastic limit of steel before the development of deformations up to 1.5% in it can be represented as:

$$\varepsilon_s \cdot 10^3 = \frac{K_0}{A-\sigma_s} - D + \frac{\sigma_s(1.9\cdot 10^5 - E_s)}{190\cdot E_s}, \quad (5)$$

where
$$K_0 = 2.931\cdot 10^{-2}(\sigma_u/\sigma_{0.2}-1)^2\sigma_{0.2}^2 + 84, \quad (6)$$

$$A = (1.186\sigma_{0.2} - 231)\cdot(\sigma_u/\sigma_{0.2} - 0.5)^2 + 0.675\cdot\sigma_{0.2}^2 + 70, \quad (7)$$

$$D = 0.01\cdot\sigma_{0.2}(1.6\cdot\sigma_u/\sigma_{0.2} - 2.126) + 0.813, \quad (8)$$

In formulas (1)-(8), the values of stress and modulus of elasticity are substituted in MPa.

Expressions (1)-(8) show that the parameters S, K, L, K_0, A, D depend on the known characteristics of the steel $\sigma_{0.2}; \sigma_u; E_s$ so they can be easily determined for each type of steel[6,7]. After substituting the numerical values of steel characteristics in expressions (2)-(4) and (6)-(8), the parameters S, K, L, K_0, A, D for each type of steel will get certain numerical values, and the formula (1) and (5) will be significantly simplified.

When prestressing or reloading the reinforcement due to the choice of plastic deformations, the values of the conditional elastic and yield stresses increase. In an extremely stretched up to the stress σ_{sp} rebar the stresses are reduced over time due to the losses from stress relaxation σ_1 and other types of losses, the total value of which is here indicated as σ_{los}. In the state of steady pre-stresses σ_{sp2} in rebar A_{sp} the selected plastic deformations will be $\varepsilon_{sp,pl} + \sigma_1/E_s$, where $\varepsilon_{sp,pl}$ - is the plastic deformation of steel when it is stressed to the level exceeding the elastic limit σ_{el}. When an external tensile force is applied, the diagram can be assumed to be linear to the point of intersection with the original curve. This point corresponds to a new increased value of the elastic limit σ'_{el}. The deformation corresponding to this stress is the following:

$$\varepsilon'_{el} = \varepsilon_{sp} + \Delta\varepsilon_s, \quad (9)$$

or according to [5]

$$\varepsilon'_{el} = \varepsilon_{sp} + 7.7\cdot 10^{-7}\sigma_{0.2}, \quad (10)$$

The new value of the elastic limit can be determined by the formula (1) with $\varepsilon_s = \varepsilon'_{el}$, and the new value of the conditional yield strength by the formula [5]:

$$\sigma'_{0.2}/\sigma_{0.2} = 1 + (\sigma_u/\sigma_{0.2} - 1)\cdot\left(\frac{0.127}{1.2214 - \sigma_{0.2}} - 0.157\right). \quad (11)$$

Let us determine the stress in a rebar when some external load effects on the element containing a pre-stretched A_{sp}, and non-stressed A_s rebar. The initial condition is the equality of incremental strain of deformations from external load in the entire rebar located at the same distance from the most compressed face[8,9]. If the distance of the reinforcement from the specified face is different (not single-row reinforcement), the relationship between the incremental strain of deformations of the pre-stretched (ε_{sp}^{N}), and non-stressed (ε_{s}^{N}) rebar is expressed as follows:

$$\frac{\varepsilon_b + \varepsilon_{sp}^{N}}{h_{op}} = \frac{\varepsilon_b + \varepsilon_{s}^{N}}{h_{os}}, \tag{12}$$

where h_{op} and h_{os} - is the distance from the most compressed face of the element to the resultant forces in the specified types of reinforcement.

In the reinforcement located within a zone stretched from an external load before applying the external stress load, the stresses will be equal: in a pre-stretched - σ_{sp2} (point A_1 in figure 1), and in non-stressed - σ_{so} (point B1).

If we take the increment of deformations from the external load in the entire reinforcement to be the same and equal ε_{s}^{N}, then the total deformations will be:

in rebar A_{sp} (point A in figure 1) - $\varepsilon_s = \varepsilon_{sp2} + \varepsilon_s^N$; (13)

in rebar A_s (point B) - $\varepsilon_{sp} = \varepsilon_s^N - \varepsilon_{so}$; (14)

The stresses corresponding to these deformations are σ_s and σ_{sd}. The increment of deformations in the reinforcement when the compressed zone of the element is destroyed is expressed by a known dependence:

$$\varepsilon_s^{N} = \frac{\varepsilon_{bu}}{1-\omega/1.1}\left(\frac{\omega}{\xi}-1\right), \tag{15}$$

and stress in elastic operation and with respect to prestress σ_{s2} - according to the formula of standards

$$\sigma_s^N = \overline{R_s}(\omega/\xi - 1) + \sigma_{s2}, \tag{16}$$

$$\overline{R_s} = \sigma_{sc,u}/(1-\omega/1.1), \tag{17}$$

Deformation included in expression (13)

$$\varepsilon_{sp2} = \varepsilon_{sp} - (\sigma_{los} - \sigma_1)/E_s, \tag{18}$$

where ε_{sp} at $\sigma_{sp} > \sigma_{el}$ is defined by the formula (5) taking $\sigma_s = \sigma_{sp}$, and at $\sigma_{sp} \leq \sigma_{el}$ deformation $\varepsilon_{sp} = \sigma_{sp}/E_s$.

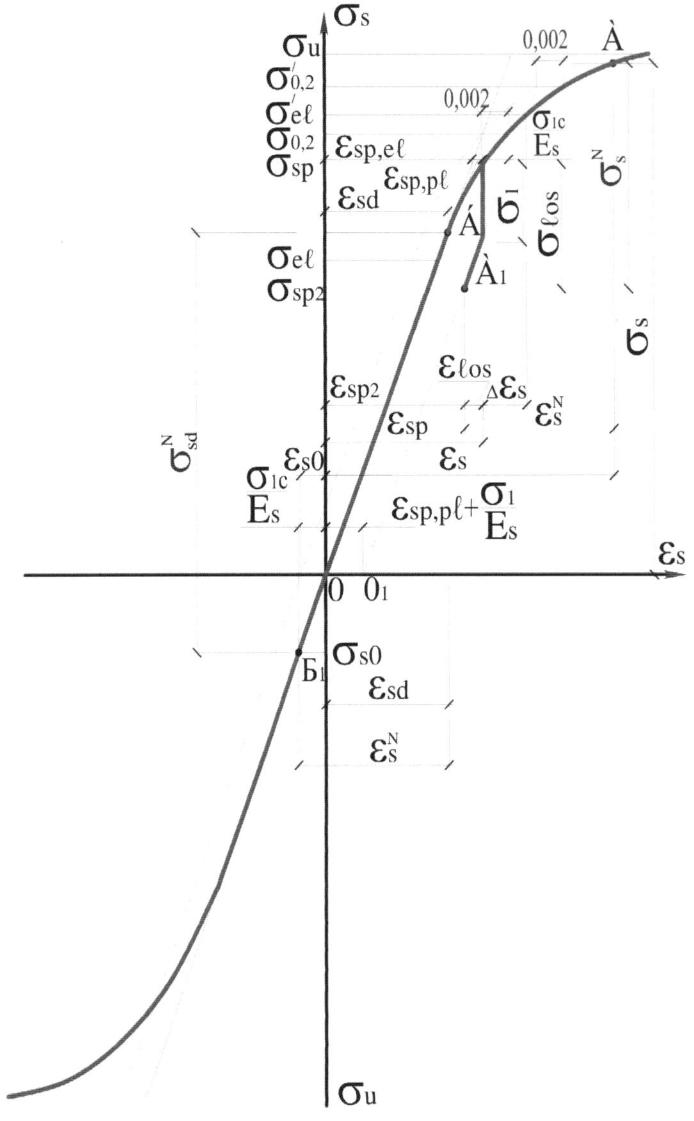

Figure1. Stresses in a prestressed and non-stressed reinforcement when it is located in a cross-section zone that is stretched from the external load.

Deformations of non-stressed reinforcement $\varepsilon_{so} = \sigma_{so} / E_s$, where σ_{so} - numerically equal to the prestress losses.

The stresses in the rebar A_{sp} after applying an external load at $\sigma_s \leq \sigma'_{el}$ is

$$\sigma_s = \left(\varepsilon_{sp2} - \varepsilon_{sp,pl} - \sigma_1 / E_s + \varepsilon_s^N\right) E_s = \left[\varepsilon_{sp} - \left(\sigma_{los} - \sigma_1\right) / E_s - \varepsilon_{sp} + \sigma_{sp} / E_s - \sigma_1 / E_s + \varepsilon_s^N\right] E_s,$$

$$\text{or } \sigma_s = \overline{R_s}\left(\omega / \xi - 1\right) + \sigma_{sp2}, \tag{19}$$

a well-known standard formula is obtained.

At $\sigma_s > \sigma'_{el}$ the total deformation of a rebar A_{sp} according to (13), (17) and (18) shall be

$$\varepsilon_s = \overline{R_s}\left(\omega / \xi - 1\right) / E_s + \varepsilon_{sp} - \left(\sigma_{los} - \sigma_1\right) / E_s, \tag{20}$$

where ε_{sp} is determined by the formula (5) $\sigma_s = \sigma_{sp}$.

The stress corresponding to the strain ε_s is determined by the formula (1).

In a non-stressed reinforcement A_s the total deformation according to (14), (15), (17) is equal to

$$\varepsilon_{sd} = \overline{R}_s(\omega/\xi - 1)/E_s - \sigma_{so}/E_s, \quad (21)$$

Corresponding stresses at $\sigma_{sp} \le \sigma_{el}$ are equal to

$$\sigma_{sd} = \overline{R}_s(\omega/\xi - 1) - \sigma_{so}, \quad (22)$$

For non-elastic work ($\sigma_{sd} > \sigma_{el}$) they are determined by the formula (1), for deformations by (21).

In reinforcement located in a zone compressed from external load (figure 2), compressive stresses from the external load corresponding to the deformations ε_s^N are added to the steady-state stresses σ'_{sp2} (point A1) and σ'_{so} (point B1).

In the rebar A'_{sp} compressive stresses are added to the tensile stresses, so it works elastically and the total stresses (point A) are determined by the formula (19). Note that for a rebar located in the compressed zone, the first term of the right-hand side of this expression is negative and its highest value is equal to $\sigma_{sc} = -\sigma_{sc,u} + \sigma_{sp2}$.

In a non-stressed rebar A'_s deformations ε'_{sd} (point B) are determined by the formula (21), and the stresses at $\sigma'_{sp} \le \sigma_{el}$ - by the formula (22). At $\sigma'_{sp} \le \sigma_{el}$ stresses σ'_{sp} are determined by the formula (1).

In a rebar of the compressed cross-section zone, the stress at the destruction of the element reaches the calculated resistance at x/a' ≥2, where a' - is the distance from the compressed face to the axis of the longitudinal reinforcement[9,10]. This condition is usually always met in compressed elements.

In case of its violation in the intervals 1< x/a' <2 the deformations in a rebar change according to the linear law:

$$\varepsilon_{sc} = \varepsilon_{bu}(\xi h_0/a' - 1) \le \varepsilon_{bu}, \quad (23)$$

The reinforcement stresses in a specified range will be:

in the rebar A'_{sp} $\quad \sigma_{sc} = \sigma_{sp2} - \sigma_{sc,u}(\xi h_0/a_p' - 1)$ (24)

in the rebar A'_s the stresses are equal to: $\sigma'_{sd} = -\sigma'_{su} - \sigma_{sc,u}(\xi h_0/a_s' - 1)$ (25)

In formulas (24)-(25) a'_p and a'_s - is the distance from the longitudinal axes of the rebar A'_{sp} and A'_s to the most compressed face.

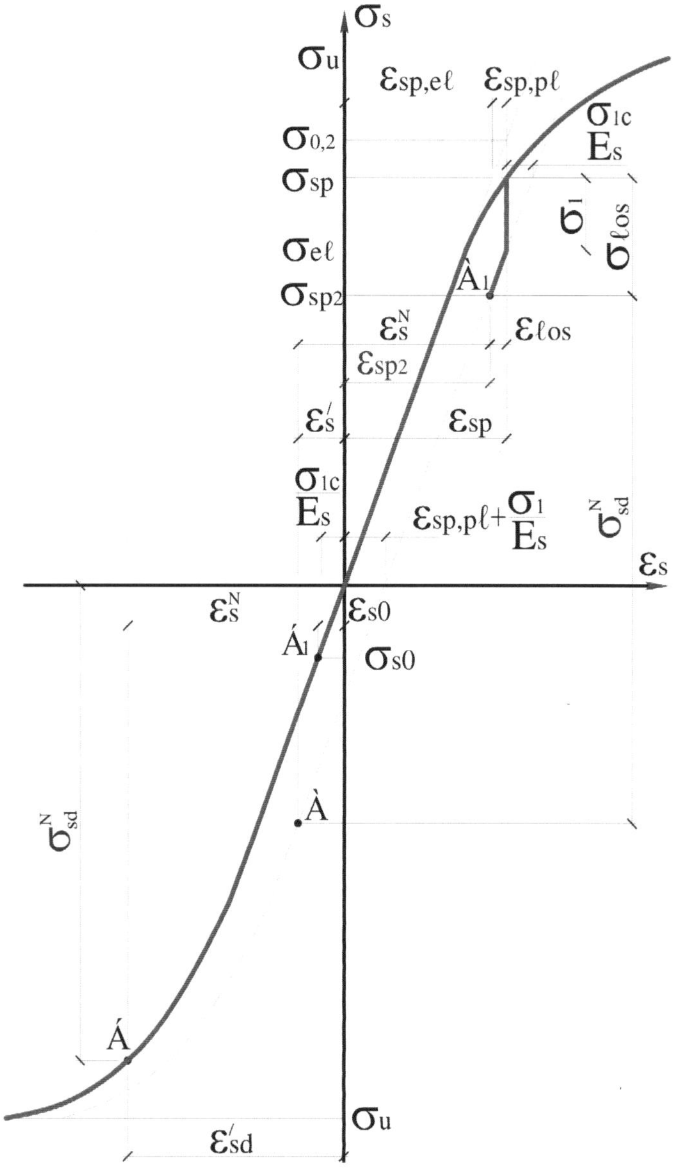

Figure2. Stresses in a prestressed and non-stressed rebar when it is located in a cross-section zone compressed from external load.

The given expressions for determining a reinforcement stress in rebars are used in the stress equilibrium equations. The first of them – the equation of moments relative to the line of action of the resultant external forces - is used to determine the height of the compressed cross-section zone and the stresses in the rebar of the stretched (or less compressed) cross-section zone[6,7]. The second - the equation of moments relative to the axis passing through the point of application of the resultant force in the rebar of the stretched zone - is used to determine the bearing capacity of the element.

When the compressed stress element is destroyed, the reinforcement stresses in the most compressed zone with a sufficiently large value of its height ($x \geq 2a'$) will obtain the maximum increment $\sigma_{sc,u}$ and will be:

in pre-stretched $A_{s(2)}$ -

$$\sigma_{s(2)} = \sigma_{sc,u} - \sigma_{s2(2)} \tag{26}$$

in non-stressed $A_{s(4)}$ -

$$\sigma_{s(4)} = \sigma_{sc,u} - \sigma_{s2(4)} \tag{27}$$

In expression (27) the stresses $\sigma_{s2(4)}$ are compressive and are entered with the sign (-), so the total stresses $\sigma_{s(4)}$ increase.

The equation of moments relative to the line of action of an external force will take the form:

$$\sigma_{s(1)}S_{(1)} + \sigma_{s(3)}S_{(3)} = \gamma_{Rb}R_b bx(e + 0.5x - h_{oz}) + (\sigma_{sc,u} - \sigma_{s2(4)})S_{(4)} + (\sigma_{sc,u} - \sigma_{s2(2)})S_{(2)}, \tag{28}$$

where: $S_{(i)} = A_{s(i)}(e + a_z - a_{(i)})$ at $i = 1;3.$

$S_{(i)} = A_{s(i)}(e_2 + a_{(i)})$ at $i = 2;4.$

γ_{Rb} is a coefficient of concrete working conditions, taking into account the previous stress history.

For elastic operation of the reinforcement of the stretched zone, the expressions for the stresses according to (16) and (17) are represented as:

$$\sigma_{s(1)} = \left[\overline{R_s}\omega h_{0(1)} + x\left(\sigma_{s2(1)} - \overline{R_s}\right)\right]/x, \tag{29}$$

$$\sigma_{s(3)} = \left[\overline{R_s}\omega h_{0(3)} + x\left(\sigma_{s2(3)} - \overline{R_s}\right)\right]/x, \tag{30}$$

Substituting (29)-(30) in (28), after the transformations we get the cubic equation

$$Ax^3 + Bx^2 + Cx + Д = 0, \tag{31}$$

where: $\mathrm{A} = 0.5\mathrm{R}_b b$ (32)

$$\mathrm{B} = \gamma_{Rb}\mathrm{R}_b b(e - h_{oz}) \tag{33}$$

$$C = SL - SK \tag{34}$$

$$SL = \sigma_{sc,u}(S_{(4)} + S_{(2)}) - \sigma_{s2(4)}S_{(4)} - \sigma_{s2(2)}S_{(2)}, \tag{35}$$

$$SK = \sigma_{s2(1)}S_{(1)} + \sigma_{s2(3)}S_{(3)} - \overline{R_s}(S_{(1)} + S_{(3)}), \tag{36}$$

$$D = -\overline{R_s}\omega(S_{(1)}h_{0(1)} + S_{(3)}h_{0(3)}), \tag{37}$$

Having determined from (28) the value of x, and from (29)-(30) $\sigma_{s(1)}$, $\sigma_{s(3)}$, the latter are compared with the new values of the elastic limits of steel σ'_{el}. If at least one of the stresses exceeds σ'_{el} a new stress value is determined. To do this the deformation $\varepsilon_{s(i)}$ is determined by formula (20), and according to it the stress $\sigma_{s(i)}$ by formula (1). After determining the stresses $\sigma_{s(1)}$ and $\sigma_{s(3)}$ and substituting them into equation (28), the latter is reduced to the square equation:

$$Ax^2 + Bx + C = 0, \tag{38}$$

from which a new value of x, after which the values of $\sigma_{s(1)}$ and $\sigma_{s(3)}$ are again determined.

The carrying capacity of an element is determined at $x \geq 2a'$ and $x \leq 0$ (with substitution x=0) by the formula:

$$N = 1/e\left[R_b \gamma_{Rb} x b\left(h_{oz} - 0.5x\right) + A_{s(2)}\left(\sigma_{sc,u} - \sigma_{s2(2)}\right)\left(h_{oz} - a_{(2)}\right) + A_{s(4)}\left(\sigma_{sc,u} - \sigma_{s2(4)}\right)\left(h_{oz} - a_{(4)}\right)\right], \quad (39)$$

and at $0 > x > 2a'$ - by the formula:

$$N = 1/e(R_b \gamma_{Rb} x b\left(h_{oz} - 0.5x\right) + A_{s(2)}\left[\sigma_{sc,u}\left(x/a_{(2)} - 1\right) - \sigma_{s2(2)}\right]\left(h_{oz} - a_{(2)}\right) + \\ + A_{s(4)}\left[\sigma_{sc,u}\left(x/a_{(4)} - 1\right) - \sigma_{s2(4)}\right]\left(h_{oz} - a_{(2)}\right) \quad (40)$$

3. Conclusion

The above given method allows you to refuse determining the calculation case and the coefficient that takes into account the work of the reinforcement beyond the conditional yield point, moreover, this method for calculating the strength of compressed elements with mixed reinforcement is represented according to the calculation program on a computer.

4. References

[1] Baykov V N 1980 Stress-strain behaviour of reinforced concrete elements based on generalized experimental dependences of physical and mechanical characteristics of concrete and reinforcement *Questions of strength, deformability and crack resistance of reinforced concrete* **8** (Rostov-on-Don)

[2] Gushcha Y P 1983 About accounting for non-elastic deformations of concrete and rebar in calculation of reinforced concrete structures*Improvement of structural forms, methods of calculation and design of reinforced concrete structures* (Moscow: Scientific Research Institute of Concrete and Reinforced Concrete) pp 11-8

[3] Madatyan S A1980*Technology of reinforcement tension and load-bearing capacity of reinforced concrete structures* (Moscow: Stroizdat) p196

[4] Mailyan R L and Mekerov B A 1983 Method of accounting for the effect of pre-stressing when calculating the strength of reinforced concrete elements *Concrete and reinforced concrete***9** 28-30

[5] Mekerov B A and Mailyan R L 1982 Analytical descriptions for the tensile diagram of high-strength reinforcing steel *New types of reinforcement and its welding.* (Collection of reports of the All-Union meeting in Volgograd: Moscow) pp 166-71

[6] Ryabukhin A and Matsiy S 2013 In-situ measurement of the anchor-pile displacement in the geotechnical conditions of Sochi *Proceedings of the 5th International Young Geotechnical Engineers' Conference (IYGEC 2013)* (France. Paris :Ecole des PontsParisTech) pp 576–8

[7] Ryabukhin A K, Lesnoy V A and Kalashaov D Y 2019 Study of the stability of soil slopes and downhills with the use of fiberglass dowels *IOP Conf. Ser.: Mat. Sc. and Eng.* **698**(7) 077003

[8] Bezuglova E. and Matsiy S 2013 Engineering and geological grounds of landslide protection reliability of structures *Landslide Science and Practice: Risk Assessment, Management and Mitigation* **6** 709-14

[9] Bezuglova E, Matsiy S and Podtelkov V. 2016 Landslide risk management at transport facilities *Landslides and Engineered Slopes. Experience, Theory and Practice* **2**, 2016. 405-9

[10] Matsiy S I and Sukhlyaeva L A 2019 Antimud flow protection with rigid thorough structures/ *Geotechnics Fundamentals and Applications in Construction: New Materials, Structures, Technologies and Calculations* (Proc. of the Int. Conf. on Geotechnics Fundamentals and Applications in Construction: New Materials, Structures, Technologies and Calculations, GFAC) pp 178-81

IOP Conf. Series: Materials Science and Engineering **889** (2020) 012027 doi:10.1088/1757-899X/889/1/012027

Investigation on bond performance between different types of FRP -reinforcedrebars and concrete

A I Komarov[1], N V Makarova[1,2] and V G Tsuprik[1]

[1] Engineering School, Far Eastern Federal University, 8 Sukhanova St., Vladivostok 690090, Russia
[2] Institute of Automation and Control Processes Far Eastern Branch of the Russian Academy of Sciences, 5, Radio st., Vladivostok, 690041, Russia

The corresponding author's e-mail address: maknat@bk.ru

Abstract. In this paper the test results of a study on the bond behaviour of FRP bars to fine-grained concrete are presented. Five series of concrete cylinders reinforced with basalt-reinforced polymer (BFRP) and glass fiber-reinforced polymer (GFRP) bars are tested in direct pullout conditions. As well as two series with steel bars are tested for comparison. For each specimen, the bond failure mode, the average bond strength, the initial bond stiffness, and the load–strain relationship curves and bar relative extension were analysed. Test results demonstrate the promise of using the FRP bars as an alternative to the steel bars in reinforcing concrete elements.

1. Introduction

Recently, the use of FRP rebar has increased, due to their unique properties. Their attractiveness is caused by such qualities as durability, corrosion resistance, electromagnetically transparency, etc. [1-5]. FRP rods are used as reinforcement for concrete structures such as bridges, foundations and marine structures, where the corrosion of steel reinforcement typically led to significant deterioration and rehabilitation needs [6]. The commonly used fiber material are glass, aramid and carbon. The increasing use of FRP bars encourages utilizing new materials, such as basalt fibers [7]. An extensive experimental investigations is being conducted to evaluate their physical, mechanical and other technological characteristics [8]. Standard test methods have been developed to determine the FRP rods characteristics for the design and manufacture of reinforced concrete structures [9].

However, the adhesion between reinforcing bars and concrete is still poorly understood due to the variety of materials and surface types [10]. Many factors such as surface deformation, mechanical interlock, chemical adhesion, embedment length, diameter, environmental and loading conditions [2-9], must be considered to fully understand the bond effects.

In the most previous studies, the geometric shape of the surface and its composition were concluded to have the greatest influence on the adhesion strength. In work [11], it is was suggested to divide all FRP reinforcing bars into three groups in accordance with the method used for establishing a bond between the core of the bars and concrete. The effects of structural fibers on the bonding mechanism changes in interface between GFRP bar and concrete were studied in work [12]. It was determined that bond failure modes largely depend on the interfacial property with the rebar. The experimental results of 180 pullout tests conducted on GFRP bars embedded into high strengthconcrete blocks [13] showed that, the bond stress is inversely proportional to the embedment length and bar. Similar study [14] showed the influence of various parameters that affect bond strength such as the embedment length, type, shape, surface characteristics, and diameter of the bar. The results of an experimental study on the bond behavior of GFRP ribbed rebars with different specially designed rib geometries to the concrete are presented in work [15]. It was shown that, in normal strength concrete, the bond strength and bond-slip performance of these ribbed rebars varies with the combinations of rib spacing and rib height.

As a matter of fact, many experimental and analytical studies showed the effectiveness of FRP-based bars for reinforced concrete. Nevertheless, technological standards and design guide-lines norms cannot take into account all diversity of materials and surface shapes by different manufacturers. Therefore, the main objective of this study is to investigate the effect of different materials and surface parameters for the bond strength between FRP rebars and concrete. In addition, to understand the failure mechanism during the loading, the relative extension of the bars is measured.

2. Experimental details

2.1. Materials

Five types of rebars, manufactured in the Primorsky krai,were tested in this study. Four GFRP rebar types with different profile texture were investigated, including one type of bar containing basalt fibers in surface and one type with sand. BFRP rebars was also tested. As well as two series with steel bars were tested for comparison. Nominal diameter of all bars was 4 mm. The properties ofrebars tested in this study are presented in Table 1.

Table 1. Properties of bars.

Type	Material	Profile	Rib height(mm)	Rib spacing(mm)	Tensile strength(MPa)	Tensile modulus(GPa)	Transverse shear strength(MPa)
1	GFRP	ribbed	0.5–0.7	8	1070	54.20	221.70
2	GFRP	ribbed	0.5	7	1162	55.00	194.00
3	GFRP	ribbed	0.3	20	1016	67.82	202.00
4	Steel	ribbed	0.25	3	710	200.00	210.00
5	Steel	plain	–	–	710	200.00	210.00
6	BFRP	ribbed	0.5–0.7	15	1085	50.00	150.00
7	GFRP	sand	–	–	800	50.00	150.00

The samples were cylindrical in diameter 55 mm, height 65 mm. In the exit points of the rods, notches were made to exclude surface cracking of the concrete (Fig. 1). The concrete used for making samples was fine-grained at the age of 28 days; the uniaxial compression strength was 26.4 MPa.The bonded length L for all the tests is taken as 45±mm chosen to be sufficiently representative of rebar deformations [16].

2.2. Test method

The pullout-test developed in [17] was carried out using a universal testing machine (Shimazu AGS-X 10 kN). The concrete specimens were placed in to the rigid device, as shown in Figure 2. The loading rate was 0.1 kN/min. The values of the full pulling force of bar (inclusion of friction forces into work) were measured. In addition the relative extension of the bars was also measured during the loading by used the extensometer fixed on the bar.

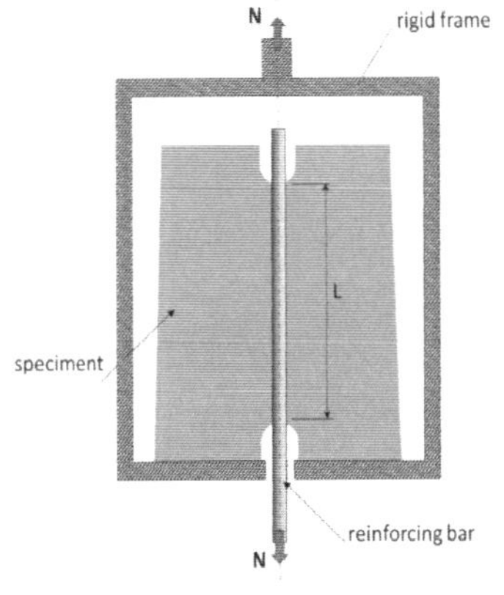

Figure 1. Schematic illustration of the principle of direct pullout testing.

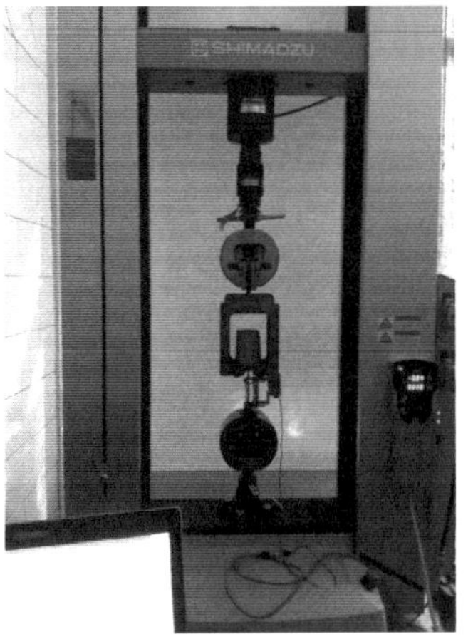

Figure 2. Testequipment.

3. Result and discussion

3.1. Bond strength

Bond strength is calculated using the value on the maximum pullout load P_{max}, assuming a uniform bond stress distribution along the embedded length *L*of the rod in the concrete. The average maximum pull-out load and bond strength of all experimental programs are presented in Table 2. The bond strength was calculated using the nominal rod diameter 4 mm.

Table 2. Average maximum bond strength of all experimental programs.

Type series	Average maximum bond stress (MPa)	Strain (mm)	Bar relative extension (%)
1	9.44±0.53	5.32±1.28	1.09±0.32
2	8.08±0.04	5.23±1.64	1.18±0.30
3	14.18±1.43	7.48±2.15	1.30±0.24
4	5.82±0.94	2.27±0.18	0.10±0.02
5	2.78±0.51	0.99±0.37	0.06±0.005
6	12.68±1.54	5.04±0.11	1.40±0.26
7	6.17±0.73	1.70±0.42	0.44±0.08

GFRP rebars with basalt ribs in surface (Type 3) show higher bond strength (14.18 MPa). BFRP rebars (Type 6) show high bond strength too (12.68 MPa). But the concrete of several specimens of this types was broken, this is characteristic of brittle fracture. The obtained average bond stress for the GFRP rebars of Type 1 and Type 2 was 9.44 MPa and 8.08 MPa respectively. For BFRP rebars with sand in surface (Type 7), the average bond stress obtained was 6.17 MPa.

Comparing the test results of the bars with the ribs on the surface showed that the main factor is the spacing of the ribs. The highest bond values were obtained on GFRP and BFRP bars with the spacing of the ribs 20 mm (Type 3) and 15 mm (Type 2). The effect of spacing is presented in the histogram shown in Figure 3. But on the basic of results, presented in [15], it was obtained that bond behavior of ribbed rebars would improve with a decrease in rib spacing.However, this conclusionneeds further more investigations to validate.

IOP Conf. Series: Materials Science and Engineering **889** (2020) 012027 doi:10.1088/1757-899X/889/1/012027

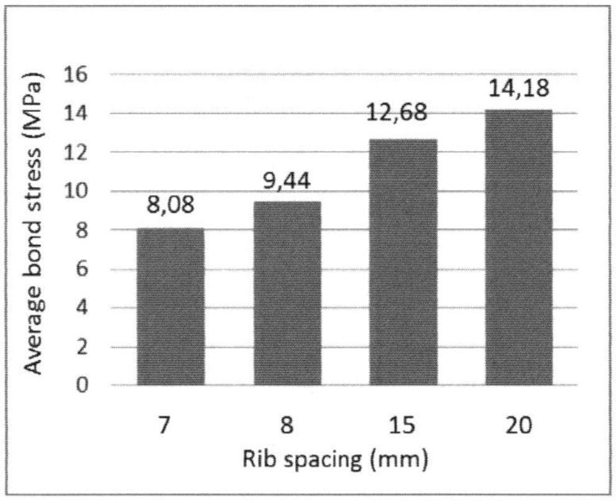

Figure 3. Effect of rib spacing on average bond stress.

Two types of steel bars were tested, having a rough (Type 4) and a plain surface (Type 5). The average bond stress obtained was 5.82 MPa for the bar with a rough surface and 2.78 MPa for the bar with a plain surface.Also, the bond strength in all types of FRP bars is greater than the corresponding steel rebars.

3.2. Load-strain curves

There are measuredfool load-straincurves (a)and local load-strain curves (b) only for the barfor every specimen of seven series. The examples of the typicalcurves are presented in figures 4–7.

The section of the curve to the peak characterizes the stiffness of the bond as the ratio between the load and the deformation. First of all, this provides a mechanism for the combined action of reinforcement and concrete in structures. The bond behaviour of alltype of rebars before peak is linear at early section. After 60% - 80% of the maximum load up to the peak, the curves become non-linear (Figure 4a, 5a, 6a, 7a). For alltypes bars, before peak load level, the relative extension strains are elastic and the curve is linear (Figure 4b, 5b, 6b, 7b).

When the load reaches their maximum value Pmax, the loose deboningzone forms between the concrete and the bar. The post-peak section of the load-strain curve characterizes the frictional shear strength in this debondingzone.After the maximum pull-out load of GFRP bar (Type 1), the debonding of the bar decreases the load to about 40% and then immediately increases of the maximum load to about 20%. The decreasing curve commencing after this is the result of frictional resistance (Figure 4,a), as well as for BFRP bar whis sand (Type 7) and ribbed steel bar (Type 4) (Figure 6,a). For these two types bars, after peak load level, the relative extension strains are not elastic and the curve is not linear (Figure 4b, 6b).

a) b)

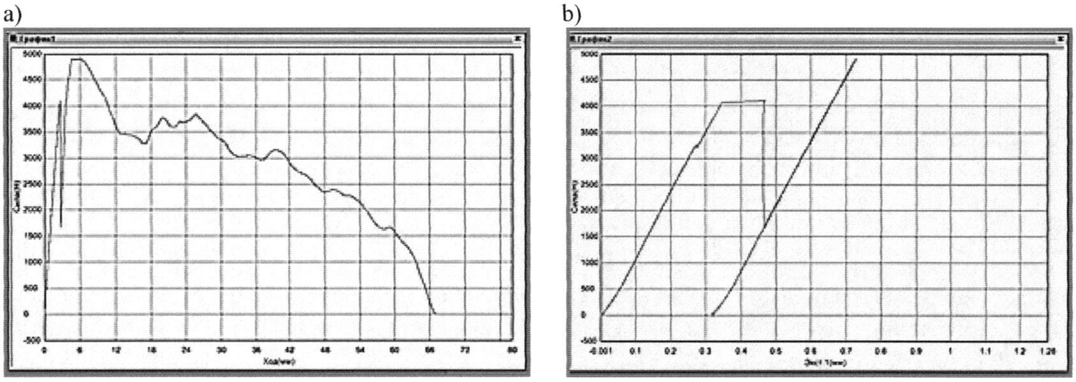

Figure 4. Specimen 1-3 (GFRP bar):a – total load-strain curve; b - local load-strain curve.

a) b)

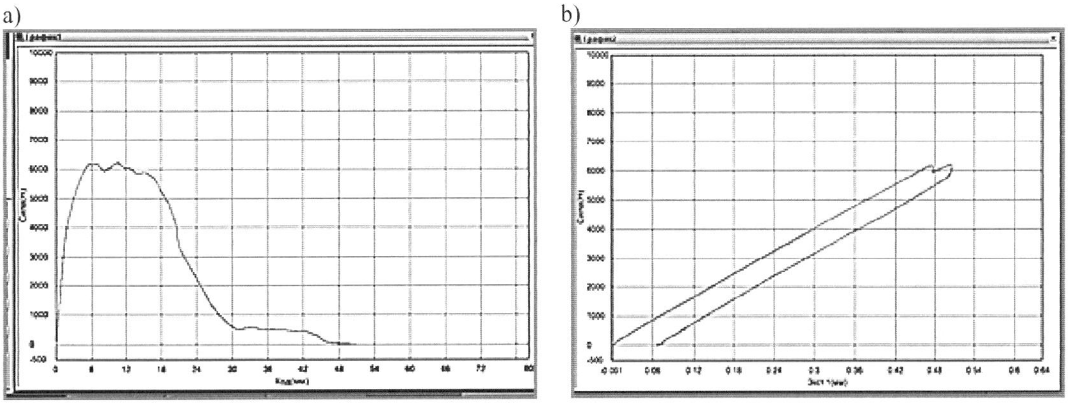

Figure 5. Specimen 3-3 (GFRP bar):a – total load-strain curve; b - local load-strain curve.

a) b)

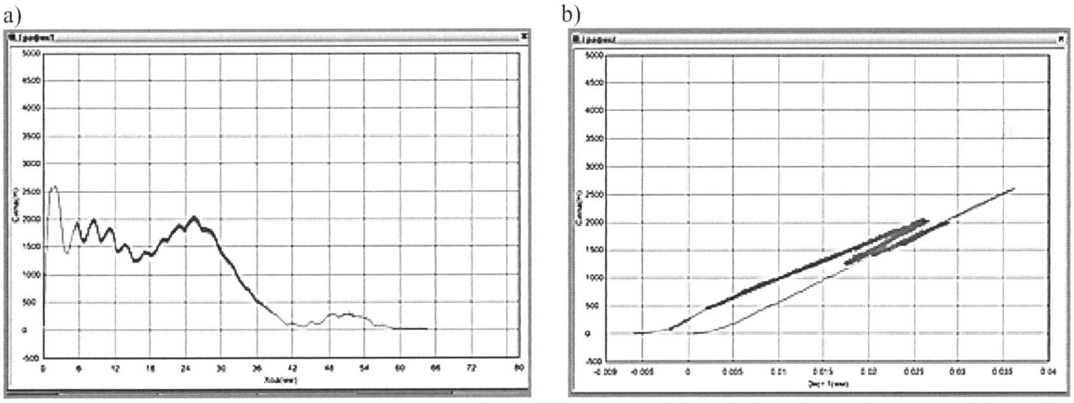

Figure 6. Specimen 4-4 (rib steel bar):a – total load-strain curve; b - local load-strain curve.

a) b)

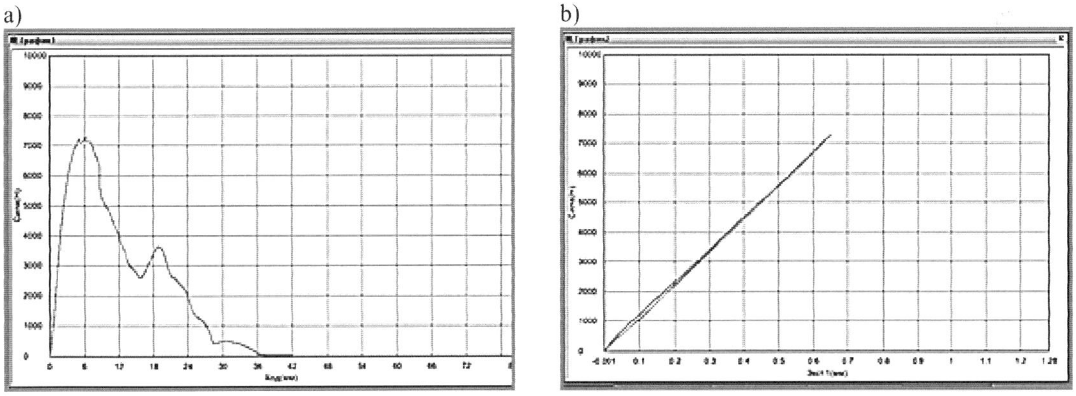

Figure 7. Specimen 6-2 (BFRP bar):a – total load-strain curve; b - local load-strain curve.

After peak load of the GFRP bar (Type 3) rapid drop of the pullout load to about 80-90% of the maximum load decreases (Figure 5,a), this is characteristic of brittle fracture with a minimal contribution of the friction force. Similarly behaviours we observed for of the BFRP bar (Type 6), moreover we noticed a second increase in the load, which was due to the redistribution of stresses after peak load along of the bar (Figure 7,a).

4. Conclusions

Test results demonstrate the promise of using the FRP bars from different manufacturers as an alternative to the steel bars in reinforcing concrete elements.It should be noted that the study of deformation process during the entire loading cycle gives important results for further investigations.

In this study, the conclusions to be drawn are that there is an adhesion and friction are the most important components of bond stress in FRP reinforcing bars. The greatest strength obtained with GFRP and BFRP rods with spacing of the ribs 15 and 20 mm. However, the analysis of the load-strain curves showed the brittle fracture mechanism; it can lead to brittle cracking of the structure during operation. Samples with low bond (but more than steel bars) strength showed the plastic nature of the deformation.

Thus, further research should be carried out directly in the concrete reinforced structures.It will contribute to introducing new development in the FRP design codes.

5. References

[1] BronnikovIV 2019 Compositereinforcement — problemsandprospectsapplications*Int. J. of Appl. ScienceandTechn.* «Integral» **3** 223(in Russian)

[2] Lapshinov A E 2015Prospects of Potential Application of Non-Metallic FRP Reinforcement in FRP-Reinforced Concrete Compressive Members as Main Longitudinal Non-Prestressed Reinforcement*Proc. of Moscow State University of Civil Engineering* **10** 96 (In Russian)

[3] Portnov G, Bakis C E, Lackey E. and Kulakov V 2013 FRP Reinforcing bars - designs and methods of manufacture (Review of Patents)*Mechanics of Composite Materials* **49**(4) 381

[4] Teplova Z S, Kiski S S and Strizhkova Y N 2014. Fiberglass reinforcement for armouring of concrete structures*Construction of Unique Buildings and Structures* **24**(9) 49 (in Russian)

[5] Behnam B and Eamon C 2013 Reliability-based design optimization of concrete flexural members reinforced with ductile FRP bars*Construction and Building Materials* **47** 942

[6] Benmokran B and Ahmed H A 2016Durability of FRP rebars in aggressive environments Proc. of 8th Int. Conference on Fibre-Reinforced Polymer (FRP) Composites in Civil Engineering (CICE2016)

[7] Elgabbas F, Cousin P, Ahmed E and Benmokrane B 2014 Development and characterization of basalt FRP reinforcing bars for concrete structuresProc. of the 7th International Conference on FRP Composites in Civil Engineering, International Institute for FRP in Construction (CICE) pp 20-2

[8] Monaldo E, Nerilli F and Vairo G. 2019 Basalt-based fiber-reinforced materials and structural applications in civil engineering*Composite Structures* **21415** 246

[9] GOST 31938-2012 Fibre-reinforced polymer bar for concrete reinforcement. General specifications (in Russian)

[10] Elgabbas F, Ahmed E A and Benmokrane B 2015 Physical and mechanical characteristics of new basalt-FRP bars for reinforcing concrete structures*Construction and Building Materials* **95** 623

[11] Wang H and Belarbi A 2010 Static and Fatigue Bond Characteristics of FRP Rebars Embedded in Fiber-reinforced Concrete*Journal of Composite Materials* **44**(13) 1605

[12] Kim B, Doh J H; Yi C K and Lee J Y 2013 Effect of structural fibers on bonding mechanism changes in interface between GFRP bar and concrete*Composites* **B 45** 768

[13] Islam S; Afefy H M; Sennah K and Azimi H 2015Bond characteristics of straight- and headed-end, ribbed-surface, GFRP bars embedded in high-strength concrete*Construction and Building Materials* **83** 283

[14] Achillides Z and Pilakoutas K 2004Bond behavior of fiber reinforced polymer bars under direct pullout conditions *J. of Composites for ConstructionASCE* **8** (2)173

[15] Hao Q; Wang Y; He Z and Ou J 2009 Bond strength of glass fiber reinforced polymer ribbed rebar in normal strength concrete*Construction and Building Materials* **23** 865

[16] Cosenza E; Manfredi G; Pecce M and Realfonzo R 1999 Bond between GFRP rebars and concrete: an experimental analysis *Proc. of the 4th Int.l Symp. on Fiber Reinforced Polymer*

IOP Conf. Series: Materials Science and Engineering **889** (2020) 012027 doi:10.1088/1757-899X/889/1/012027

for Reinforced Concrete Structures (American Concrete Inst. Detroit) pp 347-58
[17] Makarova N V and Makarov S V 2018 Femtosecond laser nanostructuring of reinforcement bars surface for improvement of its interaction with concrete *J. of Physics: Conf. Series* **1092**(1) 012082 (IOP Publishing)

IOP Conf. Series: Materials Science and Engineering **889** (2020) 012028 doi:10.1088/1757-899X/889/1/012028

Antioxidant in a model biomembrane – astaxanthin and its esters mixed with DPPC in Langmuir films

E A Kulikov[1], A A Stupnikov[1] and Y N Malakhova[1]*

[1]National Research Centre "Kurchatov Institute", 123182, Moscow, Russia

*Malakhova_YN@nrcki.ru

Abstract. Amphiphilic properties of astaxanthin molecule and its esters make it possible to include it in self-organized structures based on lipid membrane components. Astaxanthin and its mono- and diesters were isolated from the extract of the microalgae *Haematococcus Pluvialis*, purified by column chromatography, and identified by thin-layer chromatography. The absence of impurities in astaxanthin and its esters was confirmed by means of high performance liquid chromatography method with detection in the ultraviolet-visible region and mass spectrometric detection. The model systems of the cell membrane lipid bilayer — Langmuir films of mixtures of DPPC with astaxanthin and its mono- and diesters, were formed at the air-water interface and studied by a complex of methods. It was found that an increase in the amount of astaxanthin, as well as the addition of its esters to the Langmuir films of DPPC, leads to an increase of values of area per molecule at the surface pressure isotherm uprise and decrease of Langmuir film collapse pressure. The addition of astaxanthin mono- and diesters prevents the phase transition in the DPPC monolayer more strongly than unesterified astaxanthin.

1. Introduction

Astaxanthin and other carotenoids are involved in the course of biochemical processes in living organisms as antioxidants, they protect the membranes of living cells from destruction by free radicals and active oxygen forms that are produced when the skin is exposed to UV radiation from sunlight. The level of antioxidants in the body decreases with increasing oxidative stress (smoking, sunburn, aging, etc.), the source of their intake can be food, as well as dietary and biologically active additives [1], that are promising in the complex treatment of pathologies and diseases caused by aging [2–4]. It is assumed that carotenoids with a maximum absorption of 400-500 nm, including astaxanthin, perform protective functions in the human body, protecting the skin and eyes from excessive illumination of the visible part of the spectrum [4, 5], from photoaging of the skin [6]. Therefore, the development of photoprotective cosmetics and dietary supplements is relevant, astaxanthin carotenoid being one of the promising ingredients [7] that has antioxidant [8, 9], anti-inflammatory [10, 11], and wound healing [12, 13] effects. The potential application of astaxanthin is not limited to dermatology. Prospects for its use as a hepatoprotector [14], in complex therapy for the treatment of neurodegenerative diseases and brain injuries [15–17], consequences of diabetes mellitus [18], and age-related eye pathologies [19, 20] have been noted.

From the microalgae Haematococcus Pluvialis [5], containing up to 5% astaxanthin, it is extracted by organic solvents in the form of mono- and diesters [6]. It is known that the bioavailability of astaxanthin is higher than that for its esters [1, 21]. In living organisms, astaxanthin exhibits antioxidant activity both in the cytoplasm and in the lipid membrane, since it is able to integrate into it, increase its rigidity and prevent the penetration of active oxygen forms and other radicals through the membrane [2, 3, 22]. Due to the amphiphilic structure of the astaxanthin molecule, the properties of its derivatives [23], as well as mixtures with lipid components, can be studied on model biological membranes – Langmuir monolayers at the air/water interface [24-28]. The results obtained on such simplified systems can be extrapolated to biomembranes [22, 28]. The aim of this work was to study the self-assembly of mixtures of astaxanthin and its esters with DPPC in Langmuir films at the air/water interface.

2. Experimental part

Commercially available DPPC (1,2-dipalmitoyl-sn-glycero-3-phosphocholine, Sigma-Aldrich, USA), astaxanthin (Sigma-Aldrich, USA), as well as its mono - and diesters isolated from the microalgae *Haematococcus Pluvialis* extract (Sigma-Aldrich, USA) and purified by column chromatography on silica gel (mobile phase – mixture of acetone : hexane, gradient elution). Components were identified by thin-layer chromatography. On average, the microalgae *Haematococcus Pluvialis* extract contains: 75% astaxanthin monoesters, 20% astaxanthin diesters and 5% astaxanthin.

Astaxanthin and its esters were separated by high performance liquid chromatography (HPLC) with detection in UV-visible range and mass detection (Agilent 1200 series with a diode-matrix detector and 6120 Quadrupole LC/MS, Agilent Technologies Inc., USA). Astaxanthin monoesters were separated on a 150×2.1 mm reverse phase column, 5 μm particle size, Orbit-C18. Astaxanthin diesters were separated on a 250×4.6 mm column, 5 μm particle size, Eclipse-C18. The detection was performed on a UV-visible detector at a wavelength of 470 nm, as well as a mass spectrometric detector in the mode of positive ions in the range of 830–870 m/z, 870–900 m/z for astaxanthin monoesters and 1050–1150 m/z for astaxanthin diesters. The mobile phase was 50% (V/V) acetonitrile in water (Millipore, USA) (A) and 100% acetonitrile (B) for astaxanthin monoesters. The flow rate was 700 μl/min, the volume of the injected sample was 20 μl. The elution program was started with 25% A and 75% B and continued in gradient elution mode (0–12 min) to 100% B. From 12 minutes to the end it was isocratic mode of 100% (V/V) of the mobile phase B. The mobile phase was 100% (V/V) acetone : water (Millipore, USA) 90 : 10 (A) and acetone : water (Millipore, USA) 98 : 2 (B) for astaxanthin diesters. The flow rate was 800 μl/min, the volume of the injected sample was 20 μl. The elution program started with 100% A and continued in gradient elution mode (0–50 min) until 100% B.

The formation and investigation of Langmuir films properties was performed on the Langmuir Minitrough Extended (KSV, Finland) with a maximum interfacial surface area of 558 cm^2. Purified and demineralized water (Millipore, USA) with a specific resistance of 18.2 MOhm·cm (at 25 °C) was used as a subphase. The subphase was thermostated at 20 °C. The solutions of DPPC and its mixtures with 1 wt. % of astaxanthin, 1 wt. % of astaxanthin monoesters and 1 wt. % of astaxanthin diesters in chloroform with a concentration of 1 mg/ml were prepared. Langmuir films were formed after the spreading of solutions and evaporation of the solvent and were studied in the compression mode with a constant rate of 15 cm^2/min. Surface pressure was measured by the Wilhelmy method using a rough platinum plate with an accuracy of 0.1 mN/m. Surface potential was measured by the vibrating electrode method using a SPOT sensor (KSV, Finland) with an accuracy of 1 mV. The morphology of Langmuir films directly on the water surface was visualized using a BAM-300 Brewster microscope (KSV, Finland). The obtained micrographs corresponding to the interfacial surface of 100×100 μm^2 were geometrically corrected taking into account the observation at the Brewster angle (53.1°).

3. Results and discussion

Model systems of the cell membrane lipid bilayer – the Langmuir films of mixtures of DPPC with astaxanthin and its mono- and diesters were formed at the air-water interface and were studied by a complex of methods. The surface pressure vs. mean molecular area (π-A) compression isotherms and

IOP Conf. Series: Materials Science and Engineering **889** (2020) 012028 doi:10.1088/1757-899X/889/1/012028

the surface potential vs. mean molecular area (ΔU-A) compression isotherms are shown in figure 1, surface morphology of Langmuir films visualized by Brewster angle microscopy is shown in figure 2.

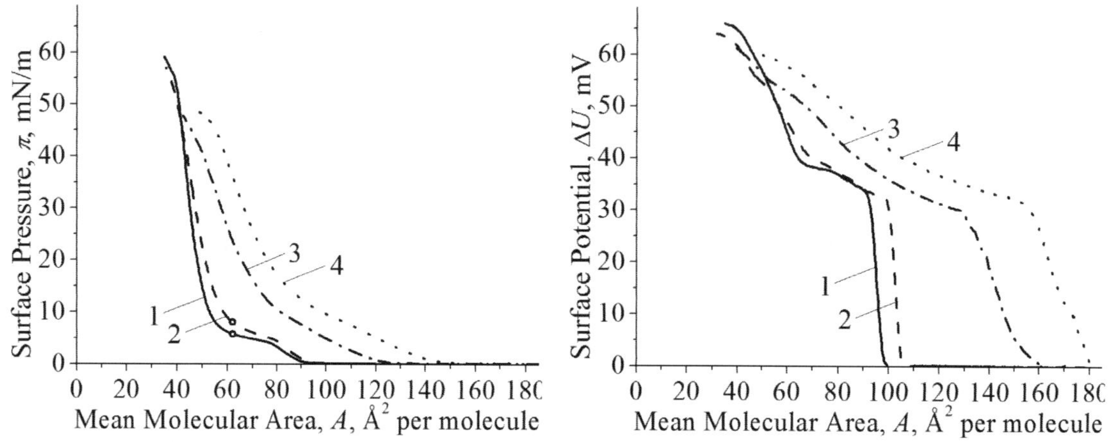

a b

Figure 1. The π-A compression isotherms (a) and the ΔU-A compression isotherms (b) of the Langmuir films of DPPC (1), DPPC mixed with 1 wt. % of astaxanthin (2), DPPC mixed with 1 wt. % of astaxanthin monoester (3), DPPC mixed with 1 wt. % of astaxanthin diester (4) at the air/water interface at 20 °C. The empty circles in plot (a) correspond to Brewster angle micrographs in figure 2.

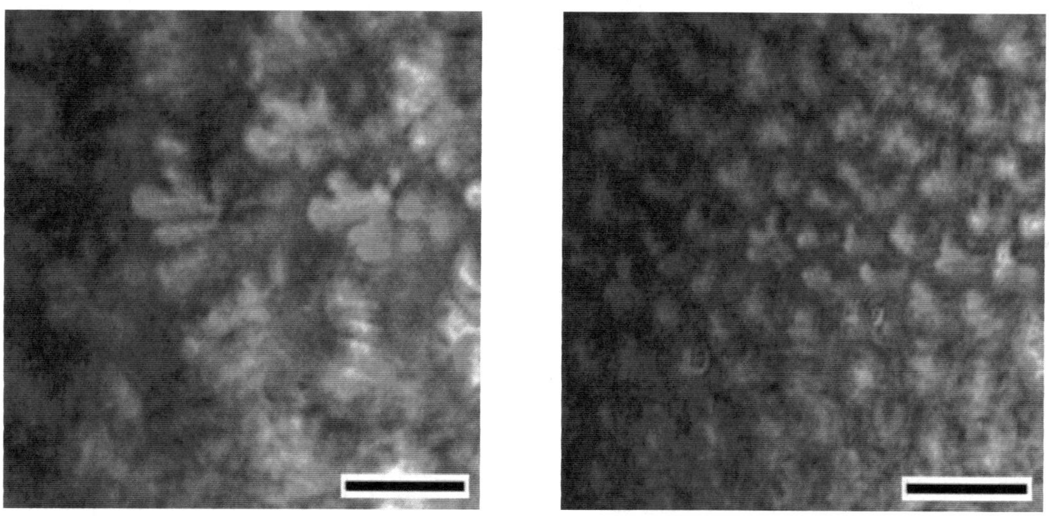

a b

Figure 2. Brewster angle micrographs for the surface of compressed Langmuir films: DPPC to a surface pressure of 5.8 mN/m (a), DPPC mixed with 1 wt. % of astaxanthin compressed to a surface pressure of 8.1 mN/m (b) at a subphase temperature of 20 °C. Scale bar 50 um.

The addition of astaxanthin, as well as its mono-and diesters in the amount of 1 wt. % of in DPPC leads to a change in the shape and parameters of the π-A and the ΔU-A compression isotherms of Langmuir films. The shape of the ΔU-A isotherms for mixed Langmuir films is characterized by the disappearance of the "plateau" section. The magnitude of the jump in the surface potential decreases to 295±5 mV both in the case of astaxanthin and its esters, and the maximum value at the point of collapse of the monolayer is reduced to 595±2 mV in the case of astaxanthin and to 555±2 mV in the case of astaxanthin esters. It was found that the mean molecular area at the beginning of growth in surface pressure increased from 90±1 to 93±1 Å^2 per molecule with the addition of astaxanthin, as well as to 122±1 and 145±1 Å^2 per molecule with the addition of astaxanthin mono- and diesters, respectively. A change in the shape of the π-A isotherms in the region of the DPPC phase transition in mixed Langmuir films was revealed. The bend at the plateau exit point was still noticeable for DPPC mixed with astaxanthin, but was no longer evident for DPPC mixed with astaxanthin esters. A significant reduction in the size of condensed phase aggregates is noticeable in the Brewster angle micrograph (figure 2b) with the addition of 1 wt. % of astaxanthin to DPPC Langmuir film compared to the pure DPPC Langmuir film (figure 2a). The bending on at the π-A isotherm associated with the end of the phase transition of the Langmuir film from the liquid expanded to the liquid condensed state is observed with increasing values of the mean molecular area (in Å^2 per molecule) during the transition from pure DPPC (57±1) to its mixtures with astaxanthin (62±1), with astaxanthin monoester (76±1), with astaxanthin diester (90±1). The surface pressure values (in mN/m) at that point also increase with the transition from pure DPPC (7.0±0.1) to its mixtures with astaxanthin (8.0±1), its mono (12.3±1) and diesters (12.5±1). Thus, the addition of mono- and diesters of astaxanthin increases the surface pressure and the mean molecular area more strongly, and therefore loosens the DPPC lipid monolayer more strongly than non-esterified astaxanthin. The stability of the Langmuir films of DPPC mixed with astaxanthin and its esters, determined by the value of the surface pressure at the collapse point, is lower compared to the Langmuir film of pure DPPC. The addition of unesterified astaxanthin reduces the stability of the Langmuir film from 55±1 mN/m at 38±0.5 Å^2 per molecule to 48.5±1 mN/m at 40±0.5 Å^2 per molecule, and to 45.0±1 mN/m at 45±1 and 55±1 Å^2 per molecule for astaxanthin mono- or diesters, respectively.

4. Conclusion

The study of astaxanthin and its esters in the model systems of the cell membrane lipid layer – Langmuir films – showed that the structure formation of DPPC from a liquid-expanded to a liquid-condensed layer is hindered both in the presence of astaxanthin and astaxanthin mono - and diesters. The shift of the surface pressure and surface potential isotherms in the direction of increasing area per mollecule, when adding astaxanthin and its esters to the Langmuir films of DPPC, was detected. The addition of astaxanthin mono-and diesters more strongly prevents the phase transition in the DPPC monolayer than non-esterified astaxanthin. As a result, the stability of Langmuir films of DPPC mixed with astaxanthin and its esters is lower than that of the Langmuir film of pure DPPC. The results obtained on model systems will be taken into account in the development of stable lipidized forms of astaxanthin and its esters with antioxidant properties able to protect the skin from active oxygen forms.

5. References

[1] Mimoun-Benarroch M, Hugot C, Rhazi L, Niamba C N and Depeint F 2016 *Bulletin UASVM Food Science and Technology* **73** 61
[2] Kidd P 2011 *Altern. Med. Rev.* **16** 355
[3] Ambati R, Phang S M, Ravi S and Aswathanarayana R 2014 *Mar. Drugs* **12** 128
[4] Guerin M, Huntley M E and Olaizola M. 2003 *Trends Biotechnol.* **21** 210
[5] Boussiba S and Vonshak A 1991 *Plant and cell Physiol.* **32** 1077
[6] Kang C D and Sim S J 2008 *Biotechnol. Lett.* **30** 441
[7] Higuera-Ciapara I, Felix-Valenzuela L and Goycoolea F M 2006 *Crit. Rev. Food Sci. Nutr.* **46** 185
[8] Weaver R J, Wang P, Hill G E and Cobine P A 2018 *J. Exp. Biol.* **221** jeb183665
[9] Kristinova V, Aaneby J, Mozuraityte R, Storro I and Rustad T 2014 *Eur. J. Lipid Sci. Technol.* **116** 857
[10] Tominaga K, Hongo N, Fujishita M, Takahashi Y and Adachi Y 2017 *J. Clin. Biochem. Nutr.* **61** 33
[11] Yoshihisa Y, Andoh T, Matsunaga K, Rehman M U, Maoka T and Shimizu T 2016 *PloS one* **11**

0152288
[12] Ritto D, Tanasawet S, Singkhorn S, Klaypradit W, Hutamekalin P, Tipmanee V and Sukketsiri W 2017 *Nutr. Res. Pract.* **11** 275
[13] Davinelli S, Nielsen M E and Scapagnini G 2018 *Nutrients* **10** 522
[14] Chiu C H, Chang C C, Lin S T, Chyau C C and Peng R Y 2016 *Int. J. Mol. Sci.* **17** 1128
[15] Grimmig B, Kim S H, Nash K, Bickford P C and Shytle R D 2017 *Geroscience* **39** 19
[16] Ito N, Saito H, Seki S, Ueda F and Asada T 2018 *J. Alzheimer's Dis.* **62** 1767
[17] Wu H, Niu H, Shao A, Wu C, Dixon B J, Zhang J, Yang S and Wang Y 2015 *Mar. Drugs* **13** 5750
[18] Roohbakhsh A, Karimi G and Iranshahi M 2017 *Biomed. Pharmacother.* **91** 31
[19] Huang J Y, Yeh P T and Hou Y C 2016 *Clin. Ophthalmol.* **10** 813
[20] Serrano G A and Narducci M 2014 *Agro Food Ind. Hi. Tech.* **25** 11
[21] Coral-Hinostroza G N, Ytrestoyl T, Ruyter B and Bjerkeng B 2004 *Comp. Biochem. Physiol. C, Pharmacol. Toxicol.* **139** 99
[22] Gruszecki W I and Strzalka K 2005 *BBA Molecular Basis of Disease* 1740 108
[23] Foss B J, Sliwka H-R, Partali V, Naess S N, Elgsaeter A, Melo T B, Naqvi K R, O'Malley S and Lockwood S F 2005 *Chem. Phys. Lipids* **135** 157
[24] Hianik T, Opstad C L, Sandorova J, Garaiova Z, Partali V and Sliwka H R 2017 *Chem. Phys. Lipids* **202** 13
[25] Ivanova T, Mircheva K, Balashev K, Panaiotov I and Boury F 2015 *Colloids Surf. B Biointerfaces* **135** 542
[26] Dopierala K and Skrzypiec M 2018 *Chem. Phys. Lipids* **213** 88
[27] Mircheva K, Gonnet M, Balashev K, Ivanova T, Boury F and Panaiotov I 2014 *Colloids Surf. A Physicochem. Eng. Asp.* **460** 209
[28] Shibata A, Kiba Y, Akati N, Fukuzawa K and Terada H 2001 *Chem. Phys. Lipids* **113** 11

Acknowledgments
The reported study was funded by RFBR according to the research project № 19-33-80004. The studies were carried out using equipment of the Resource Centers of the National Research Center "Kurchatov Institute". The authors are grateful to I.S. Kulikova for technical assistance with the HPLC measurements and to A.A. Selishcheva.

IOP Conf. Series: Materials Science and Engineering **889** (2020) 012029 doi:10.1088/1757-899X/889/1/012029

Etidronic acid adduct and bis (2-pyridyl-1,2,4-triazolyl-3) butane: properties and effect on pain sensitivity

I V Cheretaev[1*], M Yu Ravaeva[1] and E N Chuyan[1]

[1] Taurida Academy, Faculty of Biology and Chemistry, V.I. Vernadsky Crimean Federal University, 4 Academician Vernadsky Ave., Simferopol, the Republic of Crimea, Russia.

E-mail: cheretaev86@yandex.ru

Abstract. The article presents the results of experimental studies on the effect of etidrononic acid adduct and bis(2-pyridyl-1,2,4-triazolyl-3)butane (EA+BTB) on the pain sensitivity of male and female rats. This substance is created on the precursors basis, which have a pronounced analgesic activity in acute pain when injected once into male and female rats. Experiments were performed on 30 male and 30 female adult Wistar rats weighing 200-220 gr. Pain sensitivity of animals was assessed 1 hour after a single intraperitoneal injection of the studied solution substances by acute pain tests ("tail-flick" test, Randall-Sellitto test, "hot plate" test). It was found out that EA+BTB, when injected once in doses of 5 and 50 mg/kg, significantly changes the pain sensitivity of male and female rats in acute pain tests, showing gender specificity of the analgesic effect with the participation of various pain mechanisms (spinal and supraspinal) and nociception components (perceptual and mechanical). The compound EA+BTB is recommended for further in-depth preclinical tests of its analgesic activity, decoding its physiological and molecular mechanisms in order to create an improved prototype of a new generation drug based on EA+BTB.

1. Introduction

According to statistics, 9 out of 10 diseases are accompanied by pain, and a fifth of the world's population experiences unpleasant sensations connected with the chronic pain [1]. Therefore, carrying out a primary screening on laboratory animals of promising substances with the purpose of searching for new analgesics is an important task of biomedical science. Full-fledged bioscreening of new substances in laboratory rodents should include mandatory experimental studies on males and females. Numerous data, collected by various researchers over the past 30 years, have shown significant differences between the sexes in response to pain stimuli, which are manifested in the thresholds of pain sensitivity, tolerance and response to pain [2].

Currently, there are many biologically active substances with high potential in terms of possible effects on the nervous system and, in particular, analgesic activity. These compounds include bisphosphonates and 1,2,4-triazoles. These substances are interesting because they are a suitable basis for assembling complex compounds with various ligands that have specific useful properties [3-5]. Analgesic effects were found in the etidronic acid and the representatives of 1,2,4-triazole bis(2-pyridyl-1,2,4-triazolyl-3)propane), as well as in some new complex derivatives created on their basis [4-9].

The huge potential of synthetic possibilities for modifying the 1,2,4-triazole molecule opens up great prospects for the synthesis and research of the biological activity of its new derivatives with

IOP Conf. Series: Materials Science and Engineering **889** (2020) 012029 doi:10.1088/1757-899X/889/1/012029

bisphosphonates to obtain effective and safe analgesics of a new generation. One of these derivatives is an adduct of etidrononic acid and bis(2-pyridyl-1,2,4-triazolyl-3)butane (EA+BTB).

The aim of the work is to evaluate the effect of EA+BTB on the pain sensitivity of male and female rats in doses of 5 and 50 mg/kg with a single injection.

2. Material and methods

2.1. Animals

All studies on the animals were carried out in compliance with the principles set out in Directive 2010/63/EU of the European Parliament and of the Council of 22.09.2010 on the protection of animals used for scientific purposes. The experiments were performed on 30 male and 30 female adult Wistar rats weighing 200-220 gr ("FSUE Nursery of laboratory animals "Rappolovo", Saint-Petersburg, Russia). All the rats had free access to food and water. One week before the experiment, the animals with an average motor activity in the "open field" test and not exceeding 8 sec latent period of pain reaction (LPPR) in the "hot plate" test were selected. All the animals were used in experiment after one acclimation week to the housing conditions. The animals were kept in 12-hour light-dark cycle conditions in a room with a constant temperature (22 ° C) and humidity (60%).

2.2. Design of research EA+BTB on pain sensitivity, used research and statistical methods.

Pain sensitivity of rats was studied in acute pain tests ("tail-click" test, Randall-Selitto test,"hot plate" test), which are used for screening analgesic activity of various substances [2, 6-17]. The pain sensitivity of animals was evaluated after 1 hour of single intraperitoneal injection of the studied solutions of substances. Two experimental groups of males and two experimental groups of females (n=10 in each group of animals) received intraperitoneal injections of 0.2 ml EA+BTB in doses of 5 and 50 mg/kg respectively. The males (n=10) and the females (n=10) of control groups of animals received single intraperitoneal injections of 0.2 ml 0.9% NaCl solution. All the experiments to determine the threshold of pain sensitivity in rats in the listed tests were conducted from 09:00 to 15:00 h at the Department of Human and Animal Physiology and Biophysics (Faculty of Biology and Chemistry) and the Center for Collective Use of scientific equipment "Experimental physiology and Biophysics" of Taurida Academy (structural division), V.I. Vernadsky Crimean Federal University. EA+BTB was synthesized at the Department of General and Inorganic Chemistry (Faculty of Biology and Chemistry) of Taurida Academy (structural division), V.I. Vernadsky Crimean Federal University. The chemical purity of EA+BTB was at least 98 %.

In the "tail-flick" test (LE7106 Tail-flick Meter, Pan Lab Harvard Apparatus, Spain), the latent period of tail retraction (LPTR) in rats was recorded, which was determined by the time (s) of the tail flick. On the tail of each rat, sitting in the fixator, three presentations of a thermal stimulus were performed, followed by the calculation of the average value of LPTR in seconds for each animal. This test is based on a spinal flexor reflex that occurs in response to a local force on the tail by high temperature, and allows us to judge the pain sensitivity of animals mainly at the spinal level [7-12].

In the Randall-Selitto test (an experimental apparatus analgesia meter BIO-RP-R Rodent pincher, Bioseb, France), the device displays the applied force (in gr) at which the animal reacts to a gradually increasing mechanical compression of the tail – the pain threshold (PT) [13]. We performed three mechanical compressions of the tail of each rat sitting in the retainer with tongs, followed by calculating the average PT value in grams for each animal.

In the "hot plate" test (an experimental apparatus Cold and hot plate CHP, Bioseb, France), the LPPR of the animal was recorded, which was determined by the value of the time (s) of the reaction manifestation of withdrawal and licking of the limbs and (or) vocalization from the heated surface. This indicator was measured after placing the animal on a hot metal surface with a temperature of + 55 ° C. The test allows to judge the pain sensitivity of animals with the participation of supraspinal mechanisms [7-9, 14-16].

The reliability of statistical differences between the control and experimental groups with different doses of EA+BTB (5 and 50 mg/kg) was determined by using a one-way analysis of variance (ANOVA) with a posteriori Tukey test and Dunn's test of multiple comparisons.

3. Results and discussion

In the "tail-flick" test in male rats (Figure 1A), EA+BTB in a dose of 50 mg/kg caused a significant increase in LPTR by 19.5 % ($p \leq 0.05$, n=10) relatively to control (n=10). This indicates the presence of an analgesic effect of adduct EA+BTB in a dose of 50 mg/kg, implemented with the participation of the perceptual component of nociception and the spinal mechanism for regulating pain sensitivity. In female rats (n=10) in this test, EA+BTB (Figure 1A) did not significantly affect the value of LPTR.

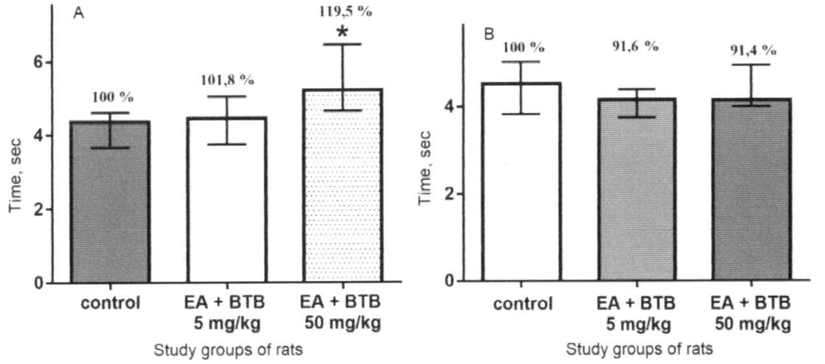

Figure 1. Latent period of tail retraction in rats (median, interquartile range from 25 to 75 %) in the "tail-flick" test in the control group and after intraperitoneal injection of adduct of ethidronic acid and bis(2-pyridyl-1,2,4-triazolyl)butane (EA+BTB) in doses of 5 and 50 mg/kg: (A) males, (B) females.

In the Randall-Sellitto test in male rats (Figure 2A, n=10), EA+BTB did not significantly change PT compared to the control (n=10). In female rats in this test, EA+BTB in a dose of 5 mg/kg (Figure 2B) significantly increased PT by 24.2% ($p \leq 0.01$, n=10) relatively to control (n=10) in response to the mechanical tail compression. This indicates the effect of EA+BTB (5 mg/kg) on the mechanical component of nociception.

In the "hot plate" test in male rats (Figure 3A), EA+BTB in a dose of 50 mg/kg significantly increased LPPR by 57.1 % ($p \leq 0.01$, n=10) compared to the control (n=10). In female rats (Figure 3B), EA+BTB in a dose of 5 mg/kg also significantly increased LPPR by 37.8 % ($p \leq 0.05$, n=10) relatively to control (n=10). Consequently, both male and female rats showed analgesic effect of EA+BTB, which is implemented with the participation of supraspinal mechanisms in the pain sensitivity regulation.

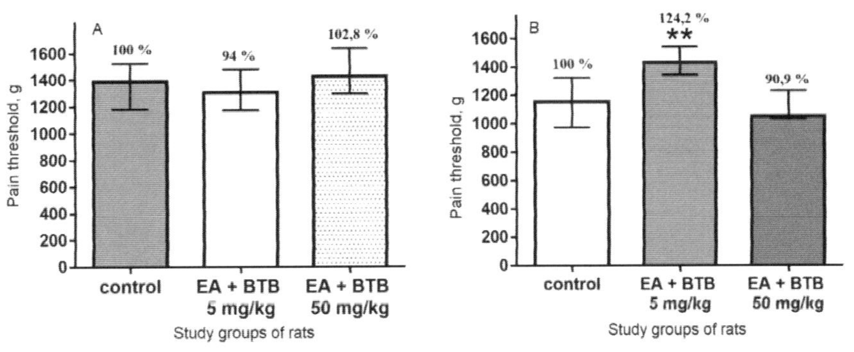

Figure 2. Pain threshold in rats (median, interquartile range from 25 to 75 %) in the Randall-Selitto test in the control group and after intraperitoneal injection of adduct of ethidronic acid and bis(2-pyridyl-1,2,4-triazolyl)butane (EA+BTB) in doses of 5 and 50 mg/kg: (A) males, (B) females.

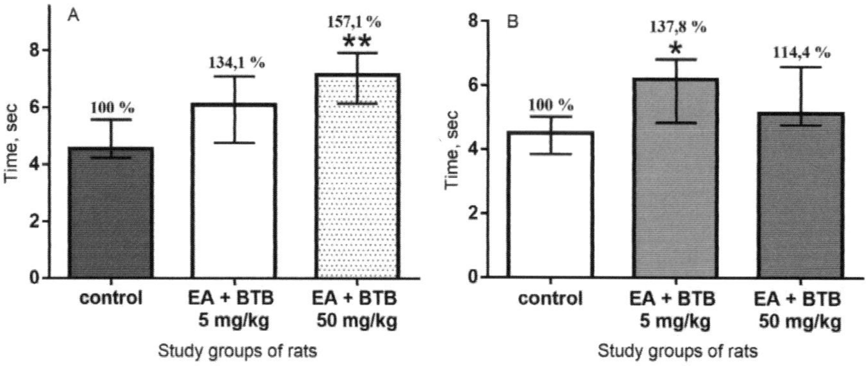

Figure 3. Latent period of pain reaction in rats (median, interquartile range from 25 to 75 %)in the "hot plate" test in the control group and after intraperitoneal injection of adduct of ethidronic acid and bis(2-pyridyl-1,2,4-triazolyl)butane (EA+BTB) in doses of 5 and 50 mg/kg: (A) males, (B) females.

Thus, both studied doses of EA+BTB, either in male or female rats in certain acute pain tests, showed an analgesic effect. However, the studies have shown that gender-specific manifestations of the analgesic effect of EA+BTB are observed. This may be conditioned by different tolerance of the receptors and structures of the nervous system to pain in male and female rats. According to the assumptions [2], male mammals, including humans, have more pronounced activation of somatosensory regions of the cortex when experiencing pain, while females have more pronounced activation of the structures responsible for the perception of the affective – emotional component of pain. Therefore, males are easier to perceive the emotional-affective component of pain, determine the source of pain irritation better and have a more clearly directed motivation to get rid of it. The authors of the review [2] believe that initially there are pronounced gender differences in the perception of pain in males and females, due to significant psychophysiological and biological differences in the functioning of the central and peripheral nervous systems. We associate the differences in the effects of different doses of EA+BTB on pain sensitivity in male and female rats exactly with these gendered features of the functioning of the rat nervous system. This is partly confirmed by the fact that BP and pain tolerance in females are lower during the mechanical compression [17]. Lower PT in females in control than in males were demonstrated in the present study in the Randall-Selitto test (Figure 2).

4. Conclusion

It was found that EA+BTB when injected once in different doses significantly changes the pain sensitivity of male and female rats in acute pain tests, showing gender specificity of the analgesic effect with the participation of various pain mechanisms and nociception components.

In the "tail-flick" test, EA+BTB in a dose of 50 mg/kg has an analgesic effect, significantly increasing LPTR in male rats. This confirms the participation of the perceptual component of nociception and the spinal mechanism of pain sensitivity regulation in the implementation of the analgesic properties of EA+BTB in a dose of 50 mg/kg in male rats. In females, the analgesic properties of EA+BTB (5 and 50 mg/kg) were not detected.

In the Randall-Sellitto test, reflecting the participation of the mechanical component of nociception in the analgesic effect of substances, in female rats EA+BTB in a dose of 5 mg/kg shows analgesic activity, and in males the effect of this substance is not detected.

In the "hot plate" test in both male and female rats, EA+BTB has an analgesic effect, which is implemented with the participation of supraspinal mechanisms for regulating pain sensitivity. This effect is gender-specific: in males, EA+BTB increases LPPR in a dose of 50 mg/kg, and in female rats – in a dose of 5 mg/kg.

The reason for the observed gender specificity of the analgesic effect of EA+BTB in male and female rats in various acute pain tests may be different tolerance of receptors and structures of the nervous system to pain, different pain perception with the participation of various nociception components in male and female rats. This is due to significant psychophysiological and biological differences in the functioning of the central and peripheral nervous systems in males and females.

The compound EA+BTB is recommended for further in-depth preclinical tests of its analgesic activity, decoding its physiological and molecular mechanisms in order to create an improved prototype of a new generation drug based on EA+BTB.

Acknowledgment

The study was carried out with the financial support of the Russian Science Foundation in the framework of the scientific project No 18-13-00024.

References

[1] Bondarenko D A, D'jachenko I A, Skobcov D I and Murashev A N 2011 *Biomedicina* **2** 84
[2] Reshetnyak V K and Reshetnyak D V 2014 *Rossiyskiy Zhurnal Bol i***3–4** 54
[3] Kuźnik A, Październiok-Holewa A, Jewula P and Kuźnik N 2020 *Eur J Pharmacol* **866** 172773
[4] Sarigol D, Uzgoren-Baran A, Tel B C, Somuncuoglu E I, Kazkayasi I, Ozadali-Sari K, Unsal-Tan O, Okay G, Ertan M and Tozkoparan B 2015 *Bioorg Med Chem* **23** 2518
[5] Thakur A, Gupta P S and Shukla P K 2016 *Int J Curr Res Aca Rev* **4** 2 277
[6] Cheretaev I V , Chuyan E N, Ravaeva M Yuand ShulginV F 2019*Int Res J***7(85)** 192
[7] Cheretaev I V, Ravaeva M Yu, Dzheldubaeva E R, Chuyan E N, Shulgin V F, Sheichmambetov N, Palaevskaya M V 2019*Scientific notes of the Crimean Federal University named after V.I. Vernadsky. Biology, Chemistry* **5 (71)** 2 162
[8] Cheretaev I V, Ravaeva M Yu, Dzheldubaeva E R, Chuyan E N, Shulgin V F, Sheichmambetov N, Palaevskaya M V2019 *Scientific notes of the Crimean Federal University named after V.I. Vernadsky. Biology, Chemistry* **5 (71)** 3 199
[9] ChuyanE, Ravaeva M, Cheretaev I, KornilenkoO, Furkovskaya A and BirukovaE 2019 *Adv Biol Sci Res*7 82
[10] Smith E S, Lewin G R 2009J Comp Physiol A Neuroethol Sens Neural Behav Physiol **195** 12 1089
[11] Gulati A, Bhalla S, Matwyshyn G, Zhang Z, Andurkar S V 2011*Pharmacology* **83** 45
[12] Xu F, Zhang B, Li T 2013*Pharmacology* **27** 4 427
[13] Randall L O, Selitto J J 1957 *Arch Int Pharmacodyn Ther* **111** 4 409
[14] Espejo E F, Mir D 1993 *Behav Brain Res* **56** 2 171
[15] Woolfe G, Macdonald A D 1944*J Pharmacol Exp Ther* **80** 3 300
[16] Rogozhin EA, Kisil O V, Cheretaev I V, Zavriev S K 2017*Antibiot Khimioter* **62** 9-10 3
[17] Riley J L, Robinson M E, Wise E A, Myers C D and Fillingim R B 1998 *Pain* **74** 2-3 181

IOP Conf. Series: Materials Science and Engineering **889** (2020) 012030 doi:10.1088/1757-899X/889/1/012030

Coordinating metal Co^{2+} and Zn^{2+} compounds with acetylsalicylic acid: the influence on the cardiorespiratory system

M Yu Ravaeva[1*], E N Chuyan[1] and I V Cheretaev[1]

[1] Taurida Academy, V.I. Vernadsky Crimean Federal University, Simferopol, Russia

E-mail: ravaevam@yandex.ru

Abstract. The paper researches the effect of cobalt ($ACCo^{2+)}$ and zinc ($ACZn^{2+}$) acetylsalicylates in doses of 5 and 10 mg/kg on the characteristics of the rats' cardiorespiratory system. It is ascertained that coordinating metal compounds with the acetylsalicylic acid have a more pronounced biological effect in comparison with a monocompound of the acetylsalicylic acid; this makes the further search for coordinating compounds' effects more prospective.

1. Introduction

Significant success in different spheres of medicine, achieved in the recent years, to a great extent is obliged to the introduction of new technologies of obtaining highly active medications into pharmaceutics. One of such technologies is complexing with metals, which is a very promising method of increase and modification of the biological activity of well-known medical substances. The research of coordinating compounds of s, p and 3d-metals with the acetylsalicylic acid, which is a traditional medical substance, is of a special interest. Different evidences of possibility of using complexes on the basis of aspirin derivatives as antineoplastic, anti-inflammatory and antimicrobial drugs have appeared recently [1-3]. It is necessary to point out that the residual of the salicylic acid is a very suitable object for complexing, and this gives new synthetic opportunities for obtaining compounds with an unusual combination of the biological activity.

According to a number of authors, the creation of complex salicylate compounds with bivalent metals is a perspective approach in this area [4]. The works of the above mentioned authors have shown that such complex derivatives, in the basis of which there are biometals (Fe, Zn, Co, Cd, etc.), surpass aspirin in anti-inflammatory effect and have less side effects.

Consequently, the creation of complex compounds with a coordinating center of Co, Zn, Fe, Mn and another range of other metals is a perspective area for the researches of biologists, chemists, and pharmaceutists. The introduction of newly synthesized chemical substances into the clinical practice is possible only providing the detailed study of their specific pharmacological activity and safety on the stage of experimental pre-clinic research. The complex approach with the use of physiological methods, comparison of the obtained results and mathematical biostatistic analysis of quantitative indicators is of great topicality.

In connection with this the aim of the current research was to identify the influence of newly synthesized compounds $ACCo^{2+}$and$ACZn^{2+}$ in doses of 5 mg/kg and 10 mg/kg on the main characteristics of the rats' cardiorespiratory system.

2. Materials and methods

2.1. Animals

All the researches on the animals were carried out according to the principles set out in Directive 2010/63/EU of the European Parliament and of the EU Council of 22.09.2010 on the protection of animals used for scientific purposes.

Healthy sexually mature male laboratory Wistar rats, weighing 180-200 gr ("FSUE "Nursery of laboratory animals "Rappolovo") and having been placed under quarantine not less than for 14 days, were selected for the experiment. The experiments were performed on 70 male rats, characterized by an average motor activity and low emotionality in the "open field" test, and which make up the majority of the population; that is why they develop the most typical reaction to the effect of different factors [5], including the tested chemical compounds.

The animals were kept under standard conditions of the vivarium at the temperature of 18–22° C on the bedding «RexofixMK 2000» (on the basis of ear shanks) with the natural 12-hour day-and-night cycle, with free access to water (State Standard 33215-2014 «The handbook of keeping and nursing laboratory animals. The regulation rules on housing equipment and procedure organization»), and to full-fledged granulated food of State Standard P-50258-92.

The research was carried out at the Center for Collective Use of Scientific Equipment "Experimental physiology and Biophysics" of the Department of Human and Animal Physiology and Biophysics of V.I. Vernadsky Crimean Federal University.

2.2. Design of research on studying the reaction of cardiorespiratory system to the injection of the tested compounds, the used research methods and statistic methods

The synthesis, the research of the composition, structure and properties of the coordinating compounds of acetylsalicylates with Co (ACCo2+) and Zn (ACZn2+) were done at the Department of General and Organic Chemistry of V. I. Vernadsky Crimean Federal University under the supervision of Professor A. N. Gusev.

The research of the biological effect of the acetylsalicylic acid, $ACCo^{2+}$and$ACZn^{2+}$ was carried out during their intraperitoneal injection into rats in doses of 5 mg/kg and 10 mg/kg. For this purpose the rats were divided into 7 groups (10 rats in each group):

Group 1 is a biological control (K, n=10), they are the animals which were injected by the physiological solution of 0.9 % in the volume of 0.2 ml;

Groups 2 – 3 are the animals which were injected by the acetylsalicylic acid in doses of 5 mg/kg and 10 mg/kg respectively (n=10), the volume of 0.2 ml;

Groups 4 – 5 are the animals which were injected by $ACCo^{2+}$in doses of 5 mg/kg and 10 mg/kg respectively (n=10), the volume of 0.2 ml;

Groups 6 – 7 are the animals which were injected by $ACZn^{2+}$in doses of 5 mg/kg and 10 mg/kg respectively (n=10), the volume of 0.2 ml.

The bioscreening was done 20 minutes later after the intraperitoneal injection in the stated concentrations as at this period their maximum concentration is observed [6]. Simultaneously the following characteristics of all the groups' animals were registered: heart rate (HR), respiratory rate (RR), systolic blood pressure (SBP) and diastolic blood pressure (DBP). The pulse pressure (PP) was calculated on the basis of the SBP and DBP indices difference.

It is necessary to point out that HR, RR, SBP, DBP and PP are important characteristics of the functional state of the whole cardiorespiratory system and cardiac performance at a definite period of time, and are ones of the main and the oldest biomarkers of the body state in general.

BP, HR and RR of the rats were registered with the help of the system NIBP200A («BiopacSystems, Inc.», USA). BP and HR were registered on the tail artery by applying the cuff on the tail base. In order to record RR the sensor was fixed on the chest area. While registering the indices, the animals were put into an individual box and were carried to the chamber Biopac with a constantly kept temperature of 33^0Cfor creation of comfortable conditions for the animal.

The record of the indices was done during 5 minutes from the moment of the sensor's signals stabilization. This time is enough for the five-fold BP measurement; HR and RR were being registered continuously. The record and the data processing were done by the computer with the help of the software program «AcqKnowledge 4.2 for MP150» (Fig. 1).

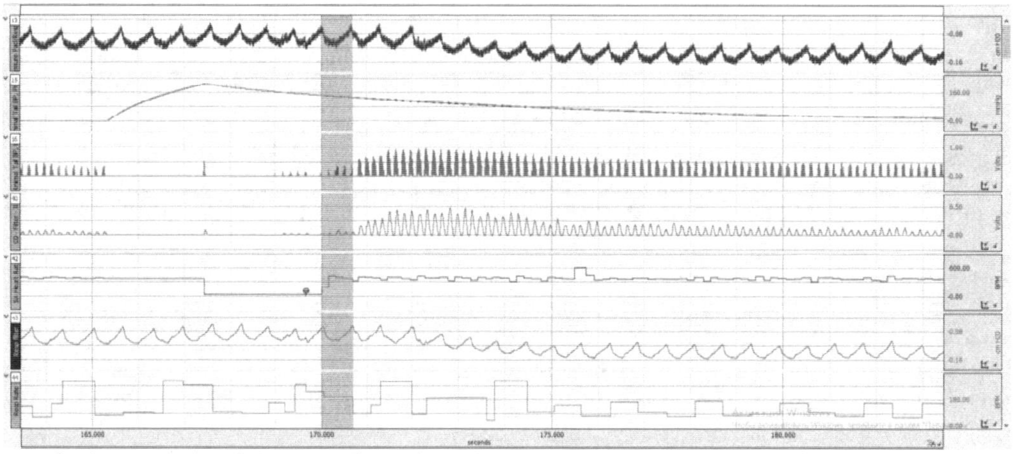

Figure 1. The example of the recording of cardiovascular system indices of the rats' BP, HR and RR under the effect of $ACZn^{2+}$in a dose of 10 mg/kg with the help of the software program «AcqKnowledge 4.2 for MP150».

The data of three repeated experiments were used for the statistical manipulation. The non-parametric statistic methods were applied as the distribution of variate values differed from the normal. The calculations, the statistical manipulation and graphic design of the obtained in the research data on the effect of the tested compounds on the physiological characteristics were done by using the program MicrosoftExcel and the program package StatSoft\STATISTICA 8. The reliability of the statistic differences between the control group (intraperitoneal injection of the physiological solution) and experimental groups with different doses of injecting the acetylsalicylic acid and acetylsalicylates Co^{2+},Zn^{2+} was assessed by Mann-Whitney test.

3. Results and discussion

During the research it was found out that the animals of the control group, after being injected by the physiological solution, did not demonstrate reliable changes of the cardiorespiratory system characteristics (Table 1). At the same time the reliable lowering of HR in animals, being injected by the tested compounds in a dose of 5 mg/kg, was registered (Table 1): at the effect of the acetylsalicylic acid by 22.0 % ($p \leq 0.05$), and at the effect of $ACCo^{2+}$and$ACZn^{2+}$by 14.1%($p \leq 0.05$) and 17.0% ($p \leq 0.05$) respectively in comparison with such in the control group. The rest characteristics of the animals' cardiorespiratory system during injecting the tested compounds in a dose of 5 mg/kg did not change reliably.

The increase of the compounds' dose up to 10 mg/kg lead to the reliable lowering of HR of the animals in all the tested groups (Table 1). Thus, during the injection of the acetylsalicylic acid HR lowered by 21.8 %($p \leq 0.05$), and during the injection of $ACCo^{2+}$and$ACZn^{2+}$by 6.9% ($p \leq 0.05$) and 14.3% ($p \leq 0.05$) in comparison with such in the control group of animals.

The characteristics of the system blood pressure and respiratory rate of animals, being injected by the acetylsalicylic acid and $ACCo^{2+}$, did not change reliably. At the same time the injection of $ACZn^{2+}$ in a dose of 10 mg/kg lead to the reliable lowering of SBP by 5.1 % ($p \leq 0.05$), rise of DBP by 14.7 %, lowering of PP by 38.3%($p \leq 0.05$) and rise of RR by 56.3% ($p \leq 0.05$) with respect to such characteristics of the animals in the control group (see Table 1).

IOP Conf. Series: Materials Science and Engineering **889** (2020) 012030 doi:10.1088/1757-899X/889/1/012030

Thus, the injection of the acetylsalicylic acid and acetylsalicylates $Co^{2+}$and Zn^{2+}lead to the changes in the laboratory animals' cardiorespiratory system functioning. At the same time, when the dose was 5 mg/kg, the reaction of the cardiorespiratory system of animals to the interjection of the tested compounds had a unidirectional character, which was shown in the negative chronotropic action. The increase in the dose up to 10 mg/kg lead to an insignificant lowering of evidence of negative chronotropic influence of all the tested compounds on the heart rate, as well as to the display of new properties of acetylsalicylate Zn^{2+}: reliable growth of DBP and RR and lowering of SBP, PP and HR.

Therefore, we may come to conclusion that introduction of the metals into the structure of the acetylsalicylic acid modifies the biological properties of the obtained coordinating compounds.

Table 1. The characteristics of the rats' cardiorespiratory system after injecting the acetylsalicylic acid and acetylsalicylates of Co^{2+} and Zn^{2+}in different concentrations.

Group		SBP, mm m.c.	DBP, mm.m.c.	PP mm.m.c.	HR, b./min	RR, r.r./min
Control (1)		118.4±1.66	74.3±1.49	44.1±1.56	468.9±9.22	98.4±4.34
The acetylsalicylic acid	5 mg/kg (2)	114.2±1.86	76.5±1.05	37.7±2.33	363.5±10.72 $p_1 \le 0.05$	98.2±3.27
	10 mg/kg (3)	117.7±1.89	79.2±1.03	38.5±1.85	366.5±10.73 $p_1 \le 0.05$	100.5±3.23
$ACCo^{2+}$	5 mg/kg (4)	114.73±4.0	72.07±1.02	42.6±1.14	403.00±7.92 $p_1 \le 0.05$	97.53±5.15
	10 mg/kg (5)	111.67±0.6	68.13±0.26 $p_3 \le 0.05$	43.5±1.31	436.87±2.75 $p_1 \le 0.05$ $p_3 \le 0.05$	95.73±1.46
$ACZn^{2+}$	5 mg/kg (6)	114.73±0.7	78.40±1.60	36.3±3.5	389.40±6.14 $p_1 \le 0.05$	99.53±2.43
	10 mg/kg (7)	112.47±0.3 $p_1 \le 0,05$ $p_3 \le 0.05$	85.27±0.38 $p_1 \le 0.05$	27.2±1.05 $p_1 \le 0.05$ $p_3 \le 0.05$	402.27±0.34 $p_1 \le 0.05$	153.87±0.46 $p_1 \le 0.05$ $p_3 \le 0.05$

Note: Mis an average arithmetic value, M±m is an error of mean, p 1-7is the confidence level of indices difference according to Mann-Whitney test concerning the corresponding animal groups; SBP is a systolic blood pressure; DBP is a diastolic blood pressure; HR is heart rate; RR is a respiratory rate

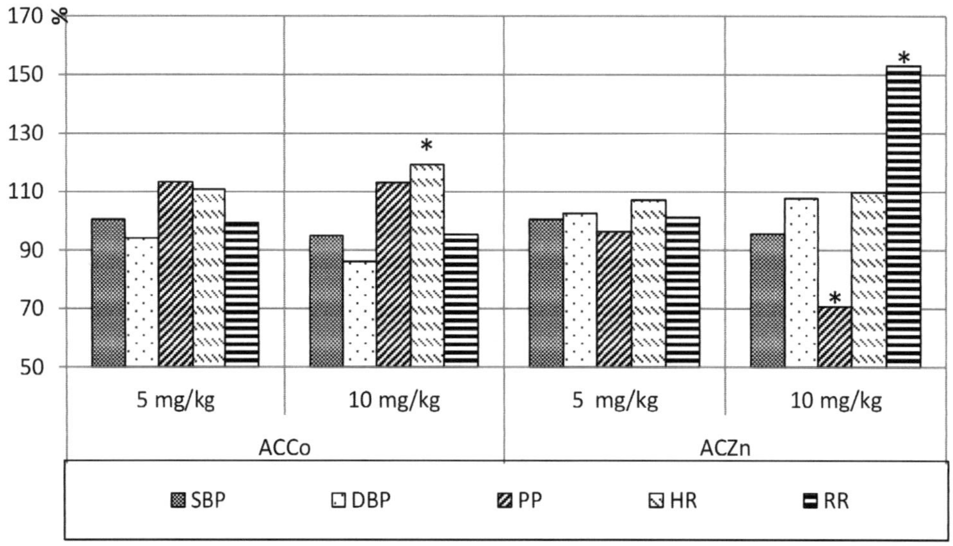

Figure 2.. The chatacteristics of the cardiorespiratory systems of animals, injected by the acetylsalicylic acid and acetylsalicylates of Co^{2+} and Zn^{2+} in concentrations of 5 mg/kg and 10 mg/kg relatively to the values in the group of animals, injected by the acetylsalicylic acid, and taken as 100 %.

Note: * is the confidence level of differences according to Mann-Whitney test relatively to the values of the indices in control; SBP is a systolic blood pressure; DBP is a diastolic blood pressure; HR is heart rate; RR is a respiratory rate.

Actually, the comparative analysis of the effectiveness of the coordinating compounds with the initial substance (the acetylsalicylic acid) showed (Fig. 2) that during the injection of the acetylsalicylates $Co^{2+}$and Zn^{2+}in a dose of 5 mg/kg the unidirectional character of the tested indices changes of the cardiorespiratory system was observed, this was proved by the absence of reliable differences in the characteristics of animals in the groups $ACCo^{2+}$and $ACZn^{2+}$ respectively to such in the group of the acetylsalicylic acid. With the increase in the dose up to 10 mg/kg $ACCo^{2+}$a reliable growth of HR by 19.2% ($p \leq 0.05$) and reliable lowering of DBP by 14% ($p \leq 0.05$)were registered, after the injection of $ACZn^{2+}$a reliable growth of HR by 53 % ($p \leq 0.05$), lowering of SBP by 5.2% ($p \leq 0.05$) and PP by 29.7 % ($p \leq 0.05$) were observed relatively to the indices of the animals in the group, injected by the acetylsalicylic acid in this dose.

It is necessary to pay attention to the differences of cardiotropic effect of the two studied derivatives of the acetylsalicylic acid – acetylsalicylates Co^{2+}, Zn^{2+} in a dose of 10 mg/kg, this is demonstrated through the differently directed influence on the pulse pressure (the tendency to the growth during the injection of $ACCo^{2+}$ and reliable lowering at the effect of $ACZn^{2+}$), significantly more pronounced growth of HR at the injection of $ACCo^{2+}$ and HR at the effect of $ACZn^{2+}$.

Thus, in the state of physiological rest of the rats because of the $ACZn^{2+}$effect in a dose of 10 mg/kg we may observe the irregularity of the heart rate (bradycardia), the increase of the breath (tachypnoea) on the background of decrease of the blood volume, thrown by the heart at the systole moment, which can be the consequence of the high intoxication of the animal's body.

4. Conclusion

The results of the current research allowed to determine that the coordinating compounds of the acetylsalicylic acid with such metals as Co and Zn not only have more pronounced biological effect in comparison with the acetylsalicylic acid, but demonstrate new properties; the most effect on the cardiorespiratory system was achieved by$ACZn^{2+}$.

The similar conformity of the pharmokinetics is observed in our previous rescarch, while studying neurotropic effects of $ACCo^2$and $ACZn^{2+}$, which showed the activated-modulating effect on the electric activity of mollusc's *Helixalbescens Rossm* neurons in comparison with the acetylsalicylic acid, and more pronounced effects of $ACCo^2$ in comparison with $ACZn^{2+}$ effects [7]. The similar results were obtained by other authors [4, 8-10], in the works of whom it is shown, that such complex derivatives, in the basis of which there are biometals (Fe, Zn, Co, Cd, etc.), surpass aspirin in the anti-inflammatory effect and have less side effects.

Thus, the results of the current rescarch allow us to claim that the salts of the acetylsalicylates, having the metal-complexingCo^{2+} and Zn^{2+}in its molecules' composition, have more pronounced and qualitatively new properties in comparison with the acetylsalicylic acid precursor's properties; this makes the further search for biological and pharmaceutical activity of these coordinating compounds more perspective, as well as it makes the technology of complexing an inexpensive and highly efficient approach to the creation of new medicines.

5. References

[1] Rubner G., Bensdorf K, Wellner A., Kircher B, Bergemann S, Ott I and Gust R 2010*J Med Chem* **53**

[2] Gilligan M M, Gartung A, Sulciner M L, Norris P C, Sukhatme V P, Bielenberg D R, Huang S, Kieran M W, Serhan C N and Panigrahy D 2019 *PNAS* **116(13)** 6292

[3] Altinoz M A, Elmaci I, Cengiz S, Emekli-Alturfan E and Ozpinar A 2018 *Chemico-Biological Interactions* **291** 29

[4] Bica K., Rijksen C., Nieuwenhuyzena M. et al. 2010 Phys. Chem **12** 2011-7.

[5] Cheretaev I V, Ravaeva M Yu, Dzheldubaeva E R, Chuyan E N, Shulgin V F, Sheichmambetov N, Palaevskaya M V 2019 Scientific notes of the Crimean Federal University named after V.I. Vernadsky. Biology, Chemistry. **5 (71)** 199

[6] LevyhA. Je.Mamchur V.I. 2015*Arterial'naja gipertenzija* **6 (44)** 57-63

[7] Cheretaev I V, Korenyuk I I, Khusainov D R, Gamma T V, Kolotilova O I and Nozdrachev A D *Neuroscience and Behavioral Physiology* 2016 **46 (6)** 644

[8] Aljohin E.K. 1999 *Sorosovskij obrazovatel'nyj zhurnal* 5 **(7)** 85-90

[9] Sokolik J., Tumova I., Blahova M. et al.2006 *ActaFacult. Farm.Univ. Comenianae* **53** 224-8

[10] Grigor'eva A.S. 2000 *Mikrojelementy v medicine* **2(1)** 17-22

Acknowledgements

The research was carried out with the financial support of the Russian Science Foundation grant № 20-33-70142 on the experimental equipment of the Center for Collective Use of Scientific Equipment "Experimental physiology and Biophysics" of the Department of human and animal physiology and Biophysics of Taurida Academy (structural division) of V. I. Vernadsky Crimean Federal University.

Improvement of high-speed steel cutting tool production technology using the method of nanostructure generation by laser ablation

Elmar Yagyaev[1*] and Seran Akimov[1]

[1]Crimean Engineering and Pedagogical University, per. Uchebnyj, 8, Simferopol, 295015, Russian Federation

*elmar1875@gmail.com

Abstract. The article reveals a problem of improving the production technology of cutting tools made of high-speed steels by the method of generating nanostructures on their surface with the help of laser ablation. The spiked microstructures obtained on the surface of the instrument are coated with nanostructures. Experimental studies to modify the surfaces of cutting tools with laser ablation method are performed in air and in liquids. As a result of the analysis of the experimental samples surface morphology we can conclude that the modification of the surface of cutting tools made of P6M5 high-speed steel with laser pulses of nanosecond duration can be used in air and in liquids as a method of increasing the resistance characteristics of a cutting tool.

1. Introduction

The operating capacity of the machining technological system mainly depends on the operating capacity of cutting tool. The product processing is accompanied by the inevitable wearing of the cutting tool. Fragmentary wearing or complete destruction of the cutting edges occurs. At the same time, cutting tool operability depends on the various surface hardening treatments used to increase the wear resistance of the contact pads [1-6].

Table 1 presents methods of increasing the cutting ability of a tool by changing the physical mechanical properties of the material [4].

Table 1. Methods of increasing the tool cutting ability

Production technologies by groups	wear-resistant coating anti-friction coating galvanic methods chemical methods chemical thermal methods physical methods mechanical methods thermomechanical methods finish grinding and sharpening method of electric spark hardening and tool extension

The solution to the problem of wearing is timely replacement or regrinding and comprehensive restoration of the cutting tool.

Replacing a cutting tool with a new one in cases of using replaceable blades is economically justified and expedient, however, when dealing with a monolithic tool, buying a new tool each time will lead to a significant increase in the cost of products.

Objective: To increase the cutting properties of a tool made of high-speed steels by nanostructuring the surface when irradiated with laser pulses of nanosecond duration.

2. Materials and Methods

Improving the wear resistance of the contact areas of the cutting tool is ensured by the application of various methods of surface hardening treatment of chemical-thermal, deformation, coating, modification of the surface layer properties of the tool and other methods [1].

An effective method of increasing the operating capacity and improving the characteristics of a cutting tool is to apply wear-resistant coatings to the working surfaces. Wear-resistant coatings reduce friction at the tool pads, increase the hardness and wear resistance of the surface layer of the tool, reduce the heat input deep into the tool, and increase heat resistance [2]. The life of the tool is increased by 4 times when processing heat-resistant steels and 5-10 times when processing structural steels.

The tool regrinding along the front and back surfaces is accompanied by the complete removal of the wear-resistant coating from the pads. After regrinding the tool, the cutting conditions are reduced by 20 - 30% compared with a tool with a wear-resistant coating. The durability period of reconditioned tools decrease by 5-10 times.

Currently, there are services for the comprehensive restoration of cutting tools with regrinding and restoration of wear-resistant coating. Resistance of reconditioned cutters and drills with restoration of coating is not inferior to the resistance of a new tool.

However, a comprehensive restoration of tools with coatings is not always feasible at an enterprise. The cost of plants for applying wear-resistant coatings is very high and not always justify its expenses for an enterprise, and besides sending to comprehensive restoration services requires a large number of cutting tools in circulation.

One of the promising methods for increasing the resistance of a cutting tool after regrinding is the generation of nanostructured coating on the surface of contact pads with laser ablation in a liquid [7-10].

Laser ablation is a method of removal (evaporation) of a substance from a surface by a laser pulse. At a low laser power, the substance evaporates or sublimates in the form of free molecules, atoms and ions [11-18]. The depth of the laser pulse action and the amount of the removed substance depend on the material, its optical properties, and also the laser power. Modern laser processing technologies make it possible to control the power and duration of a laser beam in the range from nano to femtoseconds [19-20]. Laser ablation is a well-controlled and promising process for many technological applications, which is very important for the formation of nanostructures during laser spraying.

Figure 1 shows the effect of nanosecond and femtosecond laser pulses on a metal [9, 11].

3. Result and Discussion

Experimental studies of the application of wear-resistant coatings by ablation in liquids were carried out on a MiniMarker 2-20A4 experimental setup with a ytterbium fiber laser, IPG Photonics [Figure. 2]. Maximum pulse energy – 1 mJ, working radiation wavelength – 1.064 μm, maximum power – 20 W, the pulse duration is 4-200 ns, the pulse frequency in the experiments is 100 kHz.

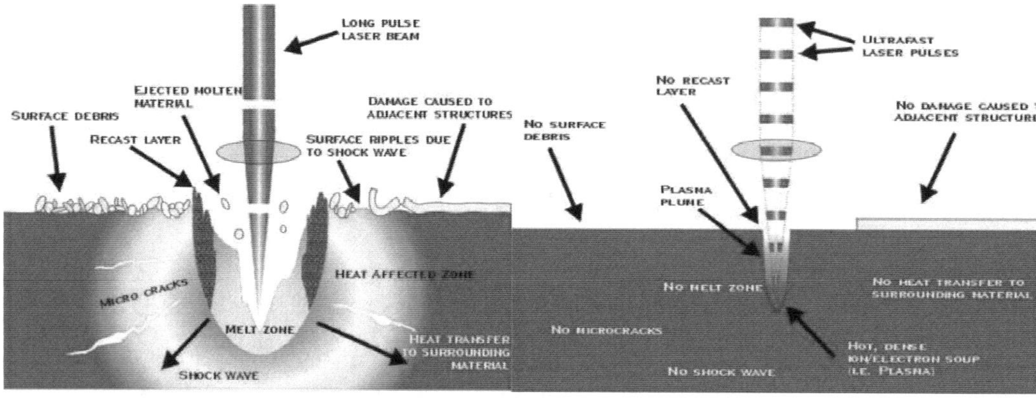

Figure 1. The effect on the metal of nanosecond and femtosecond laser pulses.

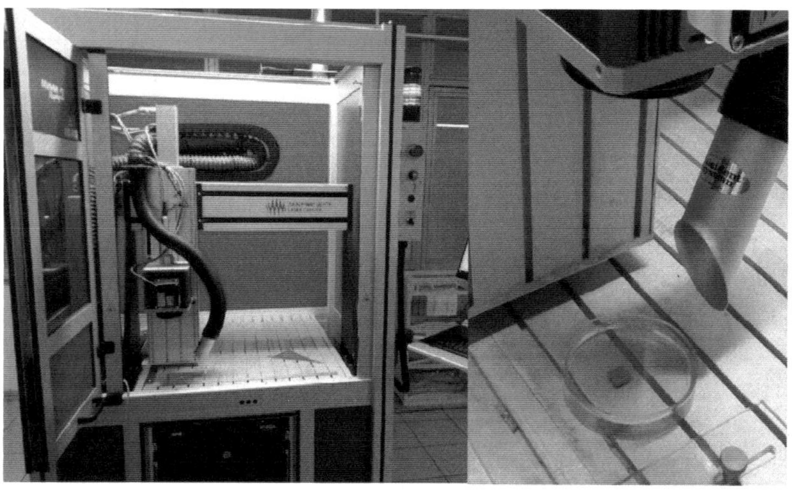

Figure 2. MiniMarker 2-20A4 experimental laser ablation unit with ytterbium fiber laser, IPG Photonics.

Table 2. Modes of laser ablation of turning tools plates

	Plate material	Pulse power P, W	Processing speed V, mm / s	Number of passes per mm.	Pulses duration, ns.	Ablation step lin / mm	Pulse frequency kHz
Insert 1	HSS	10	800	50	10	50	200
Insert 2		15	800	50	10	50	200
Insert 3	HSS	5	800	50	10	50	200
Insert 4		10	800	20	10	20	200
Insert 5	HSS	15	800	20	10	20	200
Insert 6		5	800	20	10	20	200

The results of the obtained surfaces were investigated using a 4XB metallographic microscope, with an increase in the objects under study by 250, 650 and 1250 times, and are presented in Table 3 (material of experimental plates HSS (1.3343 / S6-5-2 DIN). Table 4 presents the results of surface modification in a liquid (water).

IOP Conf. Series: Materials Science and Engineering **889** (2020) 012031 doi:10.1088/1757-899X/889/1/012031

Table 3. Surface morphology of HSS (1.3343 / S6-5-2 DIN) plates after laser ablation

	Surface morphology after laser ablation, x250 magnification	Surface morphology after laser ablation, x650 magnification	Surface morphology after laser ablation, x1250 magnification
Insert not processed			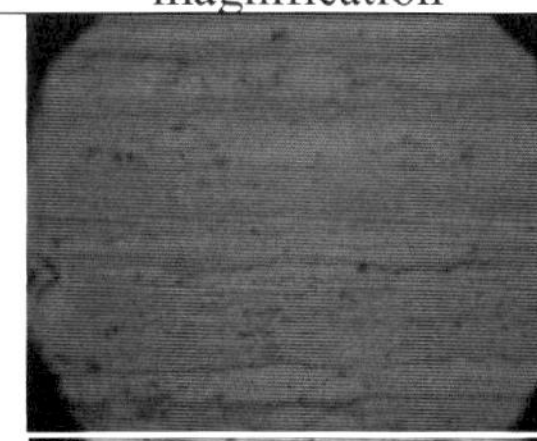
Insert 1		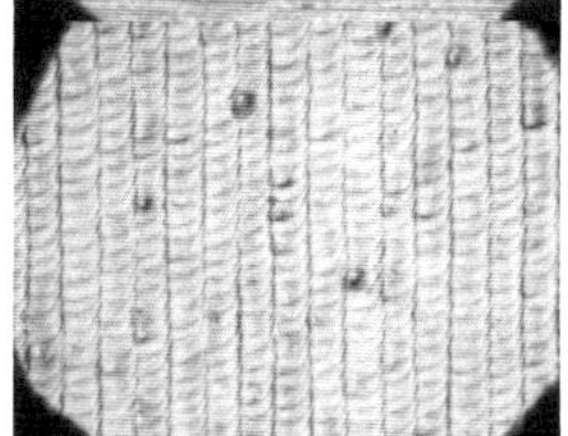	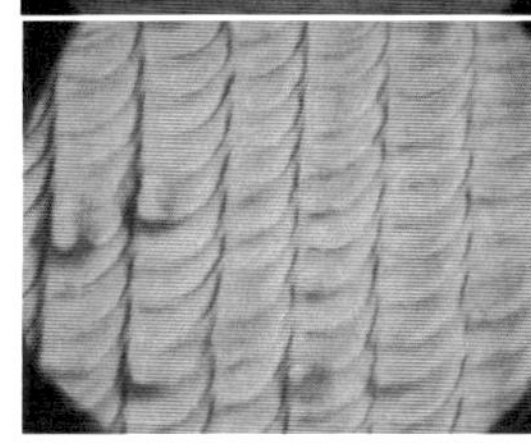
Insert 2		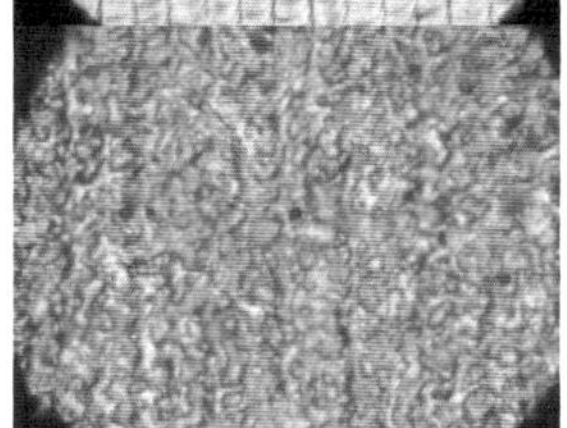	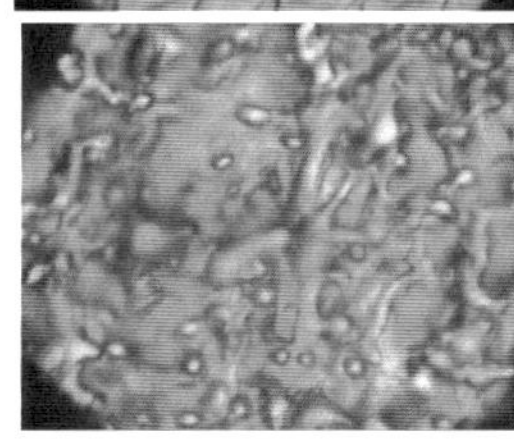
Insert 3			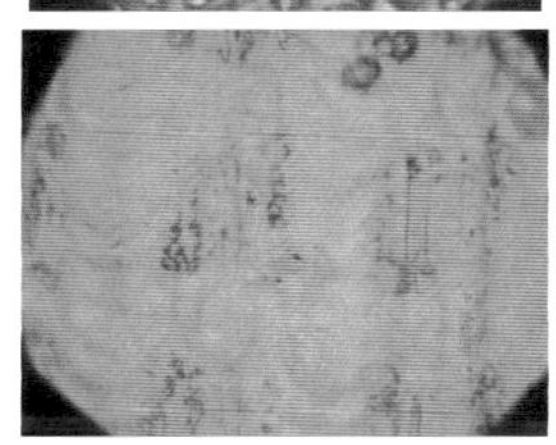
Insert 4	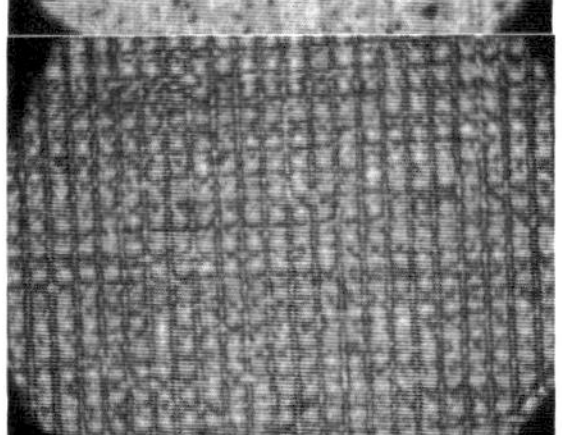	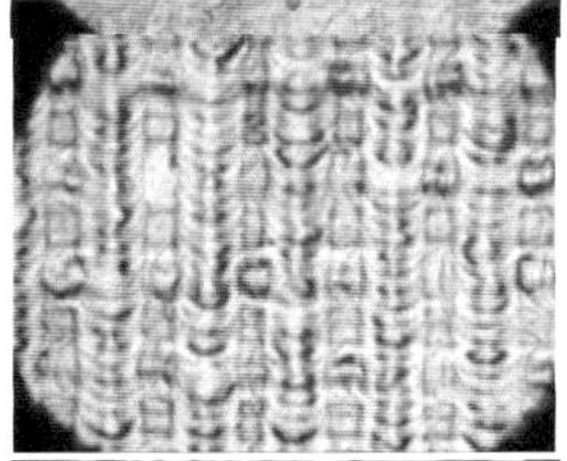	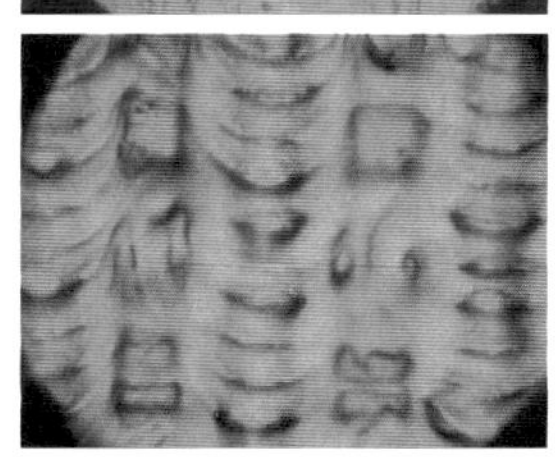
Insert 5		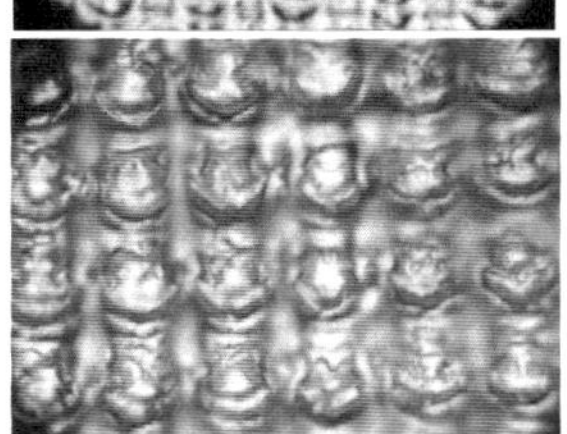	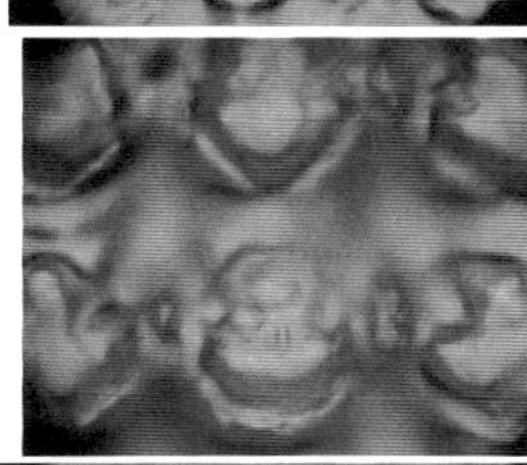
Insert 6		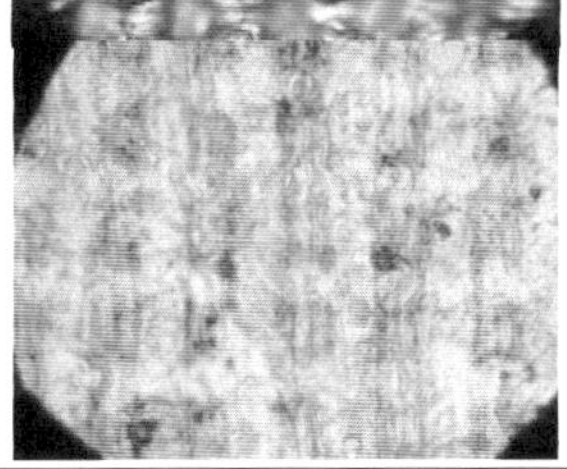	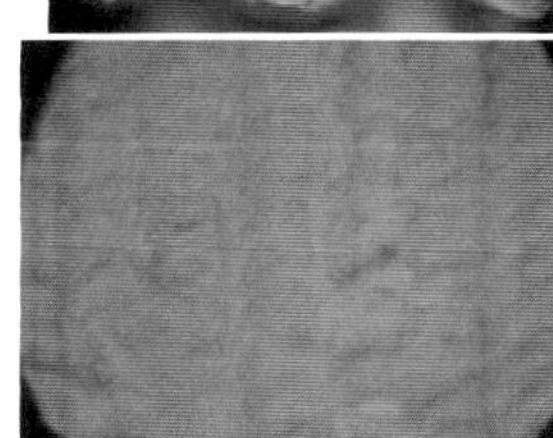

Table 4. Surface morphology of HSS (1.3343 / S6-5-2 DIN) plates after laser ablation

	Surface morphology after laser ablation, x250 magnification	Surface morphology after laser ablation, x650 magnification	Surface morphology after laser ablation, x1250 magnification
Insert 1	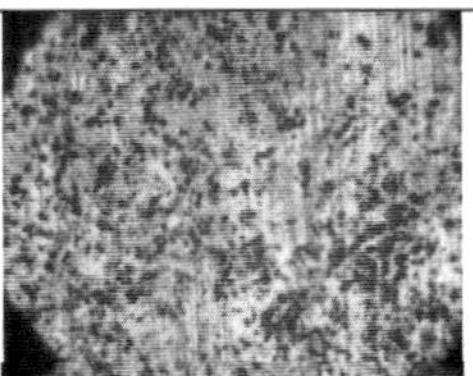	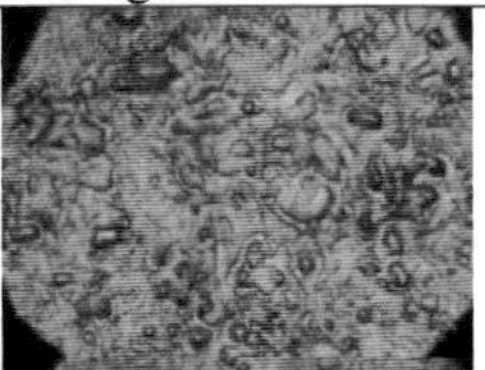	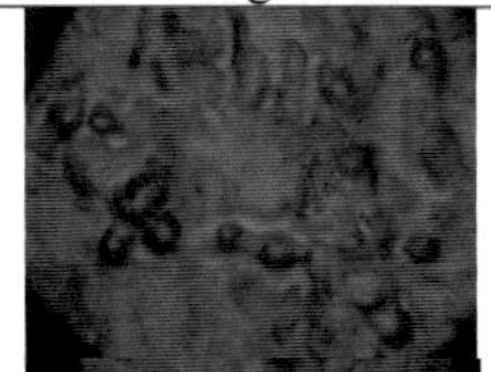
Insert 2		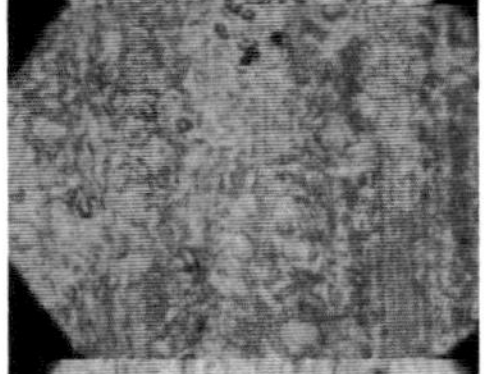	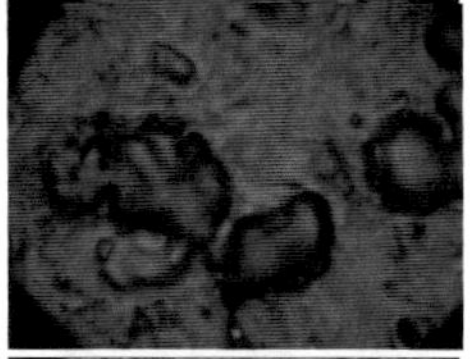
Insert 3			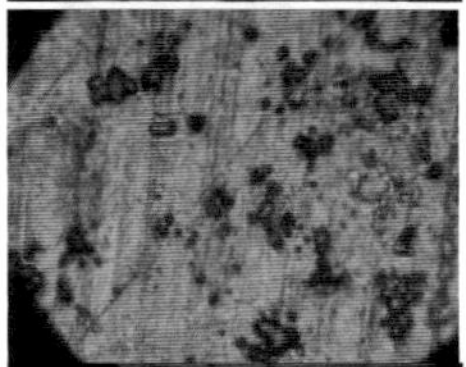
Insert 4	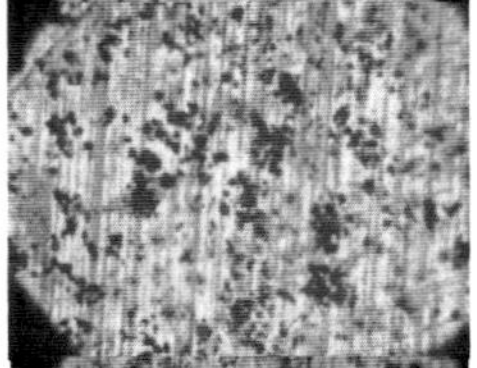	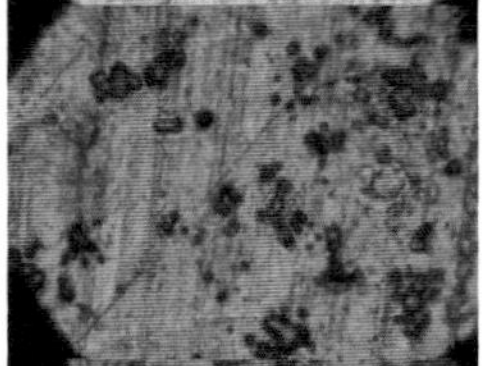	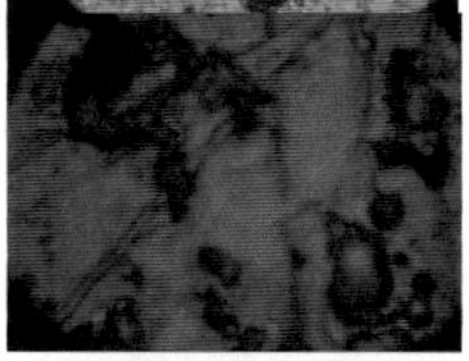
Insert 5	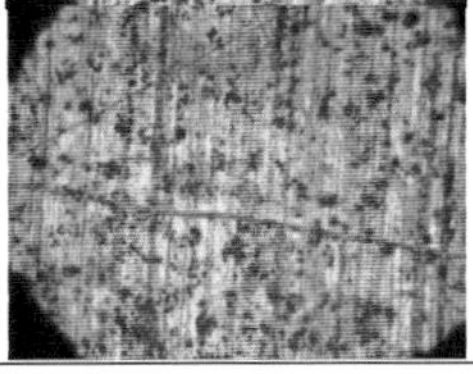	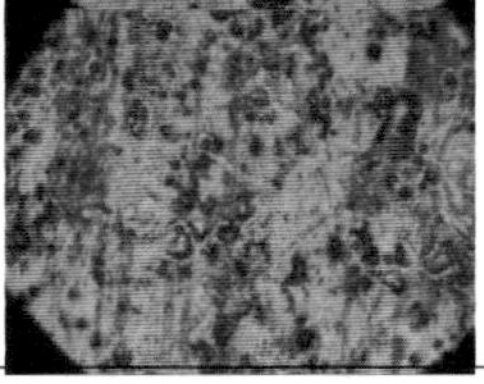	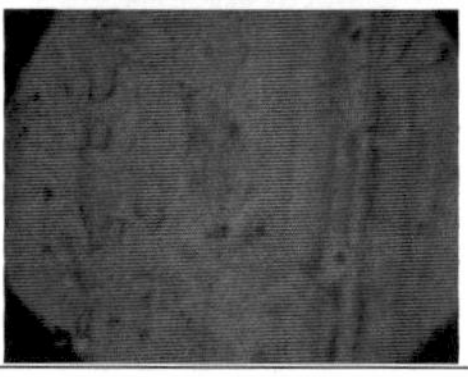

4. Conclusion

(1) As a result of the analysis of the experimental samples surface morphology, it can be concluded that modifying the surface of a cutting tool made of high-speed steel HSS (1.3343 / S6-5-2 DIN) with laser pulses of nanosecond duration in air and in liquid can be used as a method of increasing the resistance characteristics of a cutting tool.

(2) In further work, to determine the operational properties of the cutting tool plates coatings obtained by laser ablation, experimental studies of the tool life period and the treated surface quality indicators should be performed.

5. References

[1] Grigor'ev S N, Volosova M A and Borovskij V G 2005 Tekhnologicheskoe obespechenie kachestva poverhnostnogo sloya keramicheskogo instrumenta dlya povysheniya ego rabotosposobnosti pri obrabotke zakalennyh stalej *Stanki i instrument* **9** 14–19.

[2] Korotaev A D, Moshkov V Yu, Ovchinnikov S V, Pinzhin Yu P, Savostikov V M and Tyumencev A N 2005 Nanostrukturnye i nanokompozitnye sverhtverdye pokrytiya *Fizicheskaya mezomekhanika* **5** 103–6.

[3] Vereshchaka A S, Vereshchaka A A, Bublikov Yu I and Oganyan M G 2014 Nanorazmernye

mnogoslojno-kompozicionnye pokrytiya dlya povysheniya iznosostojkosti i rabotosposobnosti tverdosplavnyh koncevyh frez *Rezanie i instrumenty v tekhnologicheskih sistemah* **84** 26–42.

[4] Maksimov Yu V, Bublikov Yu I, Vereshchaka A A, Vereshchaka A S, Haustova O Yu, Kozlov L A and Hozhaev O X 2012 Nanodispersnye mnogoslojno-kompozicionnye pokrytiya dlya rezhushchih instrumentov *Izvestiya MGTU MAMI* **2(14)** 222–29.

[5] Chernobaj S P and Sablina N S 2015 Sovershenstvovanie tekhnologii izgotovleniya rezhushchego instrumenta iz bystrorezhushchih stalcj dlya obrabotki detalej letatel'nyh apparatov *Mezhdunarodnyj zhurnal eksperimental'nogo obrazovaniya* **8** 405–6.

[6] Yakubov Ch, Dzhemalyadinov R, Skakun V 2018 Improving material cutting by application of natural origin oil *MATEC Web of Conferences* 224(51):01130.

[7] Kazaryan M A, Lidskii V V and Sachkov V I 2018 On one mechanism of light ablation of nanostructures *Proceedings of SPIE - The International Society for Optical Engineering 13.* p. 1061415.

[8] Grigoryants A G, Stavertiy A Y, Bazaleeva K O, Yudina T Y, Smirnova N A, Tretyakov R S and Misyurov A I 2017 Laser surfacing of nickel-based composite war-resisting coatings reinforced with tungsten carbide *Welding International* **31** 52–7.

[9] Dang, Engracia & Chan, Cho 2002 Ultrashort pulse laser material removal *International Congress on Applications of Lasers & Electro-Optics* 21695. 10.2351/1.5065726.

[10] Shafeev G 2012 Laser Ablation in Flowing Liquid *Laser Ablation in Liquids* 549–72.

[11] Hamad A H 2016 Effects of different laser pulse regimes (nanosecond, picosecond and femtosecond) on the ablation of materials for production of nanoparticles in liquid solution *High Energy and Short Pulse Lasers* 305–25.

[12] Barmina E V 2017 Laser Alloying of Co Nanorods and Al Nanoparticles in a Liquid *ChemPhysChem* **9** 1069–73.

[13] Serkov A A, Barmina E V, Simakin A V, Kuzmin P G, Voronov V V and Shafeev G A 2015 Generation of core–shell nanoparticles Al@Ti by laser ablation in liquid for hydrogen storage *Appl. Surf. Sci.* **348** 71–4.

[14] Yan Z and Chrisey D 2012 Pulsed laser ablation in liquid for micro-nanostructure generation *Journal of Photochemistry and Photobiology C Photochemistry Reviews* **13(3)** 204–23.

[15] Kudryashov S I, Nguyen L V, Kirilenko D A, Brunkov P N, Rudenko A A, Busleev N, Saraeva I N, Nastulyavichus A A, Ionin A A and Tolordava E R 2018 Large-scale Laser Fabrication of Anti-Fouling Si Surface Nanosheet Arrays via Nanoplasmonic Ablative Self-organization in Liquid CS2 Tracked by Sulfur Dopant *ACS Appl. Nano Mater* **1** 2461–68.

[16] Streubel R, Barcikowski S, Gökce B 2016 Continuous multigram nanoparticle synthesis by high-power, high-repetition-rate ultrafast laser ablation in liquids *Opt. Lett* **41** 1486–89.

[17] Lang V, Rank A and Lasagni A F 2017 Large Area One - Step Fabrication of Three - Level Multiple - Scaled Micro and Nanostructured Nickel Sleeves for Roll - to - Roll Hot Embossing *Advanced Engineering Materials* **8** 1700126.

[18] Jiang T, Zhao Q 2020 Lung Tumors Laser Ablation. In: Pacella C, Jiang T, Mauri G. (eds) Image-guided Laser Ablation *Springer* Cham pp 107–15.

[19] Lam J 2016 Dynamical study of bubble expansion following laser ablation in liquids Applied *Physics Letters* **7** 074104.

[20] Pacella C M and Mauri G 2020 History of Laser Ablation. In: Pacella C, Jiang T, Mauri G (eds) *Image-guided Laser Ablation. Springer*, Cham pp.1–5.

IOP Conf. Series: Materials Science and Engineering **889** (2020) 012032 doi:10.1088/1757-899X/889/1/012032

Temperature dependence of $Ge_2Sb_2Te_5$ conductivity and its change during aging

Y V Vorobyov[1,*], A O Yakubov[1,2] and A V Ermachikhin[1]

[1] Ryazan State Radio Engineering University, Ryazan 390005, Russia
[2] National Research University of Electronic Technology, Zelenograd 124498, Russia

* vorobjov.y.v@rsreu.ru

Abstract. Results of studies of conductivity activation energy and its evolution during the aging process are presented for the common phase change material $Ge_2Sb_2Te_5$. The evolution of temperature dependence of conductivity during quasi-isothermal annealing are interpreted in the framework of simple Arrhenius model as the change of two corresponding parameters: activation energy and pre-exponential factor. It is shown, that the change in resistivity of the $Ge_2Sb_2Te_5$ during aging is attributed to the simultaneous change of both parameters of the Arrhenius equation, that is, activation energy and pre-exponential factor. Temperature dependencies of those parameters are shown as well, and the Meyer-Neldel rule for conductivity of $Ge_2Sb_2Te_5$ is tested on the basis of the obtained data.

1. Introduction

The modern Phase Change Memory (PCM) technology [1] relies on use of the phase state of an active material for information storage. Two states of the material, crystalline and non-crystalline (amorphous), possess strikingly different physical properties. The phase transition between them is a reversible process, which could be induced by either optical or electrical disturbance. Crystalline state, being thermodynamically preferable, virtually does not change over time. Amorphous state could also last many years provided that temperature is not sufficiently high [2].

Partial crystallization of a memory cell volume is achievable, offering opportunities for multilevel storage of information [3]. Because of great difference between electrical conductivities of both phases, resistance of a memory cell is sensitive to the fraction of crystalline phase inside the active volume. The number of different levels of resistance is limited in the first instance by the signal-to-noise ratio necessary for reliable detection. However, in practice this is not the only limitation [4]. Due to the metastable nature of amorphous materials, they are prone to the process of aging. During this process, variety of material properties is changing, and this is of great concern for the developers of PCM devices [5]. Regarding electrical PCM, the main characteristic property of the material is resistivity, and its value of the material in amorphous phase tends to gradually increase over time. This phenomenon is called "resistance drift" in the literature.

Different approaches to the study of resistance drift in phase change materials, in particular $Ge_2Sb_2Te_5$, are reported. Atomistic simulations [6] show that the predominant process happening during aging of such compound is the transformation from a structure with mainly fourfold Ge and twofold Te into the one with the majority of Ge and Te atoms in the threefold coordination. It is commonly accepted that this relaxation increases the activation energy for conduction [7]. Measurements show that thermal annealing leads to an increase of the band gap value [8,9], testifying in favour of this theory. On the other side, the modelling approach proposed in [10] suggests that the distance between trapping centres in GST decreases with increasing temperature, lowering the value of activation energy. Therefore, both time- and temperature dependence of activation energy take place in GST.

Our current study focuses on the experimental observation of how the activation energy changes with both time and temperature. Relation between those two processes could provide valuable insights on the resistance drift effect in PCM devices.

2. Methods

2.1. Measurement technique

The qualitative behaviour of the temperature dependence of conductivity in phase change materials obeys the well-known Arrhenius equation:

$$\sigma = \sigma_0 \exp\left(-\frac{E_A}{k_B T}\right), \tag{1}$$

where k_B is the Boltzmann constant, E_A is the activation energy of conductivity and σ_0 is called pre-exponential factor. Below we refer to the latter as the prefactor, for brevity.

During the process of aging, the value of conductivity gradually changes towards lower values. The obvious way to monitor these changes is the measurement of device resistance [11,12], which is inversely proportional to the conductivity value. However, (1) suggests that at a given temperature the value of conductivity is determined by two parameters, σ_0 and E_A. To study the process of resistance drift in greater detail than it is provided by resistance measurement, it is necessary to obtain temporal evolution of σ_0 and E_A during this process.

2.1.1. Time dependence.

A forementioned quantities could be obtained from temperature dependence of conductivity. However, to assign them to the process of aging at constant temperature it is necessary to obtain corresponding time dependencies in the same condition, that is, isothermally. One way to achieve this is to separate the aging process, induced by isothermal annealing, from the measurement [13]. Thus, annealing is performed sequentially in order to capture the time evolution of the resistance drift process *ex situ*. Time resolution achievable in such experiment is poor. In [14] a better technique is used allowing better time resolution by combining annealing and measurement stages in the same setup. However, initial stages of aging process still lack time resolution.

In this study we propose to measure both σ_0 and E_A simultaneously and *in situ* by means of complex temperature profile. The sample is held at *quasi*-constant temperature, that is, the temporal temperature profile is the sum of constant base level and small oscillations. The amplitude of the latter was chosen to be ±5°C to (*i*) avoid its influence on the resistance drift process and (*ii*) to provide enough data to calculation of desired quantities. The collected data is therefore a set of temperature dependencies of conductivity in the 10°C range around the base temperature. On the basis of (1), it is straightforward to define the activation energy as the slope of linear dependence $\ln\sigma = f((k_B T)^{-1})$:

$$\frac{\mathrm{d}\ln\sigma}{\mathrm{d}(k_B T)^{-1}} = -E_A. \tag{2}$$

2.1.2. Temperature dependence.

Data regarding the measurement of the temperature dependence of Arrhenius equation parameters for phase change materials is scarce, and it is not common to analyse it with respect to the resistance drift phenomena. The work [15] should be mentioned, where the values of activation energy for $Ge_2Sb_2Te_5$ are reported from the room temperature up to the melting temperature. The authors use semi-analytical approach to extract the desired quantity from the high-speed measurements of resistance, assuming prefactor σ_0 temperature-independent.

It should be noted, that the very definition of the temperature dependence of E_A and σ_0 is not straightforward. In [16] it is pointed out, that if one's approach implies the use of (1) for the *entire* temperature range, taking into account temperature dependencies of parameters leads, instead of (2), to

$$\frac{\mathrm{d}\ln\sigma}{\mathrm{d}(k_BT)^{-1}} = \frac{\mathrm{d}\ln\sigma_0}{\mathrm{d}(k_BT)^{-1}} - \frac{\mathrm{d}\,E_A}{\mathrm{d}(k_BT)^{-1}} - E_A. \tag{3}$$

Two additional terms in (3), in comparison with (2), are unknowns, and their presence does not allow deriving the $E_A(T)$ dependence from the $\sigma(T)$ dependence alone.

Because of this limitation we do not attempt to describe the full temperature range with (1) together with temperature dependent σ_0 and E_A. Instead, we assume (1) to be valid *locally* over a small range around the given temperature. This definition, therefore, (*i*) allows use of simple equation (2) to calculate the value of activation energy using an experimental $\sigma(T)$ dependence and (*ii*) is in line with our method of time-resolved measurements of E_A.

2.2. Preparation of experimental samples

Simple planar structures were used for the study. TiN/W electrodes were formed on the oxidized silicon substrates with the inter-electrode distance 1.3 mm. The $Ge_2Sb_2Te_5$ layer was deposited on the top of electrodes by means of magnetron sputtering (residual Ar pressure 0.57 Pa, power 25 W). Thickness of films was about 130 nm according to atomic-force microscope profiles.

2.3. Experimental details

Measurement of temporal dependencies was performed using Linkam HFS600E-PB4 heating stage. Resistance measurements were performed sourcing constant voltage of 10 V (Agilent E3647A) to the experimental structures and registering the current along $Ge_2Sb_2Te_5$ layer using Keithley 6485 picoammeter. Constant Ar flow was provided through the sample camera during measurements to prevent the oxidation of $Ge_2Sb_2Te_5$ at elevated temperature.

Temperature dependencies of activation energy were measured in the Janis CCS-400/204N close-cycled cryostat. The same constant voltage regime was used to measure the sample resistance; Keithley 6487 picoammeter/voltage source was used. Heating rate during measurements was limited to 2 K/min.

3. Results and Discussion

Figure 1 shows the temporal evolution of σ_0 and E_A for $Ge_2Sb_2Te_5$ thin films subjected to the annealing at 100°C. In [13] it is stated that the prefactor remains constant during aging process, suggesting that the change in activation energy alone drives the resistance drift phenomena. Authors [14], despite having identified the change of the prefactor for some compounds, also reported constant value of this quantity for $Ge_2Sb_2Te_5$. From the data in figure 1, however, it is evident, that change of σ_0 takes place in $Ge_2Sb_2Te_5$, though only at the very early stage of the process. That is, the drift of activation energy and prefactor are characterized by different time constants. Fundamentally, change of both E_A and σ_0 during aging defines the resistance drift phenomenon in phase-change materials.

In figure 2 the temperature dependence of activation energy and prefactor are shown for the sample before and after annealing at 100°C for 5 hours. Both parameters exhibit significant temperature

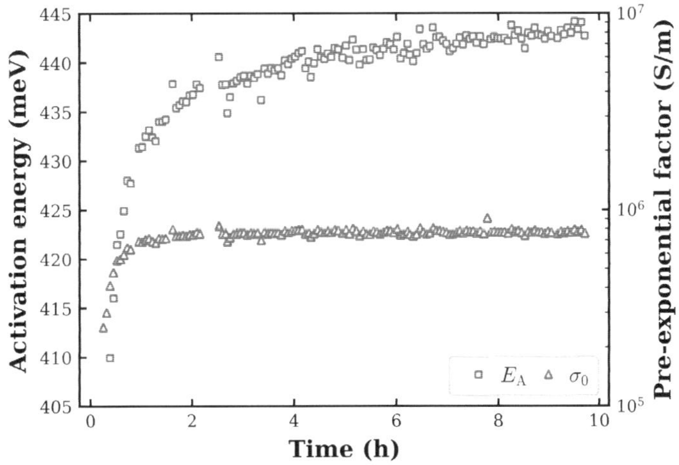

Figure 1. Evolution of the Arrhenius equation parameters for the conductivity of $Ge_2Sb_2Te_5$ under isothermal annealing.

dependence in the presented temperature range. In [15] the temperature dependence of activation energy is demonstrated for the wide temperature range and an increase of the activation energy value from 330 meV to 360 meV in 300–400 K range is reported. Our numbers are somewhat higher, but the relative difference, that is, the temperature coefficient of activation energy, is close. The difference, however, is that authors of [15] assumed the value of prefactor temperature-independent, while according to our results σ_0 increases with temperature, together with the increase of activation energy value.

The data for the initial sample show rapid increase of both E_A and σ_0 for temperatures above 60°C. We believe that the source of this feature is again the aging process. The temperature scanning during the measurement is relatively slow, and as far as at elevated temperature the drift process becomes activated, the observed increase of E_A could be attributed to time dependence of this quantity. It is not clear, if it is possible to decouple time- and temperature dependencies of E_A and σ_0 when the drift process becomes activated. However, if several undrifted samples are available, the method described

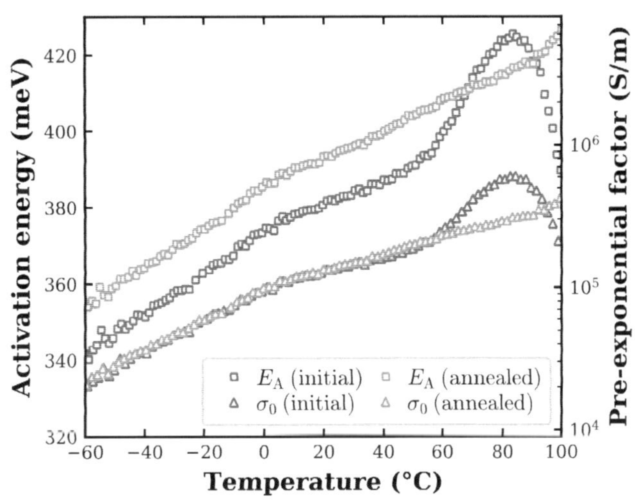

Figure 2. Temperature dependence of activation energy and pre-exponential factor before and after annealing at 100°C.

in [17] could be used utilizing different heating ramps.

The prefactor values before and after annealing practically coincide for temperatures below 60°C. From the first sight, it appears to contradict the detected change of the prefactor value during aging (figure 1). However, the figure 1 depicts values of prefactor that correspond to the elevated temperature 100°C, while the values for the same temperature in figure 2 appear to be distorted by the influence of drift process, which prevents direct comparison. Nevertheless, the fact that for lower temperatures the prefactor remains constant after annealing could still be described on the basis of the data in figure 1. It shows, that the value of prefactor saturates quickly during annealing, in comparison with the activation energy. After deposition, the samples were exposed to influence of room temperature for some time before measurements, which could already provoke the changes in σ_0 and E_A. Due to fast nature of those changes for σ_0, its value for lower temperatures could be already saturated.

The data in figure 2 clearly shows an increase of activation energy value in the whole range of temperatures after the annealing, which is in line with the results shown in figure 1. The difference ΔE_A is about 14 meV and it is spread uniformly across the whole temperature range. The drift phenomenon in the $Ge_2Sb_2Te_5$ is attributed to the band gap widening ΔE_g [8,18]. In [8] a 30 meV change after annealing at 80°C for 27 hours is detected, and it is indeed uniform from 80°C down to helium temperatures. Our results, therefore, support the theory of predominant influence of band gap widening on the resistance drift in $Ge_2Sb_2Te_5$.

Interesting to note, that the temperature dependence of σ_0 in semi-logarithmic scale clearly resembles the temperature dependence of E_A. This observation could be a manifestation of the Meyer-Neldel rule (MNR) [19] which states that the prefactor correlates with the activation energy as

$$\sigma_0 = \sigma_{00} \exp\left(\frac{E_A}{E_{MN}}\right), \tag{4}$$

where E_{MN} is the characteristic energy of MNR. Since change of temperature provides means of changing the value of E_A, the data from figure 2 could be used to test the MNR in the case of $Ge_2Sb_2Te_5$ conductivity. Figure 3 shows the $\sigma_0(E_A)$ dependence in semi-logarithmic coordinates. The dependencies are sufficiently linear and straight lines provide fits to data according to (4). The estimated value of E_{MN} is about 25 meV for both initial and annealed samples. This value is similar to those found for other chalcogenide systems [19]. Moreover, the obtained value is close to the

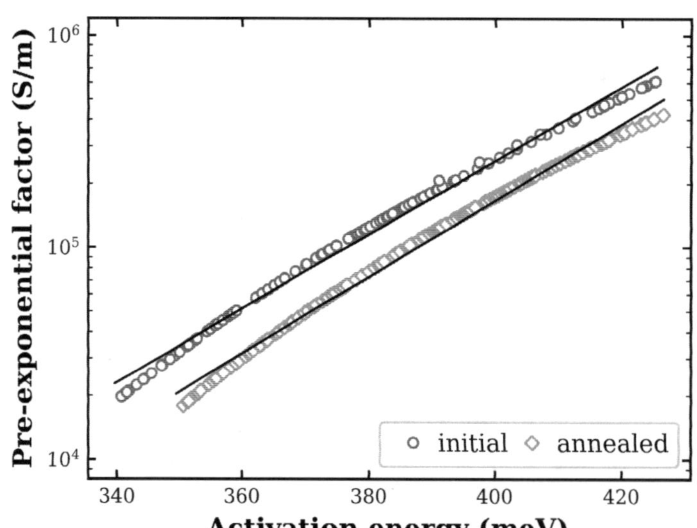

Figure 3. The test of Meyer-Neldel rule for the conductivity of $Ge_2Sb_2Te_5$.

magnitude of optical phonons in amorphous $Ge_2Sb_2Te_5$, which is in the range 15–20 meV [20]. This correlates well with the multi-excitation theory of MNR, which states that in the systems where the potential barrier for a kinetic process is large in comparison with available excitations the activation is provided by means of multiple elemental excitations.

4. Conclusions

Parameters of Arrhenius equation for conductivity are determined for the $Ge_2Sb_2Te_5$. Their evolution with both time and temperature are presented. Temporal measurements show considerable change of the activation energy and pre-exponential factor during isothermal annealing, which is attributed to the resistance drift phenomenon in phase-change materials. Temperature dependence of activation energy demonstrates a uniform shift toward higher values after annealing, which could be explained with corresponding increase of the band gap. The validity of Meyer-Neldel rule for conduction of $Ge_2Sb_2Te_5$ demonstrated leading to the value of an elemental excitation energy close to the phonon energy.

Acknowledgments

Authors acknowledge support of the Russian Science Foundation (project no. 19-73-00150).

References

[1] Noé P, Vallée C, Hippert F, Fillot F and Raty J-Y 2018 *Semicond. Sci. Technol.* **33** 013002

[2] Kozyukhin S, Vorobyov Y, Sherchenkov A, Babich A, Vishnyakov N and Boytsova O 2016 *Phys. status solidi* **213** 1831–8

[3] Stanisavljevic M, Pozidis H, Athmanathan A, Papandreou N, Mittelholzer T and Eleftheriou E 2016 *2016 IEEE 8th Int. Mem. Work. IMW 2016* 4–7

[4] Athmanathan A, Stanisavljevic M, Cheon J, Kang S, Ahn C, Yoon J, Shin M, Kim T, Papandreou N, Pozidis H and Eleftheriou E 2015 *Proc. - 2014 IEEE Asian Solid-State Circuits Conf. A-SSCC 2014* 137–40

[5] Ielmini D, Sharma D, Lavizzari S and Lacaita A L 2009 *IEEE Trans. Electron Devices* **56** 1070–7

[6] Raty J Y 2019 *Phys. Status Solidi - Rapid Res. Lett.* **13** 1–10

[7] Le Gallo M, Krebs D, Zipoli F, Salinga M and Sebastian A 2018 *Adv. Electron. Mater.* 1700627

[8] Rütten M, Kaes M, Albert A, Wuttig M and Salinga M 2015 *Sci. Rep.* **5** 1–11

[9] Krebs D, Schmidt R M, Klomfabeta J, Luckas J, Bruns G, Schlockermann C, Salinga M, Carius R and Wuttig M 2012 *J. Non. Cryst. Solids* **358** 2412–5

[10] Kaes M and Salinga M 2016 *Sci. Rep.* **6** 1–12

[11] Braga S, Cabrini A and Torelli G 2009 *Appl. Phys. Lett.* **94** 1–3

[12] Pirovano A, Lacaita A L, Pellizzer F, Kostylev S A, Benvenuti A and Bez R 2004 *IEEE Trans. Electron Devices* **51** 714–9

[13] Ielmini D, Fugazza D, Boniardi M, Montemurro G and Lacaita A L 2010 *Epcos* 7–8

[14] Wimmer M, Kaes M, Dellen C and Salinga M 2014 *Front. Phys.* **2** 1–12

[15] Muneer S, Scoggin J, Dirisaglik F, Adnane L, Cywar A, Bakan G, Cil K, Lam C, Silva H and Gokirmak A 2018 *AIP Adv.* **8** 065314

[16] Li J V., Johnston S W, Yan Y and Levi D H 2010 *Rev. Sci. Instrum.* **81** 033910

[17] Oosthoek J L M, Krebs D, Salinga M, Gravesteijn D J, Hurkx G A M and Kooi B J 2012 *J. Appl. Phys.* **112**

[18] Fantini P, Brazzelli S, Cazzini E and Mani A 2012 *Appl. Phys. Lett.* **100** 013505

[19] Shimakawa K and Abdel-Wahab F 1997 *Appl. Phys. Lett.* **70** 652–4

[20] Först M, Dekorsy T, Trappe C, Laurenzis M, Kurz H and Béchevet B 2000 *Appl. Phys. Lett.* **77** 1964–6

Hopping conduction in single crystals of the diluted magnetic semiconductor $(Zn_{1-x}Fe_x)_3As_2$ (x=0.005)

V S Zakhvalinskii[1], E A Pilyuk[1*] and T B Nikulicheva[1]

[1]Belgorod National Research University, Russia, 308015, Belgorod, 85 Pobedy str.

*pilyuk@yandex.ru

Abstract. Single crystals of the diluted magnetic semiconductor $(Zn_{1-x}Fe_x)_3As_2$ ($x = 0.005$) were obtained by modified Bridgman method. According to the results of the X-ray powder diffractometry, the material was single-phased and isomorphic and corresponded to the pure Zn_3As_2 ($x = 0.0$). The research of the electroconductivity and the magnetoresistance was carried out at the temperature range from 10 to 300 K. It was found out that the electroconductivity in the temperature range 11 ÷ 19 K corresponded to the mechanism of the Mott type variable-range hopping conductivity. The microparameters, characterizing electroconductivity $(Zn_{1-x}Fe_x)_3As_2$ ($x = 0.005$) at the temperature range of 11 ÷ 19 K, were defined.

1. Introduction

Zn_3As_2 is a semiconductor of group II_2V_5p-type conductivity, at the room temperature it has the width of the band gap$E_g \approx 1$ eV, tetragonal crystal structure and belongs to the space group $I4_1cd$ (it is so called α-phase Cd_3As_2)[1]. Solid solutions on the basis of Zn_3As_2, where a part of atoms As is changed by Mn, belong to the class of diluted magnetic semiconductors which can be applied in the heterostructures of spintroninc devices as injectors of spin-polarized charge carriers. As it is known, for a successful integration of magnetic material with nonmagnetic semiconductors in a microelectronic device, it is necessary for them to have comparable resistance values. In this case magnetic semiconductors have an indisputable advantage over metals [2]. Solid solutions of the diluted magnetic semiconductor $(Zn_{1-x}Mn_x)_3As_2$ were researched in a range of works [3-9]. We know only one work dedicated to the study of magnetic properties of a new diluted magnetic semiconductor $(Zn_{1-x}Fe_x)_3As_2$ [10], and the reference to this material in a review [11] with the reference to the work [10]. In the work [10] the single crystals $(Zn_{1-x}Fe_x)_3As_2$ were synthesized at the range of compositions $0.005 \leq x \leq 0.04$. By X-ray phase analysis it was found out that solid solutions $(Zn_{1-x}Fe_x)_3As_2$ remained homogeneous till the composition $x = 0.015$. In the diffractograms for $x \geq 0.02$ there was fixed the appearance of the second phase, identified as Fe_2As. The parameters of the crystal lattice of $(Zn_{1-x}Fe_x)_3As_2$ did not change considerably at the growth of x up to 0.02. The magnetic properties were researched by the SQUID magnetometer, the presence of the spin glass phase was ascertained, and the freezing temperatures of magnetic moments $T_f < 50$ K were determined.

In this paper we report about the research of electroconductivity and magnetoresistance in a single crystal of the solid solution of the new diluted magnetic semiconductor $(Zn_{1-x}Fe_x)_3As_2$ ($x = 0.005$), which was carried out for the first time.

2. Results and discussion

The research of the temperature dependence of single crystals of the solid soution of the new diluted magnetic semiconductor $(Zn_{1-x}Fe_x)_3As_2$ (x = 0.005) resistance was done by six-probe method. The samples were in the form of parallelipiped 2x2x5 mm, copper contacts were attached to the samples by silver conductive paste. The temperature range of measurements was from 10 to 300 K with the step of 2 K. The temperature dependence of the magnetoresistance was measured in the magnetic field of 1 T. In order to change the sample's temeparture we used the closed cycle cryostat of the model Janis CCS-350S, equipped with a thermostat and a helium compressor, 8200 Compressor, and a thermocontroller 331 of the brand Lake Shore.

Figure 1 shows the experimentally mesured dependence of the resistivity on the temperature,$\rho(T)$, of the single crystal $(Zn_{1-x}Fe_x)_3As_2$ (x = 0.005) in the zero magnetic field and in the magnetic field B = 1 T. The behaviour of the resistivity dependence on the tmperature $\rho(T)$ $(Zn_{1-x}Fe_x)_3As_2$ (x = 0.005) has a complicated character both in the zero magnetic field and in the magnetic field B = 1 T. At all the temperature range from 10 to 300 K there is a positive magnetoresistance. In the low-temperature area the dependence $\rho(T)$ has an activation character from 10 to 40 K in the zero magnetic field and from 10 to 30 K in the field B = 1 T. The further temperature increase leads to the growth of the resisitivity, which corresponds to the presence of metallic conductivity in the impurity band due to Mott transition [12].

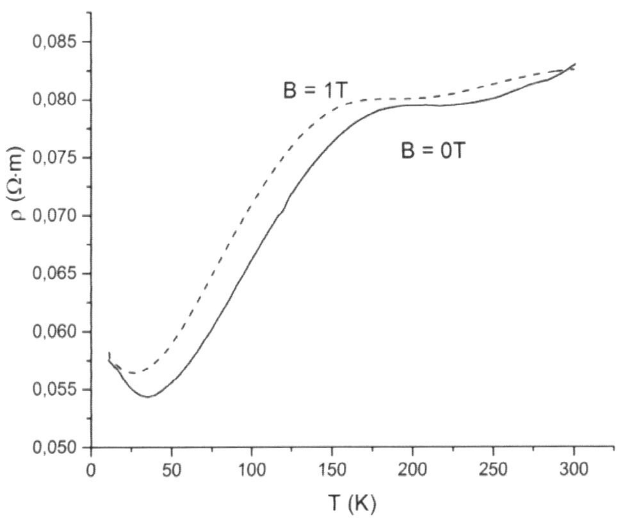

Figure 1. The dependence of ρ on T for the single crystal $(Zn_{1-x}Fe_x)_3As_2$ (x = 0.005).

The resistivity of the semiconductor at the thermal activation follows the law [13]:

$$\rho(T) = \rho_0 \exp\left[E_A \cdot (kT)^{-1}\right], \tag{1}$$

where k is Boltzmann constant; E_A is the activation energy; ρ_0 is the preexponential factor.

On the other hand, the hopping conduction in three-dimensional (3D) alloyed semiconductors can be assigned by the universal equation [14], describing the implementation of the hopping conduction with the help of different mechanisms

$$\rho(T) = DT^m \exp(T_0/T)^p, \tag{2}$$

where D is the constant; T_0 is a characteristic temperature, depending on the implemented hopping conduction mechanism, which is mathematically described by the parameter quantity p ($m - p$). The parameter p can get the following, most probable for the semiconductors alloyed by the shallow impurity, quantities: at p = 1 there is a nearest-neighbor hopping (NNH) conductivity; at p = 1/4 there is Mott variable-range hopping conductivity (Mott VRH); at p = 1/2 there is Shklovsky-Efros (SE) type VRH [15].

The quantity of the characteristic temperature T_0 in case of the implementation of one of VRH mechanisms can be presented as

$$T_{0M} = \frac{\beta_M}{kg(\mu)a^3}, \quad T_{0SE} = \frac{\beta_{SE}e^2}{\kappa ka}, \tag{3}$$

where $g(\mu)$ is the density of the localized states (DLS) near Fermi level; a is the charge carrier localization length; κ is a dielectric conductivity, $\beta_M = 21$, $\beta_{SE} = 2{,}8$ [14].

The implemented hopping conduction mechanism at low temperatures according to the equation (2) mathematically is considered to be the parameter quantity p, but, on the other hand, the power dependence T^m also influences considerably the hopping conduction mechanism. In connection with this it is suitable to rewrite the equation (2) at $B = 0$ and $B = 1$ tesla as:

$$\ln[E_a/(kT + m)] = \ln p + p \ln T_0 p + \ln(T^{-1}), \tag{4}$$

where $E_a \equiv d\ln\rho/d(kT)^{-1}$ is the local activation energy [14].

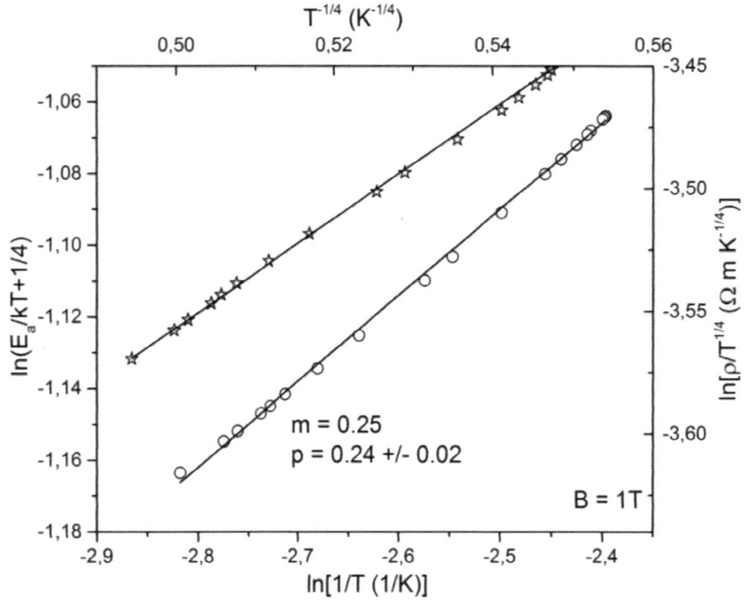

Figure 2. The dependences of $\ln(E_a/kT+1/4)$ on $\ln(1/T)$ (○) and $\ln(\rho/T^{1/4})$ on $T^{-1/4}$ (☆) at $B = 1$ tesla (temperature range 11 ÷ 19 K).

Thereby, for the definite conduction mechanism the left part of the equation (4) will be a linear function of $\ln(T^{-1})$, and p will be determined by the incline of the graph $\ln[E_A/(kT + m)]$from $\ln(T^{-1})$ (Figure 2).

For the analysis of the hopping conduction type consider $m = ¼$ in the equation (4), we get $p \approx 1/4$, which corresponds to the Mott type VRH. Mott VRH occurs in case when DLS near Fermi level is a constant and finite [12].

From the dependence $\ln(\rho/T^{1/4})$ on $T^{-1/4}$ for the sample $(Zn_{1-x}Fe_x)_3As_2$ (Figure 2) we get an extensive linear interval of low temperatures, this also allows assumption about VRH mechanism type. Quantities of T_{0M} at $B = 0$, $B = 1$ tesla can be found out from the linear sections of Figure 2 and are presented in Table 2. The activation energy E_A can be determined from the linear section of the dependence $\ln\rho$ on T^{-1} in the equation (1) at the temperature range from 11 K to 19 K, under condition that the coefficient ρ_0 faintly depends on T [14]. These values are presented in Table 1.

According to the data, obtained at $B = 0$ and in weak fields at $B = 1$ T, it is possible to find out different microscopic parameters [15].

Table 1. Coefficient ρ_0 and the activation energy E_A in $(Zn_{1-x}Fe_x)_3As_2$ ($x = 0.005$).

	ρ_0 (Ohm·m)	E_A (meV)
$B = 0$ T	1.54	0.249
$B = 1$ T	1.66	0.250

Table 2. The calculated quantities of T_{0M}, of the soft Coulomb gap in DLS near Fermi level Δ, tough gap δ, κ, a, W and g for the sample $(Zn_{1-x}Fe_x)_3As_2$ (x = 0.005).

	T_M, K	g_μ, cm^{-3} meV^{-1}	*r*, nm	*a*, nm	*κ*	Δ, meV	W, meV
$B = 0$ T	69	$2.65\cdot10^{+15}$	3.7	12	165	0.18	1.13
$B = 1$ T	25	$7.33\cdot10^{+15}$	2.9	12	180	0.04	0.53

In the sample $(Zn_{1-x}Fe_x)_3As_2$ ($x = 0.005$) the influence of Coulomb gap is minimal, Mott type conductivity is implemented at $B = 0$, $B = 1$ T. That is why DLS in the acceptor zone can be approximated by the rectangular form, the width of Coulomb gap in DLS of Fermi level is $W \approx kT_{VM}^{3/4}T_{0M}^{1/4}$ [12]. From the formula $g(\mu) \approx N_A/(2W)$ we get DLS for this type of conductivity $(g(\mu) \equiv g)$, and from the equation (3) we find *a*. The value *κ* will be got from the formula $E_A = F(K)e^2\kappa^{-1}N_A^{1/3}$ [12], where *F*(*K*) is some universal compensation function (*F*(*K*) = 0.43 [16]). Then from the formulas

$$\Delta \approx \frac{k}{2}\sqrt{T_{VM}T_{0M}}, \; g_0 = \frac{3\kappa^3(\Delta-\delta)^2}{\pi e^6}, \; g(\mu) = \frac{N_A}{2k\left(T_{VM}^3 T_{0M}\right)^{1/4}}, \tag{5}$$

we find the values Δ and *δ*. The calculated values Δ, *δ*, *κ*, *a*, *W* and *g* for the sample are presented in Table 2. The correlations between the values Δ and *W* are coordinated with the corresponding conduction mechanism at $B = 0$. The ratio $\Delta/W \sim 0{,}1$ is suitable for Mott VRH regime.

3. Conclusion

For the first time we researched the temperature dependence on electroconductivity and magnetoresistance of single crystals of the diluted magnetic semiconductor $(Zn_{1-x}Fe_x)_3As_2$ ($x = 0.005$), obtained by modified Bridgman method. According to the results of the X-ray powder diffractometry, the material was single-phased and isomorphic and corresponded to the pure Zn_3As_2 ($x = 0.0$). The research of the electroconductivity and the magnetoresistance was carried out at the temperature range from 10 to 300 K. In the low-temperature zone of the dependence *ρ*(*T*) there was an activation section from 10 to 40 K in the zero magnetic field and from 10 to 30 K in the field $B = 1$T. It was found out that the electroconductivity in the temperature range 11 ÷ 19 K corresponded to the mechanism of the Mott type variable-range hopping conductivity. The microparameters, characterizing electroconductivity $(Zn_{1-x}Fe_x)_3As_2$ ($x = 0.005$), were determined for the hopping conduction mechanism. With the growth of temperature the resistivity increases, which is characteristic of the metal conductivity, and we may presume that this is connected with the metallic conductivity appearance in the impurity zone as a result of Mott transition [12].

4. References

[1] Pietraszko A and Łukaszewicz K 1973 Thermal expansion and phase transitions of Cd_3As_2 and Zn_3As_2 *Physica Status Solidi* (a) **18(2)** 723–30, doi:10.1002/pssa.2210180234

[2] Žutić Igor, Fabian Jaroslav, and Sarma S Das 2004 Spintronics: Fundamentals and applications *Rev. Mod. Phys.***76** 323, https://doi.org/10.1103/RevModPhys.76.323
[3] Laiho R, Lashkul A V, Lisunov K G, Lähderanta E, Stamov V N and Zakhvalinskii V S 1995 Disorder in$(Zn_{1-x}Mn_x)_3As_2$ and its consequences to impurity conduction and magnetic properties *J.Phys. Cond. Matter.***7** 7629
[4] Laiho R, Lisunov K G, Stamov V N and Zakhvalinskii V 1995 Anomalous magnetoresistance of (Zn1-xMnx)$_3$As$_2$ in the region of hopping conductivity *Solid State Commun.***93** 151
[5] Laiho R, Lashkul A V, Lisunov K G, Lähderanta E, Stamov V N and Zakhvalinskii V S 1995 Vacancy-type disorder in $(Zn_{1-x}Mn_x)_3As_2$, *J.Magn. Magn.Mater.***140-144** 2019-20
[6] Laiho R, Lisunov K G, Stamov V N and Zakhvalinskii V S 1995 Magnetoresistance of $(Zn_{1-x}Mn_x)_3As_2$ in region of hopping conductivity *J. Magn. Magn. Mater.***140-144** 2021-22
[7] Laiho R, Lashkul A V, Lisunov K G, Zakhvalinskii V S and Mökinen A 1993 Low temperature conductivity of $(Zn_{1-x}Mn_x)_3As_2$ solid solutions *J. Phys. Cond.Matt.***5** 5113
[8] Lashul A V, Lähderanta E, Laiho R, Zakhvalinskii V S 1992 Freezing of magnetic moments in $(Zn_{1-x}Mn_x)_3As_2$ neer 200 K. *Phys.Rev.*B **46** (10) 6251-5
[9] Kulibatchinskii V A, Svistunov I V, Chudinov S M, Kuznetsov V D, Nateprov A N, Arushanov E K and Zakhvalinskii V S 1991 Anomalous of magnetic and galvanomagnetic properties of the diluted solid solutions $(Zn_{1-x}Mn_x)_3As_2$*Phys.Tekhn.Semicond.* **25** (12) 2201-14
[10] Laiho R, Lashkul A V, Lisunov K G, Lähderanta E, Stamov V N and Zakhvalinskii V S 1995 Disorder in $(Zn_{1-x}Mn_x)_3As_2$ and its consequences to impurity conduction and magnetic properties *J.Phys. Cond. Matter.***7** 7629
[11] Cisowski J 1997 Semimagnetic Semiconductors Based on II-V Compounds *Phys. stat. sol.* (b) **200** 311-50
[12] Mott N F and Davis E A 1979 *Electron processes in non-crystalline materials* (Oxford, Clarendon Press)
[13] Shklovskii B I and Efros A L 1984 *Electronic Properties of Doped Semiconductors* (Berlin: Springer)
[14] Laiho R, Lisunov K G, Lähderanta E and et al.2002 Variable-range hopping conductivity in $La_{1-x}Ca_xMn_{1-y}Fe_yO_3$: evidence of a complex gap in density of states near the Fermi level *J. Phys.: Condens. Matter.***14** 8043
[15] Laiho RLashkul, A V, Lisunov K G and et al. 2008 Hopping conductivity of Ni-doped p-CdSb *J. Phys.: Condens. Matter.* **20** 295204
[16] Zabrodskii A G, Andreev A G and Alekseenko M V 1992 Hopping conduction K = 0.3-series of Ge:Ga samples: saturation effect, hopping on the closest neighbors and transition to the hops with variable length *Physics and technology of semiconductors*. **26** (3) 431.

Acknowledgements
The research was carried out for account of the grant of the Russian scientific foundation (project № 19-79-00152).

IOP Conf. Series: Materials Science and Engineering **889** (2020) 012034 doi:10.1088/1757-899X/889/1/012034

Experimental evaluation of the efficiency of lubricating process media during the running-in period of a metal-cutting tool

Vladimir Skakun[1], Ruslan Dzhemalyadinov[1*], and Ilimdar Abdulkerimov[1]

[1] Department of mechanical engineering, Crimean Engineering and Pedagogical University, Simferopol, Russian Federation

* rus.dzhemalyadino@gmail.com

Abstract. The possibility of increasing the durability of a cutting insert made of high-speed steel with a wear-resistant TiN coating due to the use of lubricating process media during the running-in period of the cutting tool is considered. Based on the analysis of the data obtained, an analogy in the mechanisms of the hardening action of the refrigerant and wear-resistant coatings is presented. It is shown that the durability of a metal-cutting tool substantially depends on the initial conditions of its operation.

1. Problem statement

The development of mechanical engineering is inextricably linked with the intensification of the processes of blade machining of materials by cutting. Such processing volume accounts for up to 40% of the total labour intensity manufacturing machine parts, while the volume of dimensional processing is continuously increasing. Therefore, improving the performance of metal-cutting tools is an urgent scientific and practical task, the solution of which is related to the development and implementation in the industry of both traditional and the latest surface hardening technologies.

2. Materials and methods of research

Tools with wear-resistant coatings are increasingly used in the modern metalworking industry. The main role of the coating applied to the working surfaces of the tool is to increase the hardness, heat resistance, wear resistance, and physical and chemical passivity in relation to the processed material [1, 2]. At the same time, one of the generally accepted ways to increase the durability of a metal-cutting tool is still supplying lubricant cooling process media to the cutting zone, but due to the increase in auxiliary costs associated with transportation, storage, and subsequent disposal, the profitability of their use is significantly reduced. In addition, the widely used mineral-based industrial oils and emulsions are one of the main environmental pollutants and adversely affect the health of production personnel. Thus, in the global practice of metalworking, tendencies for the use of environmentally friendly vegetable oils and oil-based esters as process media have been identified [3–7]. The higher cost of such compounds can be offset by the implementation of an environmentally oriented minimum lubrication strategy (MMS-Minimal mengenschmierung or MQL-Minimum quantity lubrication) [8 - 11]. The essence of the method is that, with the help of modern highly efficient dosing technology, the minimum amount of lubricant is purposefully fed into the chip formation zone, where it is consumed completely and without residue during processing. The latter means a sharp reduction in the consumption of process media, and as a result, a significant reduction in the technogenic impact on the environment, which is an important step towards harmonizing production processes and nature with a reduction in the cost of manufacturing finished products.

3. Statement of the main material

The supply of a lubricating medium to the cutting zone leads to a decrease in the total contact area on the front surface of the cutting tool, and as a result to an increase in specific normal loads [12, 13]. An analysis of the obtained experimental data showed that under the conditions of using process media based on vegetable oils (with characteristic higher penetrating, lubricating, shielding, and passivating properties), this effect is enhanced. Thus, under certain operating conditions of a high-speed steel cutting tool (pressures of more than 500 MPa and temperatures of 200 ... 350 °C), we can expect hardening of its contact layers [14]. The adaptation of the working surfaces of the tool to external loading conditions is most fully manifested in the initial period of cutting - its running-in period. This process is typical for the conditions defined above and, by analogy with the running-in of rubbing pairs, it shows a significant increase in resistance to further (stationary) operating conditions. The influence of lubricating process media on the change in the contact area along the front surface and the corresponding wear on the rear surface of the cutting tool are shown in the Table 1.

Table 1. The influence of lubricating process media on the change in the contact area along the front surface and the corresponding wear on the rear surface of the cutting tool.

Dry cutting without the use of cutting fluid	**Cutting fluid**		
	Emulsion 5 %	Industrial oil	Rapeseed oil
	The method of supply and flow of cutting fluid		
	Watering the cutting fluid – 5 l/min	MQL-45 ml/h	MQL-15 ml/h
	Chip contact area		
a	b	c	d
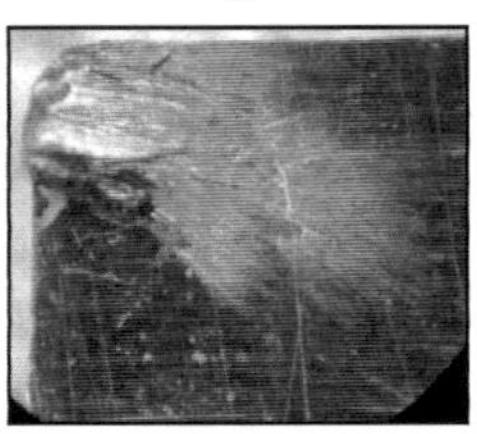	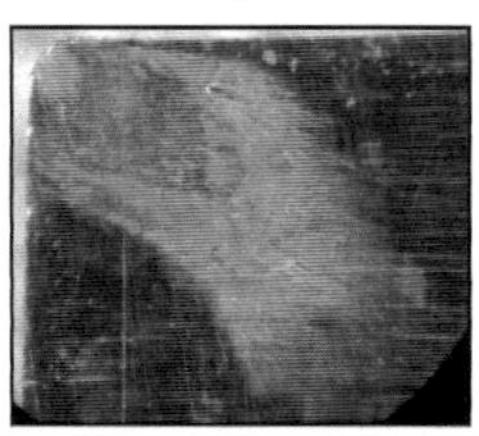	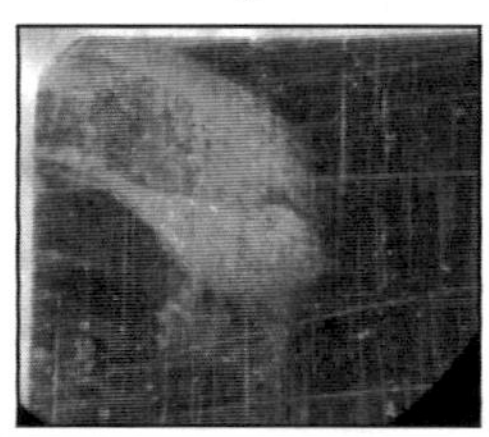	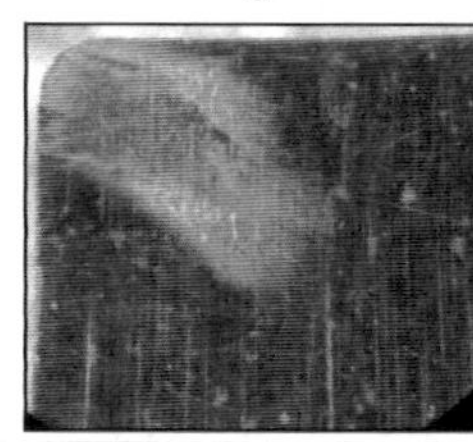
	Tool flank wear		
a	b	c	d
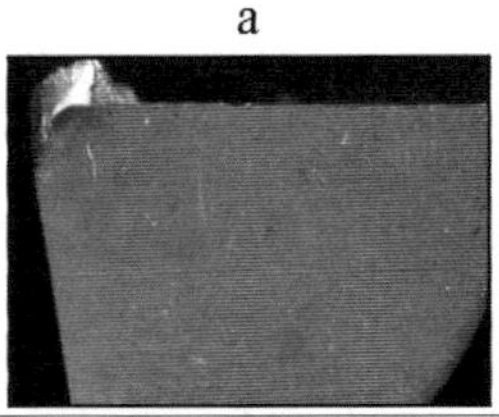	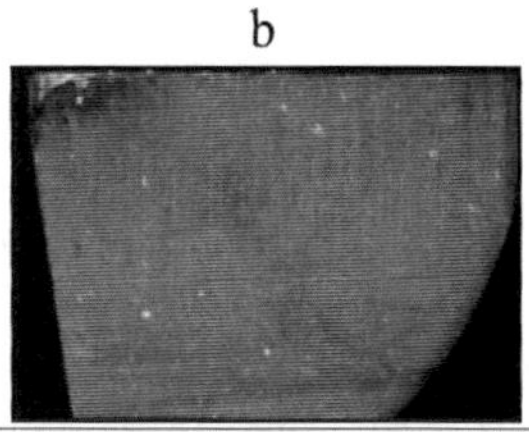	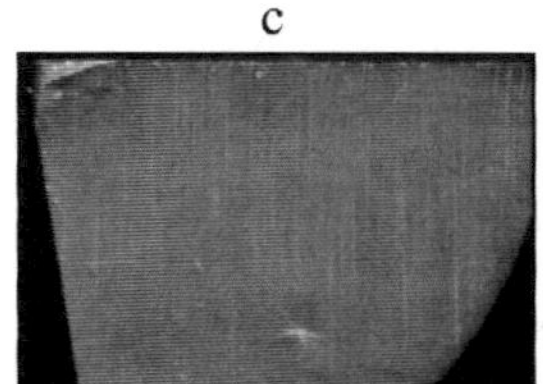	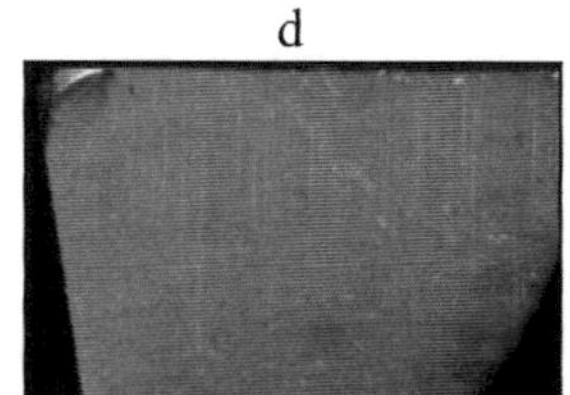

The effectiveness of applying wear-resistant coatings, in terms of the impact on the performance of the tool, is well known [15 - 17]. As a rule, after 10 ... 15 minutes of cutting, a partial or complete destruction of the wear-resistant layer occurs. However, it is still not clear why the effect of increase of durability is expressed in a much longer time even after its destruction.

When explaining this fact, the points of view of researchers diverge significantly. In some works [18], the effect of increase in the durability of a cutting tool after coating destruction is preserved due to the positive contribution of a layer of double tungsten and cobalt carbide Co3W3C (η - phase) formed at the interface with the base (substrate). In others, it is mentioned that the η - phase does not significantly affect the maintenance of the effective operation of the cutting tool after erasing of the coating [19] and possibly even worsens the characteristics of the cutting insert. There are also hypotheses explaining this circumstance by the presence in the contact area of undisturbed (stored) foci of coating or runoff of coating sections to the bottom of the crater [20].

If we consider wear-resistant coatings as a kind of intermediate process medium between the tool and the material to be processed, then we can formulate the conditions under which, using the directed supply of an external triboactive medium, it is possible to control effectively the properties of the tool material, the characteristics of contact processes, chip formation, and, as a consequence, the tool wear rate.

In most cases, the base oils in composition of such a process medium are vegetable oils, the mechanism of action of which is aimed at increasing the durability of the cutting tool, but is still not fully disclosed.

Thus, the extension of the service life of the cutting tool after wear of the coating, one way or another, is associated with a change in the properties of the main material of the tool. A detailed determination of the mechanism of this technically important phenomenon today continues to be an urgent scientific problem.

The aim of the article is an experimental evaluation of the possibility of increasing the durability of a metal-cutting tool by using oil-based process media during its running-in period.

From the point of view of the experimental research methodology, the most convenient one is the turning process, on the basis of which models that are also used on other types of machining are built.

Durability studies were carried out on a high-precision screw-cutting lathe, SAMAT 400 M model. To ensure a constant cutting speed, the Altivar 71 frequency inverter was installed on the machine. Steel 45 and chromium-nickel alloy AISI 321 were machined with ISO SPUN120308 tool turning rotary inserts made from HSSE high-speed steel with a TiN wear-resistant coating and without coating.

The technical rapeseed oil and I-20A mineral industrial oil were chosen as process media.

In these studies, a special dosing device Noga Minicool MMS (Ave. Israel) was used. A general view and principle of operation are shown in Figure 1.

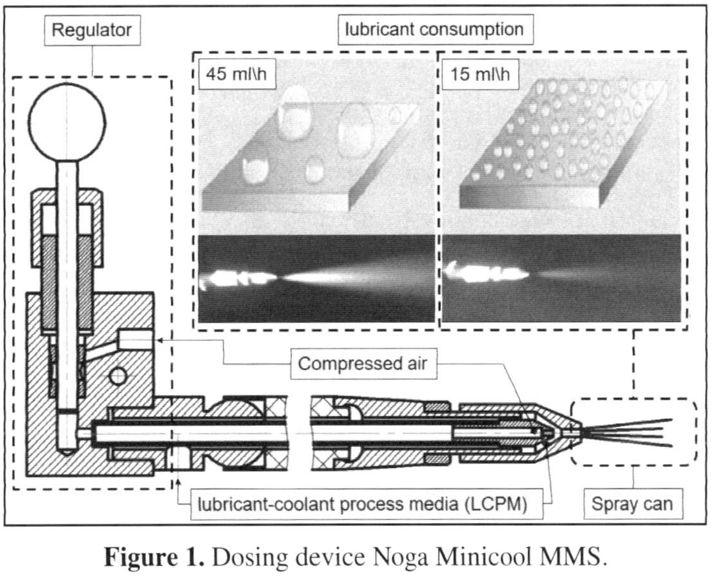

Figure 1. Dosing device Noga Minicool MMS.

The experiments were carried out at a constant cutting depth t = 0.5 mm and a supply S = 0.2 mm/rev. The cutting speed was v_C = 70 m/min for steel 45 and v_C = 30 m / min for stainless steel AISI 321.

Measurements of the wear amount of wear were made on the rear surface using an instrumental measuring microscope IMCL 150x75(2),B. The criterion was the value h_r = 0.35 mm in the normal wear zone along the rear surface of the tool.

It is commonly known that the use of process media from the point of view of wear reduction is associated with a decrease in the work of friction and temperature, trying to maintain the original properties of the tool material. The results of experimental studies presented below reveal a fundamentally new mechanism of action of process media - the possibility of hardening of the contact layers of the tool and access to wear control through this functional property.

Let's consider to which extent the influence of lubricating process media during the tool running-in period affects the durability of the metal-cutting tool during its further operation in the "dry" turning (without using process media). On the Fig. 2 and 3 summarize the results of durability tests of a cutter with a wear-resistant coating when turning steel 45 and AISI 321 after 7 minute cutting in a medium of mineral and rapeseed oils.

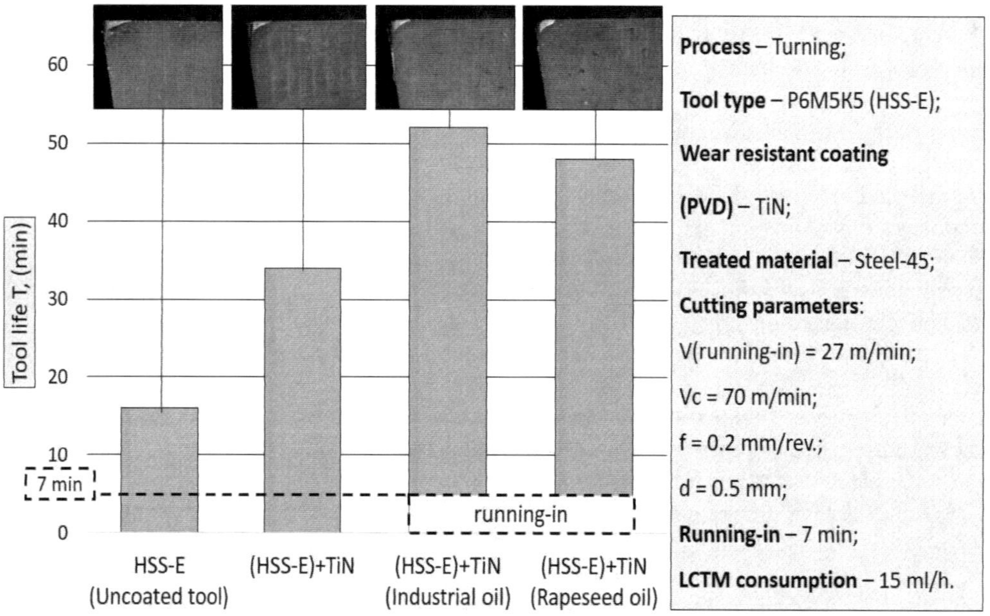

Figure 2. The durability of the cutter made of high-speed steel P6M5K5 with "dry" turning of steel 45 after preliminary running-in in the medium of mineral and rapeseed oils.

As expected, the presence of the coating significantly affects the wear rate of the cutting tool. So, when processing steel 45, the effect of increase in the durability is recorded by a twofold increase in comparison with an uncoated tool and threefold increase when processing stainless steel AISI 321. The greatest effect of wear reduction up to the accepted criterion h_r = 0.35 mm was found after preliminary 7 minute cutting in the selected process media: when processing steel 45 in a medium of mineral oil, and when processing stainless steel in a medium of technical rapeseed oil. After preliminary running-in, further operation of the cutting tool was carried out without the use of process media.

The obtained experimental data can be associated with the following circumstances. During turning, the cutting wedge is subjected to deformation loading. In the presence of a wear-resistant coating, as a result of the contact area reduction, the friction and temperature gradient over its thickness, the normal pressure on the substrate material increases, and the intensity of the heat flux decreases. This phenomenon can be likened to a decrease in cutting speed, in the limit corresponding to the range of running-in modes, in which contact layers of a hardened wear-resistant structure are formed on the surface of the tool in the presence of process media.

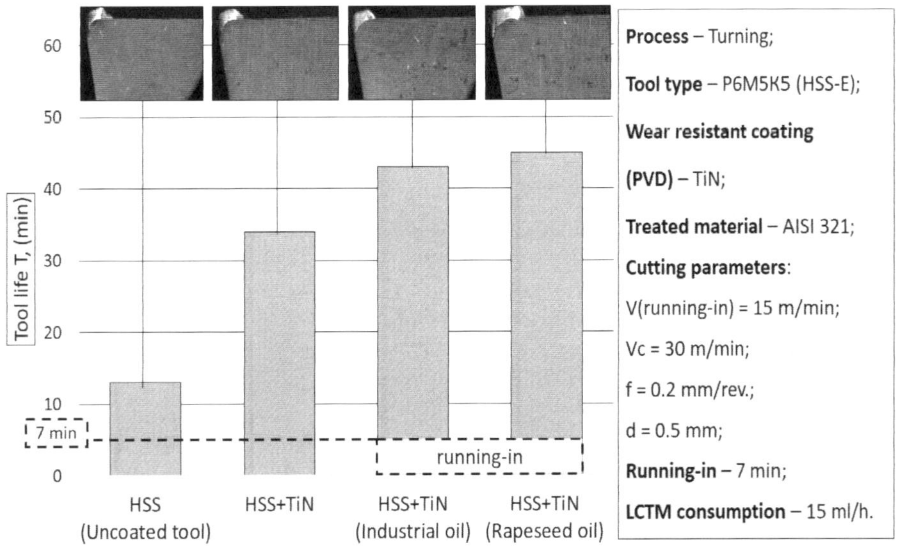

Figure 3. The durability of the cutter made of high-speed steel P6M5K5 with "dry" turning of AISI 321 after preliminary running-in in the medium of mineral and rapeseed oils.

Naturally, there should not be a full correspondence of the properties formed on the working surfaces of the tool, secondary structures during cutting in the preliminary running-in mode and substrate after the destruction of the coating. In the first case, the formation of secondary structures proceeds mainly due to deformation hardening (process media do not penetrate into plastic contact zone), in the second case, due to a combination of deformation and diffusion processes. Coating during the development of secondary structures is an active source of alloying elements and a stimulator of internal substructural mass transfer. Indeed, the diffusion activity strongly depends on deformation processes and the diffusion coefficient is directly proportional to the dislocation density. The implementation of the diffusing element from the coating into the instrumental matrix introduces additional distortion into the crystal structure, which, combined with an increasing density of dislocations, changes the parameters of the hardened layer relative to that formed under diffusion-free conditions. The formation of wear-resistant secondary structures in the surface layers of the tool by the method of preliminary running-in in the process media makes it possible to consider a wear-resistant coating as a highly effective (possibly ideal) lubricant.

4. Conclusion

Summarizing the aforesaid, it can be noted that the evaluation of the lubricating action of oil process media as a factor determining the hardening of tool contact layers in the running-in mode opens up new ways of their effective application as process media. On the one hand, taking into account a significant reduction in consumption when using dosing devices, the field of introduction of the environmentally friendly vegetable oils into the industry is expanding, on the other hand, a basis is being formed for modifying special running-in compositions, the operation of which will generally increase the durability of a metal-cutting tool.

5. References

[1] Tabakov V P and Vereshchaka A S 2013 Development of technological means for formation of multilayer composite coatings, providing increased resistance of carbide tools, for different

machining conditions *Key Engineering Materials (Precision Machining VII)* **581** 55–61
[2] Kishurov V M, Ippolitov V N, Kishurov M V and Nekrasova M Yu 2012 Influence of wear-resistant coatings on tool performance *Russian Engineering Research* **32** 182–5
[3] Yakubov Ch, Dzhemalyadinov R and Skakun V 2018 Improving material cutting by application of natural origin oil *MATEC Web of Conf.* **224** 01130
[4] Kathamore P S and Bachchhav B D 2019 Bio-based lubricant selection for metal cutting operations using MADM technique *International Journal of Mechanical and Production Engineering Research and Development* **9** 845-858
[5] Abdelrazek Ali H, Choudhury I A, Yusoff Nukman and Kazi S N 2020 Metal cutting lubricants and cutting tools: a review on the performance improvement and sustainability assessment *The International Journal of Advanced Manufacturing Technology* **106** 4221–45
[6] Aliev A 2018 Ensuring the quality of the machined holes during the hole truing and reaming on lathe machines *MATEC Web of Conferences* **224** 01122
[7] Katna R, Suhaib M and Agrawal N 2019 Nonedible vegetable oil-based cutting fluids for machining processes – a review *Materials and Manufacturing Processes* **35** 1-32
[8] Uysal A and Jawahir I S 2019 Validation of the slip-line model for serrated chip formation in orthogonal turning under dry and MQL conditions *Procedia CIRP* **82** 124-9
[9] Javaroni R L, Lopes J C, Sato B K, Sanchez L E A, Mello H J, Aguiar P R and Bianchi E C 2019 Minimum quantity of lubrication (MQL) as an eco-friendly alternative to the cutting fluids in advanced ceramics grinding *The International Journal of Advanced Manufacturing Technology* **103** 2809–19
[10] Boubekri1 N and Shaikh V 2015 Minimum quantity lubrication (MQL) in machining: Benefits and drawbacks *Journal of Industrial and Intelligent Information* **3** 205-9
[11] Sharma V S, Singh G and Sørby K 2014 A review on minimum quantity lubrication for machining processes *Materials and Manufacturing Processes* **30** 935-53
[12] Yan P, Rong Y and Wang G 2016 The effect of cutting fluids applied in metal cutting process *Proceedings of the Institution of Mechanical Engineers Part **B** Journal of Engineering Manufacture* **230** 19-37
[13] Olt J, Liyvapuu A, Madissoo M and Maksarov V V 2016 Dynamic simulation of chip formation in the process of cutting *International Journal of Materials and Product Technology* **53** 1-14
[14] Yakubov Ch 2012 *Theoretische durchdringung und nachweis der wirkungen eines umweltorientierten kühlschmierstoffeinsatzes* (Zugl.: Magdeburg, Univ.) p 157
[15] Tabakov V P and Chikhranov A V 2018 Selecting the composition of wear-resistant coatings *Russian Engineering Research* **38** 105-9
[16] Adaskin A M, Butrim V N and Vereshchaka A S 2016 Wear resistance of a coated hard-alloy tool in the machining of high-temperature chromium alloys *Russian Engineering Research* **36** 867–71
[17] Vereschaka A, Tabakov V, Grigoriev S, Sitnikov N, Milovich F, Andreev N and Bublikov J 2019 Investigation of wear mechanisms for the rake face of a cutting tool with a multilayer composite nanostructured Cr–CrN-(Ti,Cr,Al,Si)N coating in high-speed steel turning *Wear* **438–439** 17-37
[18] Ahmed R et al. 2015 Structure property relationship of suspension thermally sprayed WC-Co nanocomposite coatings *Journal of Thermal Spray Technology* **24** 357–77
[19] Smetanina K E, Andreev P V, Malekhonova N V and Lantsev E A 2019 Optimization of the phase composition of hard alloys obtained by spark plasma sintering of powders WC + 10% Co *J. Phys.: Conf. Ser.* **1347** 012064
[20] Kas'yanov S V, Kondrashov A G and Safarov D T 2017 Rapid assessment of wear-resistant tool coatings *Russian Engineering Research* **37** 969–73

IOP Conf. Series: Materials Science and Engineering **889** (2020) 012035 doi:10.1088/1757-899X/889/1/012035

Berry Phase in single crystals of the dilute magnetic semiconductor $(Cd_{0.7}Zn_{0.28}Mn_{0.02})_3As_2$

V S Zakhvalinskii[1], T B Nikulicheva[1] and E A Pilyuk[1]

[1]Belgorod National Research University, 85 Pobedy St, Belgorod 308015, Russia

e-mail: pilyuk@yandex.ru

Abstract. Investigations of transverse magnetoresistance have been performed on single crystals of diluted magnetic semiconductor $(Cd_{0.7}Zn_{0.28}Mn_{0.02})_3As_2$ (CZMA). Based on analysis of the Shubnikov-de Haas (SdH) oscillations, observed in the temperature range $T = 4.2 \div 30$ K and at transverse magnetic field $B = 0 \div 11$ T, the phase shift, β, indicating the presence of the Berry phase was determined.

1. Introduction

The unusual properties of cadmium arsenide, such as the abnormally high mobility of charge carriers, have long attracted the attention of researchers [1]. An extensive research activity on Cd_3As_2 and its solid solutions started because of the appearance of theoretical studies [2], predicting the possibility of topological properties in Cd_3As_2, and then being confirmed by experiments using angle-resolved photoemission spectroscopy (ARPES) [3]. The existence of the Dirac semimetal phase in Cd_3As_2 was experimentally confirmed using ARPES [3-5]. The properties of 3D Dirac topological semimetals in Cd_3As_2 caused an increasing interest, including the study of transport properties and the band structure of this compound [6-10]. A number of works has been devoted to the location of the 3D Dirac cones in the Brillouin zone and to refining the band structure of Cd_3As_2 [11-13]. Of particular interest is the study of the evolution of the band structure and topological properties of ternary and quaternary solid solutions based on Cd_3As_2 [14].

For single crystals of solid solutions $(Cd_{1-x}Zn_x)_3As_2$, based on the study of SdH oscillations and temperature dependences of the transverse magnetoresistance, a phase transition from Dirac semimetal (DS) to semiconductor with increasing Zn concentration above $x = 0.38$ was experimentally observed. By changing the concentration of Zn, the behavior of the temperature dependence of the resistivity changed from a metal with compositions $x = 0$–0.31 to a semiconductor behavior with $x = 0.38$–0.58 [15]. As known from the work [16], the transverse magnetoresistance was studied on non-oriented single crystals of the semimagnetic quaternary solid solution system $(Cd_{1-x-y}Zn_xMn_y)_3As_2$ ($x + y = 0.4$; $y = 0.04$ and 0.08). The effect of Mn content on the topological properties of CZMA was shown, based on analysis of SdH oscillations. In the CZMA sample with Mn concentration $y = 0.04$, a strong magnetic field dependence of the cyclotron effective mass was observed. Moreover, a phase shift close to $\beta = 0.5$ indicated the presence of the Berry phase and 3D Dirac fermions in the single crystal of CZMA with $y = 0.08$. However, an anomalous external magnetic field dependence of the cyclotron effective mass, $m_c(B) = m_c(0) + \alpha B$, was not observed. Other compositions of quaternary solid solutions of the diluted magnetic semiconductor $(Cd_{1-x-y}Zn_xMn_y)_3As_2$ also showed unusual properties. In studies of the SdH effect in single crystals of CZMA ($x + y = 0.3$), an anomalous external magnetic field dependence of the cyclotron effective mass of the charge carriers was observed for samples with a high Mn content [17, 18].

The purpose of this study was to investigate the topological properties of the diluted magnetic semiconductor $(Cd_{0.7}Zn_{0.28}Mn_{0.02})_3As_2$.

2. Experimental details

The modified Bridgeman method was used to obtain single of $(Cd_{0.7}Zn_{0.28}Mn_{0.02})_3As_2$ (CZMA), grown from stoichiometric amounts of Cd_3As_2, Zn_3As_2 and Mn_3As_2 by slow cooling, 5 °C/h, of a melt at the presence of a temperature gradient near the melting point (840°C). The composition and homogeneity of the samples were analyzed by X-ray powder diffraction and energy-dispersive X-ray spectroscopy (EDX) methods. The X-ray experiment was carried out on a diffractometer DRON-UM (FeKα - radiation, λ = 1.93604 Å, Θ- 2Θ - method). All the investigated samples $(Cd_{0.7}Zn_{0.28}Mn_{0.02})_3As_2$ demonstrated a tetragonal crystal structure (space group $P4_2/nmc$) (ICDO Card № 03-065-857) [19]. The electrodes were attached to a rectangular prism of size 1x1x5 mm^3 by soldering. Magnetoresistance measurements were made in a transverse magnetic field configuration at 0 ÷ 11 T applying six-probe method. For experiments the sample probe was inserted in a He exchange gas Dewar, where the temperature can be adjusted with an accuracy of 0.5%.

3. Results and discussion

Well-resolved single-period SdH oscillations are observed in all investigated CZMA specimens at temperatures between $4.2 \leq T \leq 30$ K. Figure 1 shows the SdH oscillations obtained at $T = 4.2$ K.

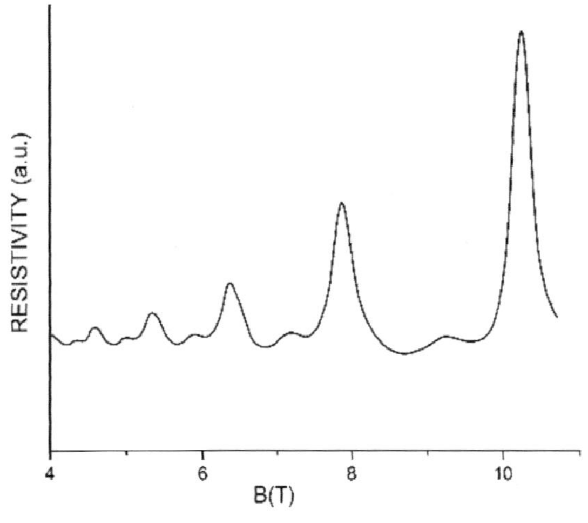

Figure 1. Oscillations of the transverse magnetoresistance obtained at $T = 4.2$ K for single crystals of $(Cd_{0.7}Zn_{0.28}Mn_{0.02})_3As_2$.

Clear linear dependence of the inverse magnetic field, $1/B_{max}$, of the maxima of the SdH oscillations on their quantum number was observed and the SdH period, P_{SdH}, did not depend on the magnetic field (Figure 2). Based on the above results of the SdH oscillations, a linear dependence of the Landau number that depends inversely on B was plotted. Fast Furrier analysis (FFA) of SdH oscillations are presented in Figure 3 for the CZMA. Well-defined peaks appear for frequencies $F_1 = 26$ T, and $F_2 = 52$ T. Note that the resolution of our FFA analysis is sufficient to calculate the number of periodic oscillations. The results of FFA analysis of the SdH oscillations in CZMA were presented by us for the first time since, in work [17], based on the study of the same series of samples in the mid-90s, the topological properties of Cd_3As_2 were not yet discovered and investigated.

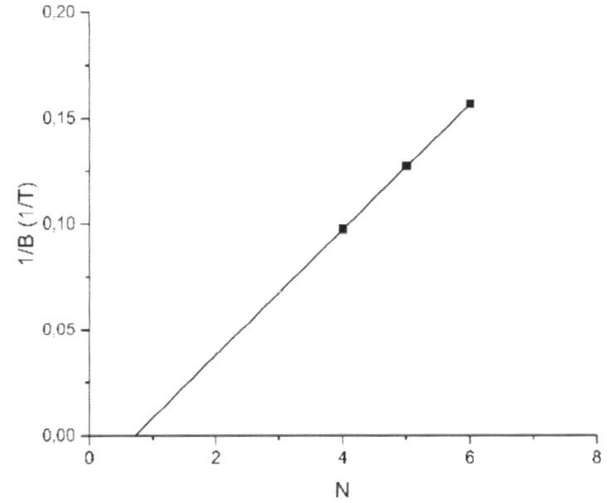

Figure 2. Landau level index plot *N* for single crystals of $(Cd_{0.7}Zn_{0.28}Mn_{0.02})_3As_2$.

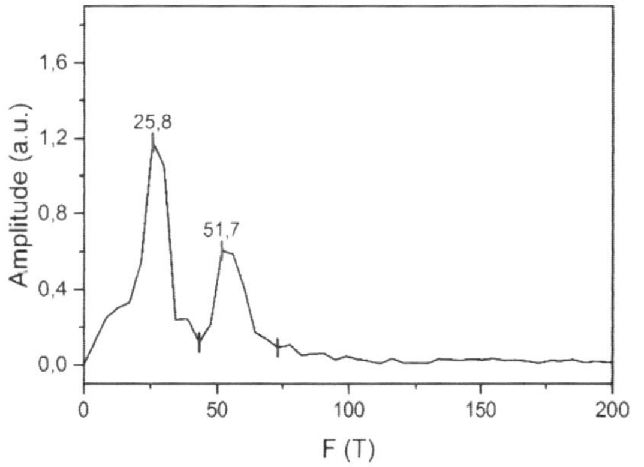

Figure 3. Results of FFT analysis of the SdH oscillations for $(Cd_{0.7}Zn_{0.28}Mn_{0.02})_3As_2$.

Around the Dirac point in momentum space, the wave function of an electron acquires a geometric Berry phase equal to π. In materials with the spin – orbit interaction and parabolic dispersion law, the Berry phase should be zero [20]. It is possible to experimentally evaluate the Berry phase from the analysis of the SdH oscillations. As follows from the work [6], the Berry phase was observed in any orientation of magnetic field and, therefore, it can be observed in the study of the oscillations of transverse magnetoresistance. The phase shift β for Dirac fermions is equal 0.5, but some deviations are possible too, such as $\beta \approx 0.45$ and 0.7 in topologic insulators $Bi_{2-x}Cu_xSe_3$ [21].

The Berry phase can be investigated in magnetic fields in which a semi-classical description of magnetic oscillations can be applied [3, 4]:

$$\rho \propto \cos\left[2\pi\left(\frac{B_F}{B}+\frac{1}{2}+\beta\right)\right],$$

where B_F is the SdH oscillation frequency, β is the Berry phase shift in the range $0 < \beta < 1$. A phase shift $\beta = 0$ (or equivalently, $\beta = 1$) corresponds a trivial case. The deviation from this value indicates a new physics with $\beta = 0.5$ that implies the existence of Dirac particles. The phase shift β for Dirac fermions is 0.5, but deviations are possible too, such as $\beta \approx 0.45$ or 0.7 for the topological isolator $Bi_{2-x}Cu_xSe_3$ [21]. Experimentally, this phase shift in the semi-classical mode can be obtained from the analysis of the SdH fan diagram, in which the sequence of the nth minimum $1/B_n$ in $\rho(B, T)$, is plotted against the Landau index N (Figure 4).

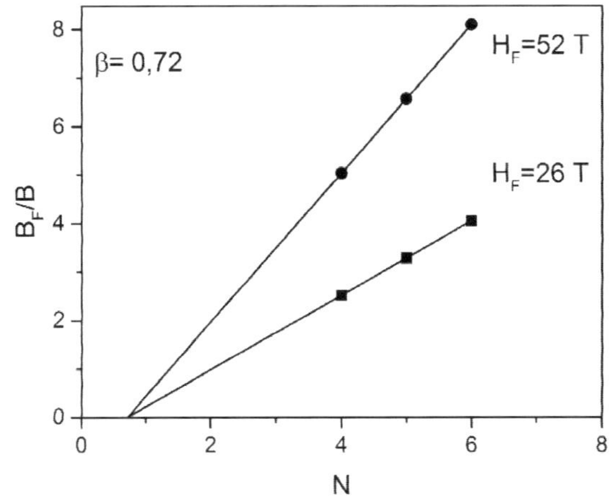

Figure 4. Landau index *N* plotted against $1/B$ for crystals of $(Cd_{0.7}Zn_{0.28}Mn_{0.02})_3As_2$.

The resulting β is close to 0.5 (Figure 4), suggesting the existence of Dirac fermions in single crystals of $(Cd_{0.7}Zn_{0.28}Mn_{0.02})_3As_2$. The existence of the Berry phase was not discussed in works [5, 6] devoted to $Cd_{1-x-y}Zn_xMn_y)_3As_2$ ($x + y = 0.3$), since the topological properties of Cd_3As_2 were not yet discovered. Finally, as mentioned above, according to the theory, the phase shift β for Dirac fermions should be equal to 0.5, however some deviations are possible.

The Berry phase can be determined by the so-called fan diagram, which is a function of Landau level index N against $1/B_{max}$. Furthermore, different studies use either the minima or maxima of the SdH oscillations. The nonzero shift of the Berry phase observed in the SdH fan diagrams (Figures. 3, 4) is associated with the vanishing mass at the Dirac point. As shown below in Table 1, the values of cyclotron masses m_c for the single crystals determined in [17], are also close to those obtained in [16] at zero magnetic field, for CZMA ($x + y = 0.4$) with $y = 0.04$ and 0.08.In order to analyze the topological properties of CZMA solid solutions $(Cd_{0.7}Zn_{0.28}Mn_{0.02})_3As_2$, we partially used the results and experimental dependences of the SdH oscillations that we reported earlier in [17]. The parameters found from the SdH oscillations of $(Cd_{0.7}Zn_{0.28}Mn_{0.02})_3As_2$ crystals [17] are given in Table 1.

Table 1. Parameters found from SdH oscillations single crystals of $(Cd_{0.7}Zn_{0.28}Mn_{0.02})_3As_2$ [17].

$n_H\times10^{18}$ (cm^{-3})	$\mu_H\times10^4$ (cm^2 V^{-1} s^{-1})	$m_c(0)/m_0$	$\alpha/m_0\times10^3$ (1/T)	T_D (K)	$T_{D\mu}$(K)
1.0	3.5	0.032	–	24	3.6

Where n_H is the Hall concentration, μ_H is the Hall mobility, m_c is cyclotron effective mass, m_0 is the electron mass, T_D is the Dingle temperature, $m_c(0)$ and α are coefficients of the linear law $m_c(B) = m_c(0) + \alpha B$, $T_{D\mu}$determines the Landau level broadening due to electron scattering on lattice defects.

A study of solid solutions based on diluted magnetic semiconductor was carried out in this paper, and is focused on the detection of the topological properties. Fan diagrams were plotted for $(Cd_{0.7}Zn_{0.28}Mn_{0.02})_3As_2$ based on the results of the fast Fourier transform (Figure 4). For $(Cd_{0.7}Zn_{0.28}Mn_{0.02})_3As_2$the phase shift β indicate the presence of the Berry phase.

4. Conclusion

In this paper we studied the topological properties in single crystals of diluted magnetic semiconductor $(Cd_{0.7}Zn_{0.28}Mn_{0.02})_3As_2$ obtained by the Bridgman method. Thus, our results are based on the analysis and observation of the SdH effect in diluted magnetic semiconductors CZMA [17]. In the presence of a magnetic field, the analysis of the SdH oscillations revealed the existence of a nontrivial Berry phase. Our results of the study of quantum transfer in solid solutions based on 3D Cd_3As_2 complement the previous ARPES and STM experiments [22]. Having a unique electronic structure, unusually high mobility and a large linear magnetoresistance at room temperature, the semimetal 3D Dirac Cd_3As_2 and its solid solutions give rise to new possibilities for future applications in electronics.

The dependence of B_F/B_N on N for $(Cd_{0.7}Zn_{0.28}Mn_{0.02})_3As_2$ shows a phase shift $\beta \approx 0.72$. A good interpretation of the results of the study of topological properties in solid solutions of CZMA based on the observation of oscillations of the transverse resistivity is complicated, since the simultaneous contribution of 2D and 3D components to the transport properties. The presence of a nonzero phase shift $\beta \neq 0$ indicates the presence of a Berry phase and confirms the presence of the topological properties in a $(Cd_{0.7}Zn_{0.28}Mn_{0.02})_3As_2$ single crystal.

5. References

[1] Arushanov E K 1992 *Progress in crystal growth and characterization of materials* **25** 131-201
[2] Wang Z, Weng H, Wu Q, Dai X and Fang Z 2013 *Physical Review B* **88** 125427
[3] Borisenko S, Gibson Q, Evtushinsky D, Zabolotnyy V, Büchner B and Cava R J 2014*Physical review letters* **113** 027603
[4] Liu Z K, Jiang J, Zhou B, Wang Z J, Zhang Y, Weng H M, Chen Y L 2014*Nature Materials* **13** 677
[5] Neupane M, Xu S-Y, Sankar R, Alidoust N, Bian G, Liu C, Hasan M Z 2014 *Nature*

Communications **51**

[6] Gelten M J, Van Es, C M, Blom F A P and Jongeneelen J W F 1980 *Solid State Communications* **33** 833

[7] Aubin M J, Rambo A and Arushanov E 1981*Physical Review B* **23** 3602

[8] Arushanov E 1980*Progress in crystal growth and characterization of materials* **3** 211

[9] Arushanov E 1992*Progress in crystal growth and characterization of materials* **25** 131

[10] Schleijpen H M A, von Ortenberg M, Gelten M J and Blom F A P 1984 *International journal of infrared and millimeter waves* **5** 171

[11] Hakl M *etal* 2018 *Physical Review B* **97** 115206

[12] Akrap A *etal* 2016 *Physical review letters* **117** 136401

[13] Conte A M, Pulci O and Bechstedt F 2017 *Scientific Reports* **7** 1.

[14] Nishihaya S, Uchida M, Nakazawa Y, Kurihara R, Akiba K, Kriener M, Miyake A, Taguchi Y, Tokunaga M and Kawasaki M 2019*Nature communications* **10** 1

[15] Lu H, Zhang X, Bian Y and Jia S 2017 *Scientific reports* **7** 1

[16] Zakhvalinskii V S, Nikulicheva T B, Lähderanta E, Shakhov M A, Nikitovskaya E A and Taran S V 2017 *Journal of Physics: Condensed Matter* **29** 455701

[17] Laiho R, Lisunov K G, Stamov V N and Zahvalinskii V S 1996*Journal of Physics and Chemistry of Solids* **57** 1

[18] Laiho R, Lähderanta E, Lisunov K G, Stamov V N and Zakhvalinskii V S 1997*Journal of Physics and Chemistry of Solids* **58** 717

[19] ICSD Database Version 2009-1 *Ref. code* 23245, 2009

[20] Mikitik G P and Sharlai Y V1999 *Physical review letters* **82** 2147

[21] Vedeneev S I, Knyazev D A, Prudkoglyad V A, Romanova T A and Sadakov A V 2015 *Journal of Experimental and Theoretical Physics* **121** 65

[22] Nishihaya S, Uchida M, Nakazawa Y, Kurihara R, Akiba K, Kriener M, Miyake A, Taguchi Y, Tokunaga Mand Kawasaki M 2019 *Nature communications* **10** 1

Acknowledgments

The work was supported by the grant of President of the Russian Federation for state support of young Russian scientists - candidates of sciences, project No. MK-238.2020.2.

AUTHOR INDEX

Abdulkerimov, Ilimdar ... 205
Abkhairova, C V ... 62
Ageeva, E.V ... 8
Akimov, Seran ... 188
Alekseeva, E ... 112
Ametov, I E ... 62
Arefiev, A S ... 98
Ashmarin, A A ... 108
Babushkina, O F ... 82
Balaev, E Yu O ... 46
Barkov, K A ... 135
Bayramukov, R A ... 56
Betsofen, S Ya ... 16, 108
Bisilov, N U ... 56
Blednova, Zh M ... 39, 46
Blyagoz, A M ... 157
Botashev, A Yu ... 56
Chekalova, Elena ... 1
Cheretaev, I V ... 177, 182
Chernousova, O V ... 135
Chernysheva, Yu V ... 104
Chornobay, S Ye ... 68, 82
Chuyan, E N ... 177, 182
Dmitrenko, D V ... 39
Dyadichev, A V ... 82
Dyadichev, V V ... 68, 82
Dyadicheva, I V ... 68
Dzhelialov, S I ... 75
Dzhemalyadinov, Ruslan ... 205
Ermachikhin, A V ... 194
Gel'manova, Z ... 27
Gololobov, G P ... 98
Gordeeva, M I ... 88
Grashkov, SA ... 12
Grushin, I A ... 16, 104
Gulyaev, V A ... 92
Gvozdeva, O N ... 20
Khalchenko, I G ... 128
Khardikov, S.V. ... 8
Khudyakova, O Yu ... 92
Knyazeva, Yu A ... 88
Kolesnikov, A V ... 82
Komarov, A I ... 165
Konakbayeva, A ... 27
Kondrat'ev, S Y ... 117
Kovalev, M ... 112
Kovalev, N M ... 142
Kretinin, K M ... 157
Kulikov, E A ... 172
Kurakin, M K ... 117
Kurapov, G V ... 46
Lebedev, M A ... 108
Lenshin, A S ... 135
Loginov, N Yu ... 92
Mailyan, D R ... 157
Makarova, N V ... 165
Malakhova, Y N ... 172
Malsugenov, R S ... 56
Melnik, A Yu ... 68
Meniuk, Dmitriy ... 32
Menyuk, S G ... 82
Mezentseva, A ... 27
Minakov, D A ... 135
Miryuk, O A ... 149
Nikulicheva, T B ... 200, 211
Novikov, A. N. ... 8
Papynov, E K ... 128
Petrov, A A ... 16
Petrov, L M ... 108
Pilyuk, E A ... 200
Pilyuk, EA ... 211
Pivovar, NA ... 12
Plomodialo, R L ... 39
Ravaeva, M Yu ... 177, 182
Savostin, D S ... 104
Seredin, P V ... 135
Serpova, M A ... 98
Shalin, A V ... 20
Shapkin, N P ... 128
Shaposhnikov, N ... 112
Shron, Leonid ... 32
Shvetsov, O V ... 117
Skakun, Vladimir ... 205
Speransky, K A ... 16
Stepushin, A S ... 20
Stupnikov, A A ... 172
Suvorov, D V ... 98
Svinoroev, Yu A ... 68
Sysoev, A A ... 12
Ternovskoy, O A ... 68
Tsuprik, V G ... 165
Volokitin, V P ... 142
Vorobyov, Y V ... 194
Voskresenskaya, I I ... 104
Yagyaev, Elmar ... 32, 188
Yakhimovich, V A ... 117

Yakubov, A O 194
Zakhvalinskii, V S 200, 211
Zhuravlev, Andrey 1
Zolotukhin, S N 142

Institute of Physics
Dirac House, Temple Back
Bristol BS1 6BE UK

ISSN: 1757-8981
ISBN 978-1-7138-3176-1

9 781713 831761

CURRAN ASSOCIATES INC.
proceedings.com